Machine-to-machine (M2M) Communications

Related titles

Handbook of Organic Materials for Optical and (Opto)Electronic Devices
(ISBN 978-0-85709-265-6)

Wearable Electronics and Photonics
(ISBN 978-1-85573-605-4)

Silicon-On-Insulator (SOI) Technology: Manufacture and Applications
(ISBN 978-0-85709-526-8)

Woodhead Publishing Series in Electronic and Optical Materials: Number 69

Machine-to-machine (M2M) Communications

Achitecture, Performance and Applications

Edited by

Carles Antón-Haro and Mischa Dohler

AMSTERDAM • BOSTON • CAMBRIDGE • HEIDELBERG
LONDON • NEW YORK • OXFORD • PARIS • SAN DIEGO
SAN FRANCISCO • SINGAPORE • SYDNEY • TOKYO
Woodhead Publishing is an imprint of Elsevier

Woodhead Publishing Limited is an imprint of Elsevier
80 High Street, Sawston, Cambridge, CB22 3HJ, UK
225 Wyman Street, Waltham, MA 02451, USA
Langford Lane, Kidlington, OX5 1GB, UK

Notice
No responsibility is assumed by the publisher for any injury and/or damage to persons or
property as a matter of products liability, negligence or otherwise, or from any use or operation
of any methods, products, instructions or ideas contained in the material herein.
Because of rapid advances in the medical sciences, in particular, independent verification of
diagnoses and drug dosages should be made.

British Library Cataloguing in Publication Data
A catalogue record for this book is available from the British Library.

Library of Congress Control Number: 2014954006

ISBN 978-1-78242-102-3 (print)
ISBN 978-1-78242-110-8 (online)

For information on all Woodhead Publishing publications
visit our website at http://store.elsevier.com/

Typeset by SPi Global
www.spi-global.com
Printed and bound in the United Kingdom

Contents

Part Three Network optimization for M2M communications 249

List of contributors

A. Alexiou   University of Piraeus, Piraeus, Greece

A. Al-Hezmi   Fraunhofer FOKUS Research Institute, Berlin, Germany

L. Alonso   Universitat Politècnica de Catalunya (UPC), Barcelona, Spain

J. Alonso-Zarate   Centre Tecnològic de Telecomunicacions de Catalunya (CTTC), Barcelona, Spain

S. Andreev   Tampere University of Technology (TUT), Tampere, Finland

C. Antón-Haro   Centre Tecnològic de Telecomunicacions de Catalunya (CTTC), Barcelona, Spain

A. Antonopoulos   Telecommunications Technological Centre of Catalonia (CTTC), Barcelona, Spain

D. Barthel   Orange, Paris, France

A. Bartoli   Universitat Politecnica de Catalunya (UPC), Barcelona, Spain

P. Bhat   Vodafone Group Services Ltd, Newbury, UK

B. Cendón   TST, Cantabria, Spain

H. Chao   Research & Innovation Center, Bell Laboratories China, Alcatel-Lucent Shanghai Bell Co. Ltd, Shanghai, China

A. Corici   Fraunhofer FOKUS Research Institute, Berlin, Germany

M. Di Renzo   CNRS–SUPELEC–Univ. Paris Sud XI, Paris, France

M. Dohler   King's College, London (KCL), London, UK; Worldsensing, Barcelona, Spain; Worldsensing, London, UK

D. Drajic   Ericsson d.o.o., Belgrade, Serbia

A. Elmangoush   Technische Universität Berlin, Berlin, Germany

F. Ennesser   Gemalto, Meudon, France

O. Galinina   Tampere University of Technology (TUT), Tampere, Finland

H. Ganem   Gemalto, Meudon, France

M. Gerasimenko   Tampere University of Technology (TUT), Tampere, Finland

A. Gotsis   University of Piraeus, Piraeus, Greece

E. Kartsakli   Technical University of Catalunya (UPC), Barcelona, Spain

Y. Koucheryavy   Tampere University of Technology (TUT), Tampere, Finland

A. Kountouris   Orange, Paris, France

S. Krco   Ericsson d.o.o., Belgrade, Serbia

A.S. Lalos   Technical University of Catalunya (UPC), Barcelona, Spain

M. Laner   Vienna University of Technology, Vienna, Austria

A. Laya   Universitat Politècnica de Catalunya (UPC), Barcelona, Spain; KTH Royal Institute of Technology (KTH), Stockholm, Sweden

T. Magedanz   Technische Universität Berlin, Berlin, Germany

J. Markendahl   KTH Royal Institute of Technology (KTH), Stockholm, Sweden

P. Martigne   Orange, Paris, France

J. Morrish   Machina Research, Reading, UK

N. Nikaein   EURECOM, Biot, France

T. Norp   TNO, Delft, The Netherlands

M. Popovic   Telekom a.d., Belgrade, Serbia

P. Popovski   Aalborg University, Aalborg, Denmark

N.K. Pratas   Aalborg University, Aalborg, Denmark

P. Svoboda   Vienna University of Technology, Vienna, Austria

S. Tennina   WEST Aquila srl, L'Aquila, Italy

H. Thomsen   Aalborg University, Aalborg, Denmark

T. Tirronen   Ericsson Research, Jorvas, Finland

C. Verikoukis   Telecommunications Technological Centre of Catalonia (CTTC), Barcelona, Spain

I. Vilajosana   Worldsensing, Barcelona, Spain

K. Wang   Centre Tecnològic de Telecomunicacions de Catalunya (CTTC), Barcelona, Spain

T. Watteyne   Dust Networks Product Group, Linear Technology, Union City, CA, USA

W. Webb   Weightless SIG, Cambridge, UK

J. Wu   Research & Innovation Center, Bell Laboratories China, Alcatel-Lucent Shanghai Bell Co. Ltd, Shanghai, China

Woodhead Publishing Series in Electronic and Optical Materials

Introduction to machine-to-machine (M2M) communications

1

C. Antón-Haro[1], M. Dohler[2]
[1]Centre Tecnològic de Telecomunicacions de Catalunya (CTTC), Barcelona, Spain; [2]King's College London (KCL), London, UK and Worldsensing, Barcelona, Spain

1.1 Introducing machine-to-machine

We have witnessed the fixed Internet emerging with virtually every computer being connected today; we are currently witnessing the emergence of the mobile Internet with the exponential explosion of smart phones, tablets, and netbooks. However, both will be dwarfed by the anticipated emergence of the Internet of Things (IoT), in which everyday objects are able to connect to the Internet, tweet, or be queried. While the impact onto economies and societies around the world is undisputed, the technologies facilitating such a ubiquitous connectivity have struggled so far and have only recently commenced to take shape.

1.1.1 Machine-to-machine and the big data opportunity

A cornerstone to this connectivity landscape is and will be machine-to-machine (M2M). M2M generally refers to information and communications technologies (ICT) able to measure, deliver, digest, and react upon information in an autonomous fashion, i.e., with no or really minimal human interaction during deployment, configuration, operation, and maintenance phases.

Flagship examples of M2M technologies are telemetry readings of status of oil and brakes of cars on the move, health state measurements of blood pressure and heartbeat of the elderly, monitoring of corrosion state of oil or gas pipelines, occupancy measurements of parking in cities, remote metering of water consumption, and real-time monitoring of critical parts of a piece of machinery.

While machines do not excel humans in writing poetry, they are definitely industry favorites when it comes to (i) repetitive jobs, like delivering water meter data once a day, and (ii) time-critical jobs with decisions taken within a few milliseconds based on the input of an enormous amount of data, like the real-time monitoring of rotating machinery parts.

M2M is all about big data, notably about (i) real-time, (ii) scalable, (iii) ubiquitous, (iv) reliable, and (v) heterogeneous big data, and thus associated opportunities. These technical properties are instrumental to the ecosystem:

Machine-to-machine (M2M) Communications. http://dx.doi.org/10.1016/B978-1-78242-102-3.00001-0

- Indeed, *real time* allows making optimal and timely decisions based on a large amount of prior collected historical data. The trend is to move away from decision making based on long-term averages to decisions based on real-time or short-term averages, making a real difference to the large amount of nonergodic industrial processes.
- *Scalability* implies that all involved stakeholders, i.e., technology providers, service companies, and finally end user in a given industry vertical, can scale up the use and deployment where and when needed without jeopardizing prior deployments. Wireless is instrumental when it comes to scalability!
- *Ubiquitous* deployment and use are important since it allows reaching this critical mass required to make technologies survive long term. A sparse or punctual use of specific M2M technologies makes sense in the bootstrapping phase of the market but is unsustainable in the long term.
- *Reliability*, often overlooked, means the industry customer gets sensor readings that can be relied on because the system had been calibrated, tested, and certified. A prominent example is the vertical of urban parking, where Google's crowdsourced platform to get parking occupancy had been discontinued since stakeholders in this space preferred to get reliable occupancy messages 24/7 from certified sensors installed at each parking spot, rather than unreliable crowdsourced data, even when offered for free.
- Finally, the big data dream can only be materialized if *heterogeneous data*, i.e., data from different verticals, are combined to give unprecedented insights into behavior, trends, and opportunities and then also *act* upon them. In fact, when referring to "big" in big data, it is less about quantity of data (i.e., large volumes of terabytes of information) but rather the quality of data (i.e., different information streams giving a new comprehension of the field, where one stream could be only a single bit).

The data being delivered by M2M systems are often referred to as the oil of the twenty-first century. It might be coincidental but the M2M data flow resembles the oil processing flow in that it is composed of data (i) upstream, (ii) mash-up, and (iii) downstream:

- *Upstream:* Here, data are being collected from sensors in the field that deliver the data wirelessly to a gateway via single-hop or mesh networking technologies. The data traverse the Internet and reach dedicated servers or cloud platforms, where they are stored and profiled for further analytics.
- *Downstream:* Here, business intelligence is applied to the data and intelligent decisions are taken. These are pushed back down to the customer, by either controlling some actuators, displaying some information/alerts in control/service platforms, or simply informing users about issues pertaining to this specific M2M application. Human–computer interfaces (HCI) or even computer–computer interfaces (beyond the typical APIs) will be instrumental in ensuring the offtake of M2M technologies.
- *Mash-up*: The data processing and business intelligent platforms are likely the most important constituent of emerging M2M systems. Data flows from machines, humans, and the Internet in general will feed intelligent analytics engines, which are able to find trends, optimize processes, and thus make the industry vertical more efficient and effective.

These three constituents are depicted in Figure 1.1.

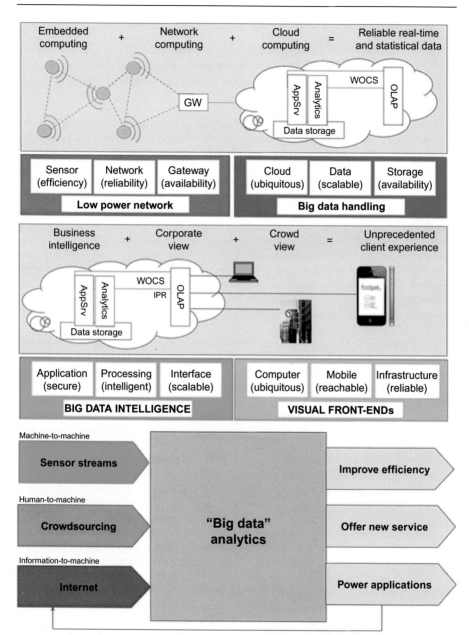

Figure 1.1 Upstream (top), downstream (middle), and mash-up (bottom) components for typical industrial M2M turnkey solutions.

1.1.2 Machine-to-machine technology landscape

As for wireless M2M technologies, the ecosystem has so far relied on ZigBee-like and 2G/3G cellular technologies; however, new players are entering the space, such as low-power Wi-Fi, Bluetooth low energy, and proprietary cellular systems. The pros and cons of these are as follows:

- *ZigBee-like technologies:* IEEE 802.15.4 and derivatives were (and still are) perceived as the holy grail for wireless sensor networking and M2M usage. Indeed, with the latest IEEE 802.15.4e amendments, it has become a very energy-efficient technology, even in the case of multi-hop mesh. However, in its very nature of providing fairly high data rates over short distances, it is against the essence of M2M, which mainly requires very low data rates over large distances. The need for frequent repeaters/gateways and skilled engineers to handle connectivity/radio planning and the lack of a global critical mass when it comes to coverage and deployment/adoption have prevented the predicted growth and will likely be the demise of the technology itself. ZigBee and derivatives have never made the jump from being a technology to being a turnkey solution, i.e., a system that is easy to use to customers whose core business is not dealing with dimensioning wireless. However, this community has achieved something that none of the others have yet: they have penetrated the control community and were able, despite clearly being technically inferior compared to any of the below systems, to become the certified choice for wireless SCADA-like systems (see, e.g., WirelessHART, ISA100.11a, and Wireless M-Bus).
- *Low-power Wi-Fi:* An interesting contender in the M2M space is emerging in the form of a well-tuned Wi-Fi system. Wi-Fi enjoys an enormous popularity in both the business-to-consumer (B2C) and B2B markets, with more than 2bn access points installed and a truly vibrant standardization ecosystem active. There is Wi-Fi coverage virtually anywhere where it is worth taking M2M data. Interestingly, it consumes significantly less energy when transmitting information [1]. This is mainly because data are transmitted in a single hop only when needed, thus requiring the radio to be switched on only at events or regular alive beacons. In contrast, ZigBee-like technologies need to listen periodically for neighbors to transmit data; this consumes in the end more energy than a simple single-hop network. It is thus to no surprise that chip manufacturing giants, like Broadcom [2], have decided to ditch the IEEE 802.15.4 stack for the more promising IEEE 802.11 stack in its gamble for the IoT and M2M market.
- *Proprietary cellular:* Cellular networks have the enormous advantage of covering large areas to reach multiple M2M devices without the need for repeaters or numerous base stations. A problem with current standardized cellular solutions (2G/3G/LTE) is the fairly slow standardization cycle and high licensing costs associated with enabling every sensor and actuator with such technology. Proprietary solutions have thus emerged, which bridge this gap. Notable players in the field are SIGFOX, Neul, Cycleo, On-Ramp, and lately also the chip giants Texas Instruments (TI) and STMicroelectronics. These technologies are, in general, fairly narrowband—thus enjoying very large link budgets—and connect hundreds if not thousands of devices to a single base station over a large area. While these solutions are technically well elaborate, they are not deemed to be future-proof since they are not standardized. Also, while coverage is guaranteed by the said systems, mobility and roaming are not supported (as of 2014).
- *Standardized cellular:* Therefore, standardized cellular M2M technologies will be a crucial constituent to the M2M landscape in the next years to come. Cellular systems enjoy worldwide coverage and interoperability, allowing for M2M devices to roam and be on the move.

In addition, interference is being handled by centralized radio resource management (RRM) algorithms, which, to a large extent, ensure that there are no coexistence, delay, or through-put problems. It is thus to no surprise that the major standardization bodies, such as 3GPP, IEEE, and ETSI, started standardizing M2M architectures and protocols.

1.1.3 Cellular M2M requirements

Before standardized cellular technologies can be used in the M2M space, the follow-ing technical requirements need to be addressed, which contrast completely the requirement on human-driven communications:

- There will be a lot of M2M nodes, i.e., by orders of magnitude more than human connections.
- More and more applications are delay-intolerant, mainly control.
- There will be little traffic per node, mainly in the uplink.
- Nodes need to run autonomously for a long time.
- Automated security and trust mechanisms need to be in place.

This book thus outlines major technical advances that allow the above requirements to be met.

1.2 The machine-to-machine market opportunity

M2M is the underlying communication network connecting sensors, actuators, machines, and objects. The market opportunities are thus directly related to the con-nectivity of the said devices. This market, in turn, is driven by larger macroeconomic market developments, which rely on the availability of the said M2M networks. Two notable examples of such larger market developments are the Internet of Things (slant toward B2C) and industrial Internet (slant toward B2B). The uptake of these para-digms will be a major boost to the M2M markets, ranging from chip manufacturer to coverage and service providers. This section provides some insights into recent market developments and associated opportunities.

1.2.1 Return-of-investment (ROI) argumentation

A major drive in the uptake of any technology is its ability to return the initial financial investment. There are three major ROI arguments to use M2M technologies:

- *Real-time instrumentation ROI:* A study by General Electric (see Figure 1.2) has identified the enormous efficiency benefits stemming from real-time instrumentation by means of M2M technologies. The study has looked at verticals like transportation, health, and manufacturing. While the study has not looked at the cost of the instrumentation, the strong ROI drive is evident.
- *Big data value ROI:* Not so well quantified as of 2014, it is understood to be an enormous value in cross correlating data from different verticals to give unique insights, which the silos would not be able to expose on their own. An example can be found in smart city

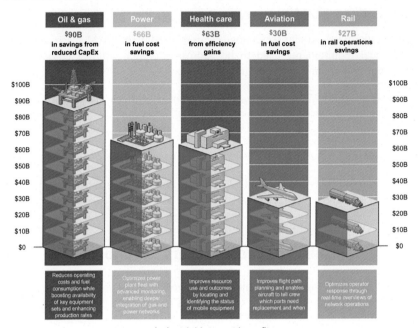

A joint venture with GE

Industrial Internet:

The power of 1%

Efficiency gains as small as 1% could have sizable benefits over 15 years when scaled up across the economic system

Industrial Internet benefits

Figure 1.2 Financial returns stemming from real-time instrumentation of various industrial verticals by means of machine-to-machine technologies [3].
Source: GE estimates/postmedia.

transportation where M2M data from traffic are cross correlated with weather and sports data, thus allowing to define viable traffic management strategies on days where congestion is likely.

- *Wireless ROI:* A fairly mundane but nonetheless important drive in rolling out wireless M2M is the fact that getting rid of cables allows achieving substantial CAPEX and OPEX gains. As indicated in Figure 1.3, while electronics and sensors become cheaper over time, human labor and cable costs (due to copper) increase. Going wireless saves cable costs (CAPEX), installation efforts (CAPEX), and system maintenance (OPEX).

The above ROI argumentations highlight that there is not only utility in using M2M technologies but also clear financial returns. It is thus to no surprise that markets commence to wake up to the opportunity, which will be scrutinized in the subsequent section.

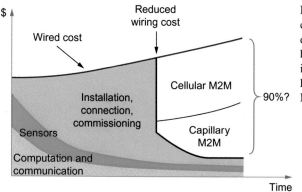

Figure 1.3 While computation and sensor costs decrease over time, the cost of human labor and cables is increasing, making a strong ROI case for using wireless M2M technologies.

1.2.2 Market overview

While M2M technologies have been so far about increasing operational efficiencies and reducing costs (see above discussion on ROI), it is now seen by industries as a true enabler for innovative services, which would have not been possible without M2M.

The M2M market is mainly business-to-business (B2B)-focused. However, due to its very nature of connectivity, it connects many previously unconnected businesses and thus acts as a catalyst for B2B2B. It is, however, still very silo-based where an M2M technology is purpose-designed for a specific market and application; and the generated data are rarely shared beyond that silo.

As shown in Figure 1.4, the major M2M service sectors are the following:

- *Energy:* Sector of utmost importance for the M2M market, which is increasingly serving the energy upstream and downstream sectors. The technology is used to improve efficiency and reduce crew cost for seismic acquisition, used in oil and gas exploration, fracking, and carbon storage. It is also instrumental in increasing the operational efficiency of the energy grid, be it along the transportation network, along the distribution network, or in the form of smart meters at the consumer premises.
- *Industrial:* M2M is used here to instrument machineries and various industrial components of high value and heavy industry manufacturing. It is instrumental for safety and for ensuring higher efficiencies.
- *Transportation:* The technology is used here to monitor traffic conditions and improve traffic flux, parking experiences, and safety, among others. It is one of the more popular markets for M2M today (2014).
- *Retail:* Anything related to optimizing the supply chain, warehouses, or stores is within the realm of M2M in this market.
- *Security and safety:* M2M is used here to provide tracking and surveillance, both fixed and mobile. The bandwidth requirements, particularly for surveillance, are often the highest among all M2M applications, which is due to frequent video transmissions.
- *Healthcare:* It is a truly emerging market with the aim to use M2M technologies to increase the ability to monitor patients in real time, among others. The main areas are clinical applications and usage in well-being (e.g., fitness monitors).

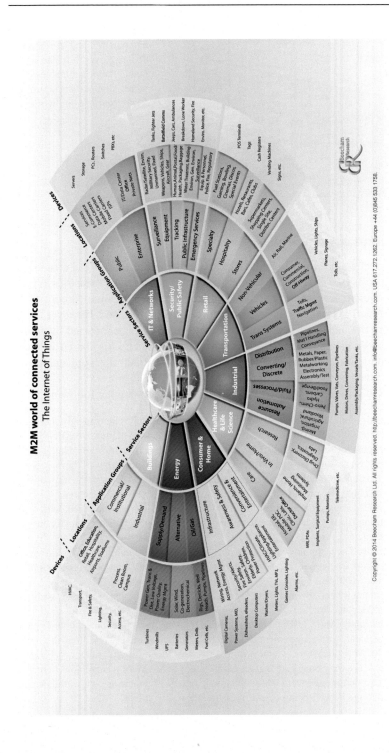

Figure 1.4 M2M market overview with different application domains [4].

- *Buildings:* M2M technology is here used to improve the efficiency of the building in terms of energy consumption, waste management, carbon footprint, etc. It is also being used to improve the well-being of the people residing in the building, be it to regulate the heating, air-conditioning, or solar flares.

Traditionally, the beneficiaries of the above markets were M2M chip manufacturers and integrators thereof. However, with growing markets and maturing technologies, new stakeholders emerge in this space. The established and emerging M2M market players, and their interaction, can be summarized as follows:

- *Component manufacturers:* At the beginning of the value chain are component manufacturers, which gear up for the manufacturing of billions of components. These pertain to the production of chips, batteries, casings, smart/SIM cards, and antennas, among others. While volumes are large in these segments, margins are often tight as technology is often standardized and the competitive edge thus often in quick time to market and price.
- *Integrators:* The ecosystem of integrators, which use off-the-shelf components to assemble products and optimize them for use in specific markets and industry verticals, is critical to a proper use and uptake of M2M technologies. Since these are often IT-originated companies selling into large and unchallenged non-IT companies (and mainly heavy industry verticals), margins are fairly large.
- *Operators:* Integrators have realized that providing industry-compliant connectivity and running M2M networks is a challenging and arduous task, which is often out of the core business of the integrators. To this end, operators are increasingly involved in M2M deployments, which handle connectivity, coverage, and availability of M2M equipment. Many emerging M2M markets require an increasing degree of mobility and support of roaming, being the core competence of traditional cellular and network operators. Providing connectivity has evolved into a commodity service, which is why margins here are fairly moderate, despite playing a major role in the uptake and growth of M2M technologies.
- *Platform providers:* Given the large amount of data involved in M2M deployments, a strong ecosystem of platform providers has emerged. General purpose horizontal platforms and market-specific platforms are emerging, where the former are more commonly found in B2B and the latter in the emerging B2C. These platforms include not only data processing and "big data" analytics but also storage capabilities (an issue often overlooked when rollout out M2M applications). We observe a strong shift to cloud platforms, as well as emerging app store-like approaches. Because the said platforms work on data volume that is not yet available in many emerging M2M markets, margins today are very unreliable; however, it is expected that—once this community is able to gather large data sets and cross correlate them across industry silos—sales margins will substantially increase.
- *Service providers:* Finally, the ecosystem of true service providers is emerging. These use technology and platforms to service certain markets. M2M is an enabler for these industries, with a unique selling point being their ability to add business intelligence and experience in process controls.

Predictions on the market volumes and addressable market sizes vary greatly for each of these players and thus are not further provided. However, we provide in Table 1.1 some historical data and predictions on the growth of M2M cellular module manufacturers by application. One observes the rapid growth and thus corroborates the market pull. Not shown here but, interestingly, the most used air interface is GSM, which is due to its technical capabilities of handling low-rate traffic in an energy-efficient

Table 1.1 Development of cellular M2M modules deployment by air interface

Application	Total cellular M2M module shipments (millions)	
	2011	2019
OEM telematics	2.85	48.31
Aftermarket telematics	5.11	38.88
Commercial telematics	4.40	13.19
AMI	6.62	23.82
Security	3.58	17.85
RMAC	5.41	14.17
ATM/POS	2.54	11.26
Vending	0.26	3.76
RID	0.04	1.82
Telehealth	0.51	11.27
Other	0.62	8.29
FWT	5.35	5.63
Industrial PDA	1.57	3.44
Total	38.85	201.68

World market, end-user basis forecast: 2011 and 2019.
Courtesy of ABI Research, June 2014.

manner. In addition, worldwide coverage is excellent, and modems are affordable. 3G UMTS technology is less energy-efficient, enjoys less coverage, and has more expensive modems. However, given that most 2G networks will be phased out shortly, the 3G M2M families will be increasingly used. LTE and LTE-A are technically very suitable to M2M but the modems, mainly due to patent costs, are very expensive to date. As for the application domains, the automotive industry logically uses cellular M2M technologies most since it requires mobility and roaming support, which are both native to cellular designs.

1.2.3 Market challenges and opportunities

To finalize the high-level market analysis of M2M, we shall highlight not only major challenges but also major opportunities.

- *Challenge of long sale cycles:* In B2B, M2M sales are exposed to long sale cycles. Between the first contact and final deal, easily, 2–4 years can pass. Given that a lot of innovative M2M solutions will come from startups or small companies, this is a major strain on their cash flow and needs to be watched out for. Once a project is assigned, there is also the danger that smaller businesses fail simply due to the large size of the projects.
- *Challenge of breaking silos:* Arguably, the largest opportunity for M2M is to capitalize on data streams, which come from different industrial silos. To bring different silos together,

however, is a major challenge, and there are ample examples from the past where this has not worked out.

- *Opportunity of big data:* Eventually, the M2M market will shift from focusing on providing connectivity to capitalizing on the actual data content, i.e., make the most of big data. This allows for the crucial shift from monitoring industries to actually optimizing them.

More in-depth market studies are provided in later chapters in this book. We move now on to some commercial and experimental examples of M2M networks.

1.3 Examples of commercial and experimental M2M network rollouts

No doubt, the importance and economic impact of M2M are on the rise. Vodafone's M2M adoption barometer [5] reveals that two-thirds out of the companies surveyed (some 600) are considering or in the process of implementing M2M solutions in their business within the next two years. The fact that major telcos such as Vodafone, Telefonica, Orange, Telenor, and Sprint have set up dedicated portals on this technology constitutes one more evidence of that. However, concerns about cost vs. potential benefit trade-offs turns out to be one of the major barriers for M2M adoption [5].

This section, on the one hand, aims to report on a number of commercial rollouts. The goal is to illustrate how M2M was successfully adopted in a variety of vertical markets. On the other, it provides a succinct description of several *pilots* and *field trials* that could turn into commercial services in a near future. Given the vast amount of commercial and experimental deployments worldwide, this section cannot but provide the reader with a small (yet, in our opinion, representative) sample of ongoing work in this area.

1.3.1 Commercial rollouts

Smart energy grids (electricity, water, and gas) have become a fertilized area for M2M rollouts. Midwest Energy, for instance, installed control units allowing farmers to monitor and control the electricity consumption of the irrigation pumps deployed in their fields [6]. On the one hand, this saves time for farmers who can avoid turning pumps on and off manually. On the other hand, this gives utilities the ability to implement demand response programs and, by doing so, prevent energy consumption peaks. Specifically, pumps become partly controlled by the utility that can inhibit their operation when electricity consumption is too high in households (e.g., due to air-conditioning system activity in hot summer afternoons). In exchange for that, farmers receive economic incentives from the utility company. For a similar agricultural irrigation scenario, Idaho Power reports reductions in the annual electricity bill in the range 12–27% for those adhering to their PeakRewards program [7]. From a technical point of view, data connectivity here is provided by a combination of *satellite and cellular* networks.

Similarly, the National Grid Energy Corporation, a company with over 3.4 million *gas* customers in the northeast of the United States, also runs a demand-side management program. Accurate meter readings with hourly consumption data (leveraging on two-way *wireless cellular* connectivity such as GSM and CDMA) make it possible for the utility to temporarily remove from the distribution system large users (e.g., factories, schools, and government buildings) as they switch to a *local* oil- or gas-based energy source. These interruptible users help mitigate the constraints imposed by the finite capacity of the generation and/or distribution system (i.e., pipelines), in particular in cold seasons. The first trial was conducted in New York City in 2004–2005 and by the end of 2007, over 2500 units had already been installed [8].

Unlike in the previous examples, utilities like Iberdrola, the first energy group in Spain, have resorted to *wired* communication technologies for smart metering infrastructure rollouts [9]. The specifications of the so-called PRIME (PoweRline Intelligent Metering Evolution) technology, which is OFDM-based, were finalized and published in 2008. Iberdrola was the first Spanish utility to pilot PRIME PLC technology. The project in Castellón, which ran until 2011, covered 100,000 metering points that were deployed in both urban and rural areas. After its successful completion, the company engaged in a mass deployment of 1,300,000 additional smart meters until 2013.

M2M connectivity can also be instrumental for the provision of sustainable energy access in developing countries where, interestingly, GSM coverage is quite extensive [10]. For instance, SharedSolar deploys microgrids connecting up to 20 families in countries like Mali, Uganda, or Haiti. Their gateways allow not only to monitor consumption in real time but also to enable payments via SMS and 2G (initially, through prepaid scratch cards). In 2012, there were 245 households connected to the Shared-Solar platform. Similarly, M-KOPA in Kenya provides solar home systems in rural areas on a pay-as-you-go basis. The system enables the customer not only to pay for energy consumption but also to finance the high initial costs that acquiring/installing a solar panel entails.

Smart cities are becoming an increasingly important vertical market for M2M, with applications ranging from air quality monitoring to security or traffic control, to name a few. Traffic congestion and parking, in particular, have become enormous problems in many cities, but fortunately, they can be partly alleviated by resorting to smart parking systems based on M2M. Typically, such systems are capable of (i) keeping track of the status of the available parking spaces, (ii) wirelessly conveying this information to a centralized server, and (iii) guiding drivers to them by means of on-street displays. Not only does this alleviate traffic congestion but also it has the side benefit of reducing greenhouse gas emissions. As an example of such systems, the city of Moscow has recently awarded a tender for the rollout of its smart parking system [11]. More innovative projects such as SFpark in San Francisco dynamically assigns parking pricing based on real-time supply and demand law [12].

Finally, M2M has also an important role to play in other sectors such as insurance or fleet control. For the former, Telefonica and Generali Seguros launched in February 2013 the so-called "Pago como conduzco" (paying the way you drive) service [13]. Data on journeys (e.g., number of kilometers, time of the day, to what extent speed limits are kept, and number of sudden braking or accelerations) are collected by a device installed in the vehicle. Such device conveys data via cellular networks to an application, which, in turn, computes the premium that allows savings of up to 40% for drivers with good habits. Customers can also monitor their ratings as drivers via a mobile app or a portal.

For fleet control scenarios, the United Kingdom-based company Trafficmaster and the Norwegian company Telenor Connexion have developed a stolen vehicle-tracking application with *global* (i.e., worldwide) footprint [14]. This can be achieved by means of a customized M2M subscriber identification module (SIM) plus a personalized roaming profile and price plan, which facilitates a cost-effective solution for this service.

1.3.2 Pilots and field trials

In a logistics context, TST Sistemas has developed an M2M system for stock management in hospitals of items such as drugs or surgery materials [15]. The system, still in a test bed phase, monitors the corresponding stock levels, relieving medical personnel (e.g., nurses) from such burden. It is based on ETSI's M2M hybrid architecture and combines a number of wireless and wired technologies, such as NFC (for devices in charge of reading the level of remaining stocks) or UTRAN, Wi-Fi, ZigBee, and Ethernet, for data transmission.

The smart grid test bed deployed in Jeju Island in South Korea goes one step beyond the DSM commercial rollouts described earlier in this section. The goal here is to move from a *closed* utility system to an *open* service platform capable of managing, in real time, millions of meters, feeder automation devices, and distributed generation assets [16]. For telco operators, the challenge lies in (i) dealing with large amounts of data in real time and (ii) providing two-way low-latency communication network guaranteeing the stability and security of the electricity grid.

In an mHealth context, Ericsson, Dialog Axiata, and Asiri Surgical Hospital conducted a study in Sri Lanka to assess, with a dedicated pilot, the feasibility of remote patient monitoring via M2M [17]. Selected patients suffering from cardiac diseases were equipped with noninvasive wireless sensor capable of monitoring vital signs. The collected sensor data were transmitted via a Bluetooth radio interface to a communication device. In turn, the device conveyed data over a 3G/GPRS network to a remote application for further analysis by medical personnel. The emphasis here was mostly on usability aspects. The study concluded that the mHealth solution was more useful for patients rather than for healthy individuals, who had concerns about carrying, e.g., bulky measurement devices (pulse oximeter and blood pressure monitor) during working hours. Users were also in favor of replacing the aforementioned communication device by a regular mobile phone.

Finally, there are also some salient examples of pilot deployments for smart cities. This includes, for instance, the large-scale M2M platform deployed in Santander, Spain [18]. From an application point of view, its target is twofold: (i) environmental monitoring, on the basis of the measurements (noise and pollution) collected by the platform, and (ii) helping drivers find a free parking space. The platform is composed of sensors, actuators, cameras, and screens to offer useful information to citizens. It is conceived as an open platform in that it allows conducting remote experiments (applications and protocols) by third parties. Still, in a context of environmental monitoring applications, a related test bed was deployed in Belgrade, Serbia [19]. The twist here was the usage of *mobile* sensors (mounted on public transportation vehicles such as buses) in order to increase the size of the monitoring area.

1.4 Machine-to-machine standards and initiatives

Market fragmentation and lack of global standards are frequent criticisms of M2M communication systems. The harmonization of technologies at a global scale is addressed in standards development organizations (SDOs), on the one hand, and industrial associations or special interest groups (SIGs), on the other. Sections 1.4.1 and 1.4.2 attempt to provide a global picture of both. Next, a number of supra-SDO and supra-SIG initiatives are succinctly described in Section 1.4.3. In addition to all of these, special attention must be paid to enhancing current M2M technologies in order to make them future-proof. Section 1.4.4 thus focuses on a number of research and innovation projects, which investigate tomorrow's M2M communication networks.

1.4.1 Standards development organizations

The *European Telecommunications Standards Institute* (ETSI) produces global ICT standards including fixed, mobile, radio, converged, broadcast, and Internet technologies. In ETSI, most of the standardization work in the M2M arena is conducted within the machine-to-machine communications technical committee (TC-M2M). This committee was established in 2009 with the mission of ensuring that M2M services deployed worldwide are interoperable. Figure 1.5 provides an overview of the M2M architecture proposed by ETSI's M2M, along with the envisaged domains, namely, application, network, and device [20].

ETSI M2M defines a service capability layer (SCL) on top of connectivity layers. Hence, for the network domain (see the above figure), it leverages on existing technologies such as 3GPP's GERAN/UTRAN/eUTRAN (i.e., 2G, 3G, or LTE networks), WiMAX, or other fixed or satellite networks. Likewise, for the M2M area network domain, it relies on the availability of short-range communication technologies such as Wi-Fi, ZigBee, or power-line communications (PLCs), to name a few. ETSI released in 2012 and 2013 the first and second versions, respectively, of its M2M standard. The core specs in this suite are those describing service requirements, the functional architecture, and communication interfaces (central column in Figure 1.6).

The activities conducted by ETSI's Smart Card Platform Technical Committee (TC-SCP) are very relevant to M2M communications too. To give an example, the availability of embedded and remotely programmable subscriber identity modules (SIM), a topic under the umbrella of this committee, is instrumental for the successful deployment of M2M networks. For 3GPP technology-related developments, ETSI-SCP

M2M domains

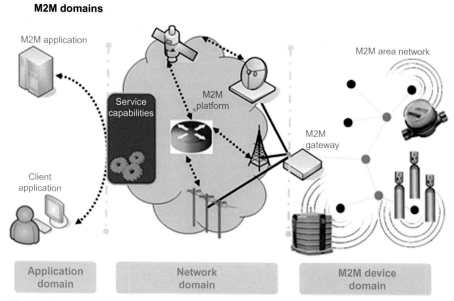

Figure 1.5 Overview of M2M domains as standardized by ETSI [20] (© European Telecommunications Standards Institute 2011. Further use, modification, copy, and distribution are strictly prohibited).

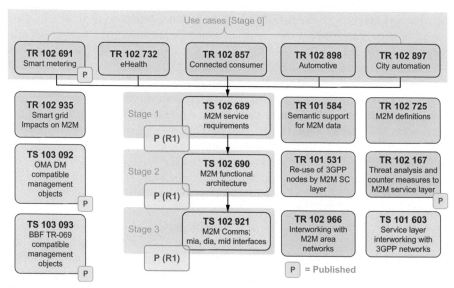

Figure 1.6 Specifications in ETSI's M2M standard [20] (© European Telecommunications Standards Institute 2011. Further use, modification, copy, and distribution are strictly prohibited).

actively collaborates with GSMA (Global System for Mobile Communications Association)'s Smart Card Application Group (SCAG), as well as 3GPP's Core Network and Terminals working group 6 (CT6, Smart Card Application Aspects).

The *3rd Generation Partnership Project* (3GPP) is responsible for the development and maintenance of the Global System for Mobile Communications (GSM), the Universal Mobile Telecommunications System (UMTS) and its Long-Term Evolution (LTE) and beyond, and the Internet Protocol Multimedia Subsystem (IMS) specifications. Hence, 3GPP is a key standardization body as far as delivering M2M communications over *cellular* (wide-area) networks or, in the 3GPP jargon, machine-type communications (MTC) is concerned. Differently from ETSI, 3GPP deals with specific systems and protocols. A high number of working groups (WGs) under, e.g., the Technical Specification Groups for Radio Access Network (TSG-RAN) or Service and Systems Aspects (TSG-SA), contribute very actively to the work on MTC-related optimizations for LTE networks. The prioritization of topics and activities is discussed in work items such as the one on "System Improvements for Machine-Type Communication" (SIMTC) of Release 11.

The *Institute of Electrical and Electronics Engineers* (IEEE) has introduced several enhancements to its air interface for broadband *wireless access* systems (i.e., IEEE 802.16) in order to more effectively support M2M applications. To that aim, two amendments (IEEE 802.16p and 802.16.1b) were developed by the IEEE 802.16's machine-to-machine (M2M) Task Group. Complementarily, the IEEE 802.15 Working Group for Wireless Personal Area Networks develops standards for *short-range* wireless networks composed of, e.g., personal digital assistants (PDAs), cell phones, and, in general, mobile and computing devices. In particular, Task Group 4 investigates low data rate solutions with multimonth to multiyear battery life and very low complexity. The e, g, and k amendments to IEEE 802.15.4 are particularly relevant to M2M communications and are reaching maturity as of 2014. In more details, the IEEE 802.15 Task Group 4e is aimed to define a MAC amendment to the existing standard 802.15.4-2006. According to its foundational chart, the intent of this amendment is "to enhance and add functionality to the 802.15.4-2006 MAC to better support the industrial markets and permit compatibility with modifications being proposed within the Chinese WPAN." The role of IEEE 802.15 Smart Utility Networks (SUN) Task Group 4g is to create a PHY amendment to 802.15.4 to provide a global standard that facilitates very large-scale process control applications such as the utility smart-grid network capable of supporting large, geographically diverse networks with minimal infrastructure and potentially millions of fixed end points. Last but not least, the IEEE 802.15.4k amendment addresses applications such as critical infrastructure monitoring.

Within the *International Telecommunication Union*–Telecommunication Standardization Sector (ITU-T), the Focus Group on the M2M service layer, established in 2012 and terminated in 2013, was responsible for studying the requirements and specifications for a common M2M service layer (including architecture, protocols, and API aspects). The strategy was to reuse to the largest extent possible what has already been specified by other SDOs. Yet such service layer is aimed to support different application domains, such as eHealth, smart grids, and industrial automation;

the focus of the group was on eHealth (e.g., remote patient monitoring and ambient assisted living).

According to its foundational charter, the TR-50 Smart Device Communications Engineering Committee of the *Telecommunications Industry Association* (TIA) is in charge of developing *interface standards* for the communication of events and information between M2M systems and smart devices, applications, or networks. In this context, TR-50 is developing a smart device communication framework that is *agnostic* to the underlying transport and access networks (both wired and wireless) and to the vertical application domain by means of a suitable application programming interface (API). The IEEE 802.15.6TM-2012 standard is optimized to serve TIA's TR-50 committee around body area networking. TR-50 also fosters collaboration and pursues coordination with other SDOs. As an example, it supports (and hosts) the machine-to-machine standardization task force (MSTF) of the Global Standards Collaboration (GSC). Besides, TIA is a founding member of the oneM2M initiative (see Section 1.4.3).

Finally, the (ongoing or past) work conducted by a number of working groups within the *Internet Engineering Task Force* (IETF) is also relevant to M2M systems. Mostly, it addresses networking aspects in the so-called capillary networks, that is, beyond the M2M gateway in ETSI's architecture. This includes, but is not limited to, the ROLL (Routing over Low-Power and Lossy networks), CoRE (Constrained REstful Environments), MEXT (Mobility EXTensions for IPv6), 6LoWPAN (IPv6 over Low-Power WPAN), and 6TSCH (IPV6 over IEEE 802.1.5.4e Time Slotted Channel Hopping) working groups.

1.4.2   Industrial associations and special interest groups (SIGs)

The activities carried out in the aforementioned SDOs are nicely complemented by the efforts made by a number of industrial associations and SIGs (which are often fed into the above-discussed standardization bodies). Since the number of industrial associations is quite high, we will focus on a particularly relevant subset only.

The *Global System for Mobile Communications Association* (GSMA) gathers around 800 mobile operators worldwide and more than 200 companies in the broader mobile ecosystem (e.g., handset makers, software companies, media, and entertainment). Within GSMA, the work conducted in the Smart Card Application Group (SCAG), which is aimed to promote smart card adoption and identification of mobile operator requirements, and favor functionality/quality enhancements (form factor, embedded versions, over-the-air reprogramming, etc.), is of notable importance for players in the M2M arena.

As for *Weightless SIG*, it promotes the adoption of an open standard for cellular M2M communications. To date, over 1500 organizations have joined the Weightless SIG. Its specification, for which version 1.0 is already available [21], is specifically tailored for the operation of M2M networks in white space spectrum (TV bands, UHF). Some salient features and design requirements include the optimization for low-cost hardware, extended coverage (to reach, e.g., metering devices in home basements or in remote places), ultralow-power operation to enhance network lifetime, and secure and guaranteed message delivery.

Finally, the *Alliance for Telecommunications Industry Solutions* (ATIS) launched its M2M committee in July 2012. It aims to define the elements of a common service layer leveraging on network capabilities and the requirements for the interfaces toward the application and transport layers. ATIS is one of the founding members of the oneM2M alliance (see Section 1.4.3).

1.4.3   Global industrial M2M initiatives

In an attempt to stimulate global harmonization and avoid duplication of activities between SDOs, we have recently witnessed the advent of supra-SDO and supra-SIG initiatives. This includes, for instance, the *oneM2M Partnership Project,* which was formed in July 2012 with the support of ETSI, TIA, ATIS, CCSA, TTA, ARIB, and TTC as founding members. Inter alia, oneM2M pursues [22] the following: (i) the definition of a common service layer allowing an independent view of end-to-end services; (ii) the design of open/standard interfaces, APIs, and protocols; and (iii) the facilitation of interoperability, this including test and conformance specifications.

Likewise, the goal of the machine-to-machine Standardization Task Force (MSTF) of the *Global Standards Collaboration* (GSC), a group of major SDOs centered on the International Telecommunication Union, is to "facilitate global coordination and harmonization in the area of M2M standardization by reaching out to a broad range of participants in the field and openly sharing relevant M2M information."

To close this section, in Table 1.2, we summarize the main players (standardization bodies, associations, and SIGs) and the specific working groups within those organizations working toward the standardization of M2M communications.

1.4.4   International innovation projects on M2M

Here, we provide an overview of several research and innovation projects in the field of M2M. Whereas the scope and size of those projects varies, the work of the projects listed here is/was mostly funded by the European Commission (EC).

The EXALTED project (EXpAnding LTE for Devices, 2010–2013) turns out to be one of the flagship projects in this area [23]. Funded under the umbrella of EC's Seventh Framework Program (FP7), EXALTED counted with the participation of the main industrial stakeholders, from both the manufacturers (Sierra Wireless, Ericsson, Alcatel-Lucent, and Gemalto) and mobile network operators (e.g., Vodafone) sides. The main goal was to develop a cost-, spectrum-, and energy-efficient radio access technology for M2M applications, the so-called LTE-M overlay, adapted to coexist within a high-capacity LTE network. Special attention was paid to scalability issues (to, e.g., avoid congestion in the random access procedures) and cost aspects, to ensure affordability of LTE M2M modules. EXALTED partners had a very active involvement in standardization bodies and SIGs such as 3GPP, ETSI, and the GSMA association.

Table 1.2 Main organizations and specific working groups in the M2M arena

Standards development organization/ association/SIG	Main relevant working group(s), committees, amendments
ETSI	M2M, SCP
3GPP	TSG-RAN, TSG-SA, TSG-CT (and WGs thereof)
IEEE	802.15.4g, 802.15.4k, 802.15.4e, 802.16p, 802.16.1b
ITU-T	Focus Group M2M
TIA	TR-50
IETF	ROLL, CoRE, MEXT, 6LoWPAN, 6TSCH
GSMA	SCAG
Weightless SIG	PHY, MAC, security, regulation
ATIS	M2M committee
oneM2M	Requirements, architecture, security, management
GSC	MSTF

The scope of the 5GNOW (5th generation non-orthogonal waveforms for asynchronous signalling) and LOLA (Achieving LOw-LAtency in Wireless Communications) projects, on the contrary, is much narrower than that of EXALTED. In terms of topics, however, they nicely complement one another. 5GNOW, on the one hand, investigates alternatives to the bulky physical layer procedures necessary to ensure strict synchronism in LTE(-A) networks [24]. This is particularly relevant for machine-type communications that typically combine small data payloads with a large number of quasi-simultaneous random accesses. LOLA [25], on the other hand, focuses on physical and MAC layer techniques aimed at achieving low-latency transmission in cellular (LTE and LTE-A) and wireless mesh networks. The project targeted two main scenarios: M2M applications in mobile environments (with sensors being attached, e.g., to trains or vehicles) and gaming services.

On a different theme, the BETaaS (Building the Environment for the Things as a Service) and VITRO (Virtualised distributed plaTfoRms of smart Objects) projects addressed network virtualization and service platform aspects. More specifically, BETaaS proposes to build a platform to run M2M applications on top of services residing in a *local* (vs. remote) cloud of gateways composed of, e.g., smart phone and home routers [26]. Particular attention was paid to scalability, security, and dependability issues. Interoperability with the main network architectures being standardized, namely, those of ETSI M2M and IoT-A, was of utmost importance. As for the VITRO project [27], the emphasis here was in virtualization mechanisms for sensor networks. Ultimately, the goal was to facilitate the federation of heterogeneous large-scale networks of objects and devices by providing them with an homogeneous abstract appearance.

As for projects funded by the National Science Foundation, "Synergy: Resilient Wireless Sensor-Actuator Networks" [28] investigates, inter alia, how to reduce bit rates in real-time control systems by resorting to quantization and event-triggered feedback. Research is validated in a multirobot test bed composed of a number of unmanned ground vehicles where M2M communication hardware/software is seamlessly integrated with the multirobot control architecture.

1.5 Book rationale and overview

Machine-to-machine (M2M) communication is essentially composed of three key ingredients: (i) a wireless end *device*, (ii) an infrastructure-based or infrastructureless wireless carrier *network*, and (iii) the *back end server* network. This edited M2M book is focused on the networking part of the architecture, dwelling on the technology required to connect the device domain with the application domain. The aim of this book is thus to provide a detailed technical insight into latest key aspects of M2M networks, with application to some chosen market verticals like smart cities and eHealth. The book digs into the technical details of the technology required to provide such applications. We deal with security, capillary and cellular access methods, and latest developments in standardization bodies. We also aim to identify challenges and open issues, thus making the material presented in this book useful for industry and inspiring for research.

The book is organized in four distinctive parts. Part One (Architectures and Standards for M2M communications) starts with a chapter devoted to M2M architectures currently being standardized and the important key issues related to these standards (Chapter 2). While the previous chapter addresses high-level architectural elements and their interaction, the subsequent three chapters deal with specific systems and protocols. Specifically, Chapter 3 (Overview of 3GPP machine-type communication standardization) focuses on the cellular M2M, notably 3GPP, developments. It encompasses, among other aspects, a review of key MTC feature enhancements, which have been elaborated to date. Departing from wide-area network approaches, Chapter 4 focuses on low-power mesh technology as an enabler of short-range M2M networks. The chapter, on the one hand, provides an overview of current standards in this area and, on the other, discusses a number of current challenges (e.g., reliability and power efficiency) and future ones of low-power mesh solutions. Given the above chapters, it has become clear that each of the M2M technologies will have its market share. Interoperation between these heterogeneous sets of technologies is hence paramount. The focus of Chapter 5 (Interworking technologies for machine-to-machine communications) is hence to embark onto some of the important issues related to interoperability between cellular, Wi-Fi, and ZigBee-based M2M networks. Complementarily, Chapter 6 revolves around how the TV white space spectrum can be used by M2M technologies. To close Part One, the chapter entitled "Supporting machine-to-machine (M2M) communications in long-term evolution (LTE) networks" discusses in detail the most significant challenges that 3GPP faces in supporting M2M communications over LTE: handling large number of devices, attaining very low-energy consumption, ensuring low-complexity protocols to bring down the cost of devices, and extending coverage to facilitate single-hop transmissions.

Part Two (Access, scheduling, mobility and security protocols) deals with a number of issues mostly in relation to the lower layers of M2M communication networks.

First, in Chapter 8 (Traffic models for M2M communications: types and applications) the authors introduce traffic models that are useful for M2M design. It encompasses a review of contributions available in the literature complemented with results from the experimental work conducted by the authors. The two subsequent chapters review the random access procedure in LTE-A networks, describe RAN/core network overload issues and the solutions proposed to prevent and/or resolve overload in the RACH control channel, and assess the performance of several packet scheduling strategies suitable for M2M data traffic. Mobility support is an important issue in future M2M systems since a large market share of current cellular M2M systems is expected to be within the automotive industry or eHealth. To this end, Chapter 11 (Mobility management for M2M communications) summarizes the challenges and issues related to supporting mobility. The focus is on cellular systems and the access network thereof. Security issues are discussed in the last two chapters of Part Two where the authors provide, on the one hand, a security taxonomy for M2M communications in capillary networks along with valuable information for the design of reliable embedded communication systems and, on the other, an overview of security, trust, and provisioning schemes for M2M devices and services.

The next block of chapters, gathered in Part Three (Network optimization for M2M communications), revolves around a number of optimizations needed in order to efficiently run M2M networks (vs. traditional broadband data networks). Handling large groups of devices requires optimizations in subscription management and billing. Thus, in "Group-based optimization of large groups of devices in M2M communications networks" (Chapter 14), the authors describe what group-based optimization cellular operators are currently providing a number of standardization efforts being conducted in the 3GPP, along with other optimizations that can be foreseen in the future. Next, in "Optimizing power saving in cellular networks for machine-to-machine (M2M) communications," the authors analyze power-saving issues from both a device and network operation points of view. This chapter describes the challenges, outlines solutions, and discusses performance results. Chapter 16 goes into further details and discuses other approaches to improve energy efficiency from a user equipment (i.e., device) perspective. This includes, for instance, discontinuous reception mechanisms. Complementarily, an extensive analysis on the access delay experienced by M2M devices (and some aspects on energy efficiency too) is conducted in Chapter 17 (Energy and delay performance of machine-type communications (MTC) in long-term evolution-advanced (LTE-A)).

Finally, in Part Four (Business models and applications), we start by providing some rationale of the emergence of the M2M market, with emphasis on business opportunities and business models and addressable markets and their growth (Chapter 18). It will allow decision makers to understand the potential, benefits, and ROI of using and deploying M2M solutions. The last chapters address two M2M verticals, namely, *smart cities*, one of the quickest emerging markets where, given the enormous amount of problems in urban areas, M2M are seen as a facilitator for improving living conditions, mainly due to the ability to gather useful information 24/7 in a reliable manner, and, finally, *eHealth*, a slowly forming market that, given its heterogeneous nature (people at home, at work, on the move, etc.), requires very tailored technologies to facilitate the said market needs.

Acknowledgment

This work is partially supported by the FP7 project NEWCOM# (318306) funded by the European Commission.

References

[1] S. Tozlu, Feasibility of Wi-Fi enabled sensors for Internet of Things, in: 7th Wireless Communications and Mobile Computing Conference (IWCMC), 2011, pp. 291–296. http://dx.doi.org/10.1109/IWCMC.2011.5982548.

[2] Broadcom, Zigbee and Z-wave are out. Broadcom's new chips bet on Bluetooth and Wi-Fi for IoT, 2013. Available from: http://gigaom.com/2013/05/29/zigbee-and-z-wave-are-out-broadcoms-new-chips-bet-on-bluetooth-and-wi-fi-for-iot/ (accessed 15.10.13).

[3] General Electric, New industrial Internet service technologies from GE could eliminate $150 billion in waste, 2012. Available from: http://www.gereports.com/post/74545148994/ge-invests-in-pivotal-to-accelerate-new-analytic (accessed 15.06.14).

[4] Beecham Research (2013). Available from: http://www.m2mforum.it/wp-content/uploads/BRL_M2MForum14May2013i.pdf (accessed 15.10.13).

[5] Vodafone, The M2M adoption barometer 2013, White Paper, 2013. Available from: https://m2m.vodafone.com/insight_news/2013-06-26-the-m2m-adoption-barometer-2013.jsp.

[6] Midwest, Turnkey project for Midwest energy, 2010. Available form: https://www.m2mcomm.com/projects/turnkey-project-for-midwest-energy/index.html (accessed 04.09.13).

[7] Idaho, Idaho power peaks reward program, 2009. Available from: https://www.m2mcomm.com/projects/idaho-power/index.html (accessed 04.09.13).

[8] National, M2M communications helps to optimize the delivery of natural gas in New York city during periods of high demand or short supply, 2007. Available from: https://www.m2mcomm.com/projects/national-grid/index.html (accessed 04.09.13).

[9] Iberdrola, Iberdrola in Spain: smart rollout of PRIME technology, 2012. Available from: http://www.landisgyr.com/webfoo/wp-content/uploads/2012/11/D000044642-_SuccessStory-Iberdrola-2012_b_en.pdf.

[10] GSMA, Sustainable energy & water access through M2M connectivity, mobile enabled community services report, 2013. Available from: http://www.gsma.com/mobilefordevelopment/wp-content/uploads/2013/01/Sustainable-Energy-and-Water-Access-through-M2M-Connectivity.pdf.

[11] Worldsensing, Worldsensing & partners win Moscow Smart Parking Tender, Press Release, 2013. Available from: http://www.worldsensing.com/index.php?option=com_k2&view=itemlist&layout=category&task=category&id=1&Itemid=4&lang=en (accessed 04.09.13).

[12] SFpark, Mission bay parking planning, 2012. Available from: http://sfpark.org/how-it-works/mission-bay-parking-planning (accessed 04.09.13).

[13] Telefonica, Generali Seguros and Telefónica launch Pago como conduzco, Press Release, 2013. Available from: http://blog.digital.telefonica.com/?press-release=generali-seguros-and-telefonica-launch-pago-como-conduzco (accessed 04.09.13).

[14] Telenor, Trafficmaster goes global with Telenor Connexion, Press Release, 2013. Available from: http://www.telenorconnexion.com/news/trafficmaster-goes-global-with-telenor-connexion (accessed 04.09.13).

[15] J. Rico, B. Cendón, J. Lanza, J. Valiño, Bringing IoT to hospital logistics systems—demonstrating the concept, in: IEEE Wireless Communications and Networking Conference (WCNC), 2012.

[16] GSMA, South Korea: Jeju Island smart grid test-bed developing next generation utility networks, GSMA smart cities report, 2012a. Available from: http://www.gsma.com/con nectedliving/wp-content/uploads/2012/09/cl_jeju_09_121.pdf.

[17] GSMA, Remote monitoring as an mHealth solution in Sri Lanka, GSMA mHealth report, 2012b. Available from: http://www.gsma.com/connectedliving/wp-content/uploads/ 2012/10/cl_mhealth_dialog_lores_09_12.pdf.

[18] Santander, Smart Santander Project, ICT-257992, 2011. Available from: http://www. smartsantander.eu (accessed 04.09.13).

[19] Belgrade, Smart City project in Serbia for environmental monitoring by Public Transportation, 2011. Available from: http://www.libelium.com/smart_city_environmental_param eters_public_transportation_waspmote (accessed 04.09.13).

[20] ETSI-M2M. Machine-to-Machine communications (M2M); Functional architecture, Technical Specification, TS 102 690, 2011. Available from: http://www.etsi.org/ deliver/etsi_ts/102600_102699/102690/01.02.01_60/ts_102690v010201p.pdf.

[21] Weightless, Weightless Specification 1.0, Weightless SIG, 2013.

[22] oneM2M, One M2M alliance, 2012. Available from: http://www.onem2m.org.

[23] EXALTED, EXpAnding LTE for devices, ICT- 258512, 2010. Available from: http:// www.ict-exalted.eu (accessed 04.09.13).

[24] 5GNOW. 5th generation non-orthogonal waveforms for asynchronous signalling, ICT-318555, 2012. Available from: http://www.5gnow.eu (accessed 04.09.13).

[25] LOLA, Achieving low-latency in wireless communications, ICT-248993, 2010. Available from: http://www.ict-lola.eu (accessed 04.09.13).

[26] BETAAS. Building the environment for the things as a service, ICT- 317674, 2012. Available from: http://www.betaas.eu (accessed 04.09.13).

[27] VITRO, Virtualised distributed plaTfoRms of smart objects, ICT- 257245, 2010. Available from: http://www.vitro-fp7.eu (accessed 04.09.13).

[28] Synergy, Synergy: resilient wireless sensor-actuator networks, 2012. Available from: http://www.nsf.gov/awardsearch/showAward?AWD_ID=1239222&Histori calAwards=false (accessed 04.09.13).

Part One

Architectures and standards

Overview of ETSI machine-to-machine and oneM2M architectures

2

P. Martigne
Orange, Paris, France

2.1 Introduction

This chapter provides the reader with some clues as to how and why machine-to-machine (M2M) standards have been elaborated. Not only does it present some parts of the M2M standards history, but also it intends to give some up-to-date (Q2 2014) information on the present and future trends for these standards.

The first section tells about the foundation of the M2M standard in the European Telecommunication Standards Institute (ETSI), which was a noticeably big step toward an end-to-end view of a generic M2M architecture offering enablers to any type of M2M application. The section explains how this standardization effort, which was then undertaken in parallel by standardization bodies located in other regions of the world, led to the creation of a more worldwide standardization partnership, called *oneM2M*, in order to converge to one global international M2M standard.

It is followed by a section dedicated to giving some technical details on this standardized generic M2M architecture, including identification of key reference points for APIs (application programming interfaces). The latter, which are also described in the second section, are major enablers for third-party application providers to use the M2M architecture for more added value services compared to "siloed" architectures.

Then, the third section aims at showing the practical use of the M2M standard by specific M2M domains. In particular, it illustrates, as one among other possible examples, how the smart home domain can leverage the enablers from the generic M2M standard as a federating infrastructure for all the smart home stakeholders.

The chapter ends with some personal thoughts on the future trends of M2M standardization, among which we can note a clear evolution toward the standardization and for the emerging IoT (Internet of Things).

2.2 Need and rationale for M2M standards

This section advocates the need for standardization that came up before 2010 and details how M2M standards activity emerged from the initial fragmented M2M market landscape. We particularly focus on the creation of ETSI M2M Technical Committee as the

Machine-to-machine (M2M) Communications. http://dx.doi.org/10.1016/B978-1-78242-102-3.00002-2

first known initiative to deal with an end-to-end architecture as a common framework for any M2M application. Evolution to oneM2M standardization partnership, which encompasses international standardization bodies' activities about M2M (ETSI M2M, CCSA, TIA TR-50, TTA, ATIS, ARIB, and TTC) constitutes a major step towards an unified standardized framework for the M2M/IOT ecosystem. We also mention some of the use cases that helped structure these generic M2M specifications.

When looking back to the early 2000s, the only M2M systems that were deployed relied on requirements that were defined (and redefined) on a case-by-case basis, without any particular attention paid to possible longer-term use cases nor to any design of a possible converging infrastructure that would have been leveraged from various use cases. Indeed, the M2M market was in an increasing development phase, but due to lack of standardization in this field, it was dominated by vertical solutions and/or proprietary solutions. This situation, where each applicative domain was developing one tailored solution per business case, is depicted as the "siloes" situation, where there are nearly as many siloes as use cases; this is illustrated on the left side of Figure 2.1.

This led to heterogeneous solutions reusing existing connectivity networks that had been previously designed for other purposes, with the risk of saturating and misusing these networks when the M2M would come to its mature phase in terms of number of connected devices. Moreover, this was not an optimized configuration in terms of midterm evolution of services proposed to the final customer, because it was just designed to answer the immediate service needs in the shortest time interval.

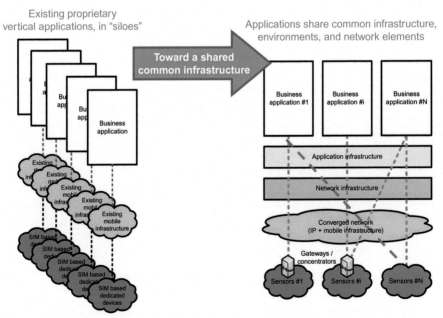

Figure 2.1 From the existing "siloes" situation toward a generic standardized applicative framework.

In 2008, the ETSI decided to circumvent this situation and created an ad hoc group dedicated to the study of standardization gaps to help the M2M market be developed in a more fruitful and sustainable way, forecasting the possible development of cross domain services between the various applicative domains.

After a 6-month in-depth analysis, this ad hoc group submitted its report to the ETSI board that advocated the need for a standardization group that would specify a generic "horizontal" M2M architecture as a common tool for all M2M application providers.

Consequently, the ETSI M2M Technical Committee was created, to provide overall standardization that would enable various business scenarios and use a common architectural framework in order to accelerate the development of the M2M market. At that period of time, indeed, the M2M value chain was beginning to evolve from single-client vertical solutions to the end-to-end provision of value-added services. This evolution needed to be supported by an appropriate standardization, to provide enablers for such an end-to-end view beyond the existing individual vertical solutions, while integrating some of the existing components. This trend is illustrated on the right side in Figure 2.1.

ETSI M2M [1] worked on listing the requirements that rose from the various M2M applicative domains, cataloged in use case technical reports. The intention was not to elaborate one technical report for each possible M2M applicative domain, but rather to have a representative set of concrete examples from where requirements could be translated into generic requirements that could then be applied to any M2M applicative domain.

These generic requirements are listed in the ETSI M2M Service Requirements Technical Specification [2], which was the first ETSI M2M TS [3] published (in August 2010 for Release 1), followed by the functional architecture specification [4], Release 1 published in October 2011, and the specification on the key reference points related to this architecture [5], Release 1 published in February 2012. Since then, ETSI M2M standard Release 2 has been published, providing a mature specification, with additional functionalities compared to Release 1 such as charging feature. The list of ETSI M2M published deliverables is shown in Figure 2.2.

But some other standardization bodies in the world also began the same kind of standardization, and it became crucial to make all these standardization activities coherent at the international level for the benefit of worldwide interoperability of the next future M2M system deployments.

As a result of the voluntary combined action of seven big standardization organizations (SDOs) of the world involved in M2M "horizontal" standards, namely, ARIB, ATIS, CCSA, ETSI, TIA, TTA, and TTC, a unique standardization partnership for M2M was created in 2012, called oneM2M, of which the funding SDOs were defined as "Partners Type 1." These partners, represented by members from various companies (telecom operators, manufacturers, integrators, device manufacturers, chipset vendors, etc.) were later joined by collaborating "Partners Type 2" (PT2) SDOs, namely, Broadband Forum [6], Continua Health Alliance [7], Home Gateway Initiative (HGI), and OMA (Open Mobile Alliance) [8], to bring any existing technical solution they had standardized to the oneM2M community so as to possibly complement the oneM2M core specifications. PT1 members and PT2 partners are thus building together the oneM2M standard (Figure 2.3); as a matter of fact, such a synergy

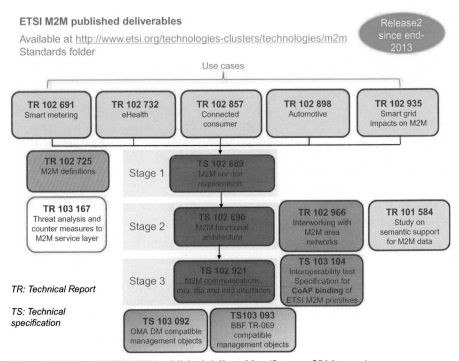

ETSI M2M published deliverables

Available at http://www.etsi.org/technologies-clusters/technologies/m2m
Standards folder

Release2 since end-2013

Use cases

| TR 102 691 Smart metering | TR 102 732 eHealth | TR 102 857 Connected consumer | TR 102 898 Automotive | TR 102 935 Smart grid impacts on M2M |

TR 102 725 M2M definitions | Stage 1 | TS 102 689 M2M service requirements

TR 103 167 Threat analysis and counter measures to M2M service layer | Stage 2 | TS 102 690 M2M functional architecture | TR 102 966 Interworking with M2M area networks | TR 101 584 Study on semantic support for M2M data

TR: Technical Report

Stage 3 | TS 102 921 M2M communications; mia, dia and mid interfaces | TS 103 104 Interoperability test Specification for **CoAP binding** of ETSI M2M primitives

TS: Technical specification

TS 103 092 OMA DM compatible management objects | TS103 093 BBF TR-069 compatible management objects

Figure 2.2 List of ETSI M2M published deliverables (January 2014 status).

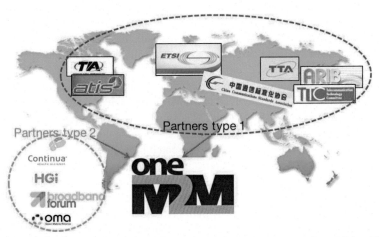

Partners type 2

Partners type 1

one M2M

Figure 2.3 oneM2M standardization partnership (PT1, funding organizations; PT2, additional contributing organizations).

between all these organizations had been the challenging motivation for the creation of oneM2M partnership as a place where all verticals would be encouraged to contribute and bring pieces into the whole coherent oneM2M puzzle.

The technical content from the funding SDOs was forwarded to oneM2M, where the main architectural design concepts from ETSI M2M were reused. These are described in the next section.

As for the continuing activity of ETSI M2M, renamed "SmartM2M" to reflect its new terms of reference, it is focussing on Smart City and also on Smart Appliances domains, for example, with the task to answer European market particular needs, also including the adaptation of the future oneM2M standard for its adoption in Europe.

2.3   Standardized M2M architecture

This section includes some detailed explanation about key components of M2M architectures that are standardized, based on published specifications from ETSI M2M, and ongoing specifications from oneM2M. Description of functional architecture and APIs is part of this section. A particular mention on the need for abstracting the heterogeneous underlying (connectivity) technologies will be made, explaining the driving overall concept of a standardized service layer that constitutes the starting point for ETSI M2M/oneM2M standardization activity.

As M2M applications are more and more popular due to the availability of mature technologies embedded in more and more daily used devices, there are an increasing number of applications to be considered under the "M2M" domain, which is now often meant as also incorporating the IoT domain.

This leads to the consideration of many various and heterogeneous area networks being part of the M2M system, each of them being particularly well designed for one or few more application domain(s). For example, a network that connects devices for the home monitoring domain has different requirements, in terms of, for example, coverage, robustness, and privacy, than a network used for devices dedicated to outdoor environmental monitoring.

ETSI M2M has considered this first type of network that they called "M2M area networks" as one main element in the global functional architecture describing an M2M system, as illustrated at the bottom of Figure 2.4. It includes local networks of connected devices, sensors, and actuators, also called "objects." M2M gateways can be used to aggregate data from these heterogeneous area networks and devices.

The issue that naturally arises is how to have these heterogeneous networks be usable by applications coming from diverse application providers, without requiring these applications to be "preaware" of what connectivity will enable them to run. By considering an upper layer above the transport layer, ETSI M2M specifies a way to deal with data notwithstanding the heterogeneity of connecting technologies between the objects. Thus, the standard relies on the specification of a service capability layer (SCL) that is expected to be implemented on devices and/or gateways in

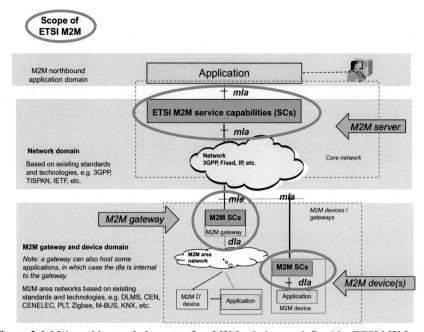

Figure 2.4 Main architectural elements of an M2M solution, as defined by ETSI M2M. Adapted from Technical Specification 102 690 Functional Architecture specification. *Source*: ETSI.

the device and gateway domain and on platforms in the network domain, as a set of generic common functionalities that the M2M system will expose to any M2M application.

The "service layer" can be defined as a conceptual layer within a network service provider architecture, which provides middleware serving third-party value-added services and applications. The service layer provides capability servers owned by a, for example, telecommunication network service provider, accessed through open and secure APIs by application layer servers owned by, for example, third-party content providers. The service layer also provides an interface to core networks at a lower resource layer (also named "control layer" and "transport layer"—the transport layer is also referred to as the access layer in some architectures).

The high-level architectural view in Figure 2.4 again shows the parts specified by ETSI M2M in the overall picture of a M2M system; clearly, the SCL that realizes the "M2M service capabilities" (SCs) is at the heart of the ETSI M2M architecture specification.

Whether the application is embedded on a device or available from the cloud, it is expected to similarly access the functionalities provided by the SCLs, that is, via APIs that are specified in a homogeneous and unified way, independent from the application and from the access network.

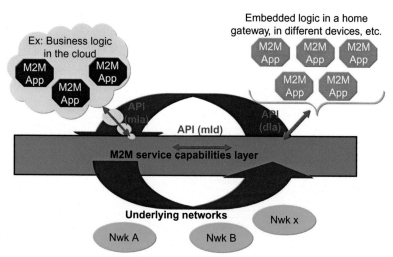

Figure 2.5 SCL-centric high-level view of the ETSI M2M standard.

Figure 2.5 provides a high-level illustration of the ETSI M2M concept considering the service layer level to abstract the underlying access technologies and taking care of any kind of M2M applications. It shows the SCL-core solution, with the three main reference points defined by ETSI M2M, namely, the mIa and dIa, to expose the SCs to the applications, and the mId, to exchange information between SCLs.

As a mnemonic for the names of these interfaces, the reader can notice that "*mI*[x]" stands for *m*ain *I*nterface (with relation to the network domain side) and "*dI*[x]" for *d*evice *I*nterface (with respect to the device and gateway domain side), then x = *a* (in mI*a* and dI*a*) when the interface is for the communication with an *a*pplication and x = *d* (in mI*d* and dI*d*) when the interface is for the communication with an SCL in the *d*evice domain side.

The driving idea was to specify a unified way to leverage these "SCs" from the application side. How to structure and address the SCL in a flexible way was guided by the RESTful style [9] choice, corresponding to an IT-driven trend that fitted well for the ETSI M2M goal of storing and sharing data across different applications: a REST-based API does not require the client to know anything about the structure of the API; it lets the server to provide whatever information the client needs to interact with the service. All entities (e.g., SCLs, applications, and associated data such as access rights or subscriptions) are represented as resources that are uniquely addressable via their URI (Uniform Resource Identifier).

The ETSI M2M standard organizes these resources into a tree structure, shown in Figure 2.6, with a <sclBase> root, which represents the instantiation of a SCL on the network domain, on the gateway domain, or on the device domain side. The collections that stand under the root each contain an attribute that in turn contains all the references to the children. This allows the URLs of the children to be retrieved using partial addressing. A collection resource that can contain multiple types of child resources shall have such an attribute per type of child.

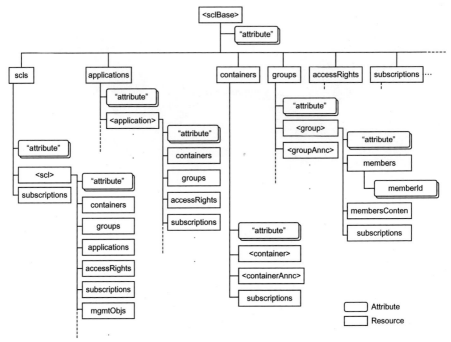

Figure 2.6 ETSI M2M resource tree presented at ETSI M2M Workshop 2011.

The main resource tree contains the following collections:

– scls, which consists in the collection of Remote SCLs that have registered and of SCLs that the < sclbase > is registered to. For example, when considering an SCL on a gateway (noted GSCL), the scls collection under the < gsclbase > of this gateway will contain the list of NSCLs (SCLs implemented on the network domain side) in connection with this GSCL.
– applications, for local applications that have registered with the < sclbase >.
– containers, for the list of the containers created by local or remote entities for the storage of data.
– groups, for the collection of groups that are created and do not have a containment relation with a specific entity.
– accessRights, as the collection of accessRights that are created and which do not have a containment relation with a specific entity.
– subscriptions, as the collection of 0–*n* < subscription > resources that represent an individual subscription.

Also, the discovery resource is included under the < sclbase > to enable a discovery on < sclbase >.

ETSI M2M standard specifies that a resource representation shall be an XML document with the name of the resource type as its root element. Each resource type shall have a representation that contains all its attributes as a sequence of XML elements. As for the content type, it shall use the XML, JSON, EXI, or Fast Infoset format.

Operations on a resource, based on the four "CRUD" verbs (create, retrieve, update, and delete) defined by the RESTful style, are supported by means of primitives that constitute the communication on the mIa, dIa, and mId APIs specified in ETSI M2M TS 102 921. Then, the ETSI M2M standard specifies the mapping of these primitives to the well-known http (Hypertext Transport Protocol) (Normative Annex C of *ETSI M2M TS 102 921*) for the transmission and to the CoAP (Constrained Application Protocol) (Normative Annex D of *ETSI M2M TS 102 921*) particularly fitted for use by devices having, for example, memory and CPU (central processing unit) limitations.

Figure 2.7 recalls the different possible implementation scenarios expected from the ETSI M2M standard, which can be leveraged by any M2M applicative domain.

The M2M service layer standardization is now in the hands of oneM2M standardization partnership, presented in Section 2.1, which is exploiting the same REST-based concept as ETSI M2M standard and is specifying the same kind of interfaces between the service layer and the applications. Yet, the nomenclature in oneM2M standard varies from the one in ETSI M2M, with the mIa and dIa interfaces being now called Mca (for M2M communication with applications), mId being replaced by Mcc (for M2M communication between common service entities (CSEs)), and CSEs replacing the SCLs. oneM2M additionally introduces the Mcn reference point that designates communication flows between a CSE and the underlying network services entity (NSE). This reference point enables a CSE to use the services (other than transport and connectivity services) provided by the underlying NSE, for example, 3GPP MTC (machine-type communication). At the date of March 2014, oneM2M

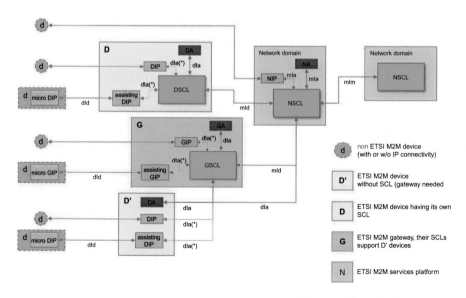

Figure 2.7 Possible deployment scenarios from ETSI M2M standard.
Source: ETSI.

Reference	Title	Global scope	Status
TR-0001	oneM2M Use Cases collection	M2M use cases from various segments of the industry, in order to anticipate the requirements for oneM2M system	✓ Oct2013
TS-0002	oneM2M Requirements	Specification of the M2M requirements that structure the oneM2M solution	✓ Oct2013
TR-0002	Part1: Analysis of the architectures proposed for consideration by oneM2M	Synthetic description of the architectures transfered to oneM2M by the founding SDOs (oneM2M Partners Type 1), to ensure a common understanding of existing M2M architectural approaches.	✓ Aug2013
TR-0003	Part2: Study for the merging of architectures proposed for consideration by oneM2M	Evaluation of existing M2M-related Architecture work: comparison of functional entities, of reference points, and very global clues on architecture styles	✓ Aug2013
TR-0006	Study of management capability enablement technologies for oneM2M consideration	Describe the state-of-the-art technologies (e.g., OMA DM, BBF TR069,...) relevant to oneM2M management capabilities. Analyze these technologies to match with the oneM2M requirements on management aspects.	✓ Dec2013

Figure 2.8 Published oneM2M deliverables as of March 2014.

standard is ongoing, with some first deliverables already published, as shown in Figure 2.8.

Next expected key specifications are oneM2M Architecture TS-0001 and oneM2M Core Protocols TS-0004, announced for autumn of 2014.

2.4 Using M2M standards for "vertical" domains, the example of the smart home

The M2M architecture described in the previous section is designed as a transverse architecture, with a service layer approach that is generic enough to be used by any kind of M2M applications. Then, for real implementation, it was felt necessary to take some distance with the objective of a generic standard and to come back to particular application domain constraint reality. This section is dedicated to ongoing standardization work in HGI [10] industrial organization targeting a smart home gateway-centric reference architecture that can federate the smart home application developers, manufacturers, operators, and service providers. The needs for an abstraction layer and for semantics in this particular smart home domain make it cowork with the ETSI M2M/oneM2M standardization bodies.

The previous section demonstrated how the M2M service layer standard elaborated initially by ETSI M2M and now by oneM2M, fitted to any M2M applicative domain. In particular, M2M enables a diversity of promising services in home surroundings.

Some exemplary applications include smart grid (demand and response use cases interacting with energy-consuming home appliances), healthcare, home automation, and security (e.g., smoke detection or intrusion detection use cases), which involve different types of actors, yet requiring some similar types of functionalities

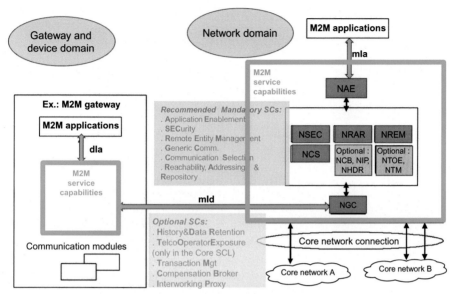

Figure 2.9 Common framework of functionalities proposed to the different M2M application providers.

(authentication, identification, access rights, device management, store and forward data, charging records, etc.) for running their applications.

Figure 2.9 recalls some examples of functionalities (named "service capabilities" in ETSI M2M or "common service functions" in oneM2M terminology) from the M2M standard that can be exploited by various M2M application providers, as most of these functionalities are out of their core business scope and application providers can then focus on their core development. In this figure, the SCs are listed on the network domain side, which gives acronyms with an "N" prefix to designate these SCs on the NSCL (NSEC, NRAR, etc.). The same SCs can be found on the gateway SCL with a G prefix instead of the N (GSEC, GRAR, etc.) or on the device SCL (DSEC, DRAR, etc.).

When looking at the smart home example, to allow for applications to be shared among different service providers, a minimum set of functionalities are foreseen to be provided via standardized APIs, such as the following:

- Authentication
- Access rights
- Application subscription to some events from devices at home
- Notification to applications when events occur
- Information on what software configuration is needed for correct running of the application (s) in the smart home environment
- Provisioning hardware and applications

These are typically a subset of the generic SCs provided by the ETSI M2M framework shown just above. When looking at how this can be instantiated on a home gateway

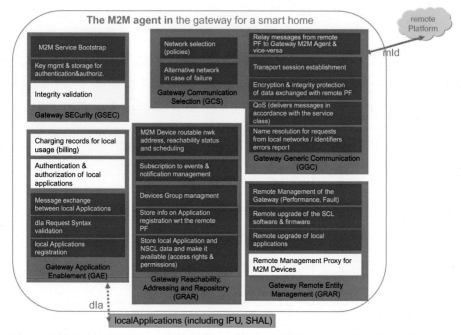

Figure 2.10 Examples of some ETSI M2M service capabilities that could be instantiated on a home gateway GSCL.

service layer, some minimum capabilities from ETSI M2M can be particularly noted as proposed in Figure 2.10 (functionalities considered as essential for the smart home case are written in the blue boxes, whereas functionalities with the lowest priority for the smart home are put in white boxes). This instantiation is called "M2M Agent" in this figure.

The use of ETSI M2M local API (dIa) allows legacy local networks to access the M2M agent (in a way that is further described later in this section), while the use of ETSI M2M API out from the HG toward the cloud (mId) allows us to address third-party applications through the mIa cloud API.

By looking even more concretely at the global smart home service architecture, depicted in Figure 2.11, it is composed of three main parts:

1. The "in-home" part, inside the house, deals with local applications, embedded in devices and/or a gateway; this is where we can find the device manufacturers and home automation specialists (e.g., for monitoring the shutters or for monitoring the lighting) providing ad hoc smart home subsystems.
2. The border part between the house and the external world that may consist in interfacing a (smart) home gateway with a remote service platform.
3. The "cloud" part, where remote applications can target home devices by interfacing with the service platform that will forward the remote application parameters up to the addressed home devices.

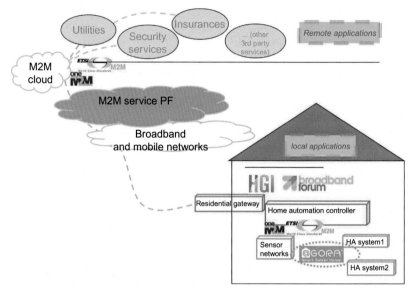

Figure 2.11 Smart home overall architecture.

While some standards are already quoted on this figure, they are not intended to be the only ones that the smart home domain can look at. Rather, it underlines which organizations are today particularly seeking for federating discussions around generic tools for the smart home that are independent from the executive environment and from connectivity technologies chosen for the deployment of smart home services.

From the smart home perspective, the in-home part has specific requirements when considering the presence of possible resource-limited devices and also in terms of interoperability, which is particularly important to gain confidence of the end user who would be more keen to invest in sustainable solutions with a warranty of having them evolved in accordance to the evolution of his own future needs.

A variety of devices present in the house are accompanied by a variety of wireless connectivity technologies that are standardized or proprietary. In order to enhance the smart home experience for the end user, HGI realized that there was a need to at least unify, for example, some installation and pairing mechanisms among these various in-home wireless technologies. HGI realized that the most efficient way to promote the minimum set of requirements that could be expected from any of these technologies would be to elaborate a requirement specification that would be circulated among these preidentified technology alliances and get their feedback about how their technology could meet these requirements or not. HGI has published this specification [11] and intends to use the feedbacks from the different alliances to make references to those technologies that best fit the smart home requirements cataloged in RD-039.

Then, with the vision to federate the smart home application developers, service providers, gateway operators, gateway manufacturers, and device manufacturers around concrete enablers toward an efficient and interoperable way of providing smart home applications to the end user, HGI and BBF are coworking to a device template, which intends to provide a tool to derive coherent device data models. The HGI GWD-042 guideline working document aims at providing a means to specify the devices, their operations, and eventually a common template enabling all smart home stakeholders to share common definitions of the structure and meaning of data. In particular, main objectives for delivering the GWD-042 document are to

– explain the type of information considered useful for software developers and end users to be able to access, for all kinds of smart home use cases, appliances, and devices;
– describe a worldwide process of collecting and modeling information, which would facilitate smart home deployments;
– discuss specific formats ("templates") to facilitate the collection process.

This technical work is expected to be taken into account by the global initiative initiated by the European Commission DG CONNECT on specifying semantics enablers for interoperability of smart appliances.

Another approach in federating smart home actors is to address the services interoperability issue in a house by getting the different owners of home automation subsystems that agree on a common way to exchange events among all their subsystems via a "homebus," which stands as a logical bus on which abstracted events from any of the home automation subsystems can circulate and be used by any other HA subsystem connected to this homebus. In France, this approach is the one undertaken by the "AGORA des réseaux domiciliaires," presented to the workshop organized by the EC DG CONNECT [12]. Similar approaches were chosen in other European countries (e.g., Energy@home in Italy and EEBus in Germany) in which the European Commission DG CONNECT wants to leverage toward a convergent technical solution that the TNO Institute (located in the Netherlands) is in charge of recommending with the support of all the stakeholders already involved in such developments. The context, objectives, and proposed way forward for this study are documented by TNO [13].

HGI also federates the smart home gateway actors around a smart home abstraction layer at the gateway level so that the application developer will not have to adapt to each specific technology for proposing its application to be implemented in the house infrastructure. In HGI smart home reference architecture requirements working document HGI RWD-036 (publication expected before the end of 2014), key reference points are identified, as shown in Figure 2.12, including these RPs that enable the application developer to access to this abstraction layer in a unified way.

Not only HGI but also BBF, ETSI M2M/oneM2M, and OSGi [14] are some of other SDOs targeting such an abstraction of local connectivity technologies and consequently involved in the HGI smart home reference architecture elaboration.

One typical example addressed as the first step for this abstraction layer is how a specific technology, for example, ZigBee, can be "abstracted," that is, how its commands and attributes can be mapped into a generic language, while another specific technology, for example, KNX, is also "abstracted" into this same generic language.

In its guideline to interworking with M2M area networks [15], ETSI M2M provides a possible way to do that, by exploiting the oBIX 1.1 (Open Building Information Xchange) language [16], which leverages the application/xml syntax. ETSI Technical

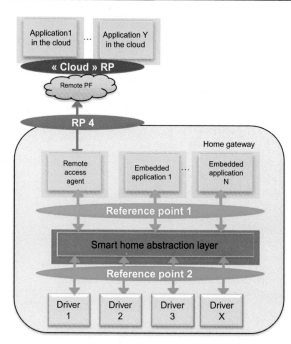

Figure 2.12 Smart home gateway-centric overview. Adapted from an HGI figure.

Report 102966 provides examples of mapping native ZigBee primitive types (Annex B.2.1 of this Technical Report), native wM-Bus primitive types and units (Annex B.3.1), and native KNX data point types (Annex B.4.1) to oBIX types and units. This mapping of multiple technology-specific semantics into one common first-level semantic convention is the first step toward getting the independence of applications with regard to the underlying area network hardware and technology.

Beyond this first level of semantic mapping, ETSI Technical Report 102966 also provides a way to discover the M2M area network structure, create an ETSI M2M resource structure representing the M2M area network structure in the ETSI M2M SCL, and manage the ETSI M2M resource structure in case the M2M area network structure changes. This is performed through the use of a dedicated application, namely, the interworking proxy unit (IPU). Figure 2.13 shows an example of how the IPU allows us to represent a ZigBee network into ETSI M2M resources that can then be manipulated through standardized ETSI M2M APIs.

A typical implementation of such an IPU would be on a home gateway GSCL (gateway service capability layer). Then, let's consider a very basic use case (not particularly interesting by itself but useful here to explain the approach), such as a lighting application in the cloud that would send a request to switch off some lights that are in the house, these lights being connected via ZigBee to the home gateway that embeds a GSCL. The request is sent from the cloud through the mIa interface to the NSCL on the remote platform where it is retargeted via the mId interface to this GSCL that has the representation of the on/off cluster of the ZB lights as resources.

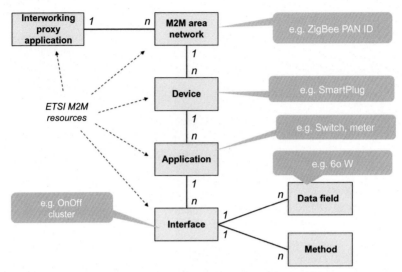

Figure 2.13 Example of ZigBee network representation into ETSI M2M resources. Adapted from ETSI Technical Report 102 966.

Here, the status of the lights is put to "off," which is mapped from the oBIX language back to the native ZB language (because the "tag" defined in Technical Report 102966 for the described semantic mapping would have indicated that the lights represented in this GSCL were using ZB technology), so that it results in a ZB command to the lamps to be switched off. In case some of the lights are connected via KNX technology to the same gateway, there will be a similar representation in this GSCL of these lamps, issued by a second IPU, with a "tag" mentioning the KNX technology this time; but the whole process would remain the same.

The Technical Report 102966 document additionally defines a profile of dIa that can be used when the support for a new M2M area network is added by means of a USB dongle, considered as a device with limited resources, which is plugged to the home gateway, for example. The mapping of the M2M area network (IPUs, M2M area networks, devices, applications, and interfaces) to the ETSI M2M resource architecture is described into two steps:

– An assisted step (in order to limit the external device resource requirements)
– A controlled step to define which REST resources may be created by the external device with little or no control of the M2M gateway (in order to enable extensibility, transparency to native protocols, and innovation) and which resources must be created by the M2M gateway itself (to remain under control of the M2M operator)

The dId interface was thus introduced in the Release 2 of the ETSI M2M standard (Normative Annex L of *ETSI M2M Technical Specification 102 921 Rel2*) to decrease the burden put on the resource-limited devices by defining the "assisting GIP" element, by reference to the classical gateway interworking proxy application defined in ETSI M2M Technical Report 102 966 to allow abstraction of M2M area network

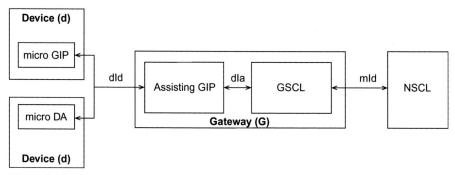

Figure 2.14 dId interface for limited resource devices.

technologies and mapping into the ETSI M2M resource structure. The "assisting GIP" takes care of mapping the data from the constrained device into the full standardized format of dIa. Figure 2.14, from Annex E of the Technical Report on Interworking with M2M area networks (*ETSI M2M Technical Report 102 966*), illustrates the approach. This enables, for example, the deployment of applications (micro DA—device application) that would be embedded in the dongle (device d) connected to a gateway we just mentioned before.

Another concept when considering a constraint device is to allow for a simplified use of the dIa interface through its extension to NSCL allowing access of M2M standardized services to devices not supporting DSCL nor GSCLs. This could be the case of constrained devices that cannot implement the full ETSI M2M protocol stack. It is also suggested to reduce the verbosity of responses to resource retrieval requests and to shorten the length of URIs and messages (Annexes M and N of *Technical Specification 102 921 Rel2*).

2.5 Conclusions and future trends for M2M standardization

The horizontal architecture approach proposed by ETSI M2M and then by the oneM2M standard is particularly relevant for cross domain interactions. The ETSI M2M "store and share" resource-based paradigm-standardized framework makes data available by the platform to any other applications through a service layer that provides sufficient abstraction to let the framework be independent from the underlying connectivity technologies and from devices. Uniformed interfaces simplify implementation and enable interoperability.

The oneM2M partnership has been formed to develop one unique M2M service layer standard with the implication of worldwide standardization bodies complemented by the crucial support from "vertical" industrial or SDOs. This allows us to ensure that the real-world needs are taken into account in such a standardization effort and that the resulting standard is useful for the industries that wish to implement it.

The key advantage in using such a common standardized framework resides in the possible sharing of some data with other actors enabling an expected evolution toward enriched applicative offers to the end users, with interoperable devices in terms of applicative capabilities, which will hopefully encourage the M2M (and also IoT) market to takeoff.

This chapter also showed the example of the smart home "vertical" looking at how to leverage such a M2M framework for its particular use cases. This led to push the ideas toward a need for a communalized way of structuring data and of correctly interpreting their meaning. This drives the standardization efforts into the semantics and ontology worlds, which are so needed by anybody who wants to benefit from cross application synergies. This was also the direction foreseen by the European Commission DG CONNECT, leading to the study to be available from TNO, with the support from all appropriate stakeholders, expected in the beginning of 2015, from which ETSI Smart M2M will be able to make a standard for the smart appliance ontology and for the mapping to an even more conceptualized data model (meta-data model) from the M2M service layer framework.

Indeed, some use cases and devices may be considered from different application domains that imply more vocabulary to be understood by the devices involved in these scenarios. This vocabulary and the consideration of relationship between entities involved are forming what we call semantics. Both concepts of abstraction and of semantic support are distinguished in Figure 2.15. These constitute fundamental elements to enable application developers to focus on the application development without prestudying each particular technology that the device running its application will use.

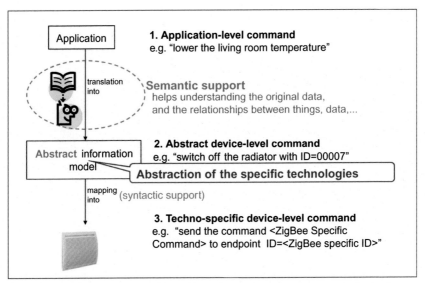

Figure 2.15 Abstraction of technologies and semantics support concepts as the basis for an interoperable M2M system open to applications developers.

It is thus anticipated that to go further in the interoperability of M2M/IoT applications, the M2M standardization activity will go more deeply into integrating semantics and ontology approaches, also considering some first directions from research projects that anticipated the need for interoperability among ontologies. oneM2M is already looking at these aspects via its working group 5 that has a dedicated item on "abstraction and semantics," in contact with, among others, the European Cluster IERC AC4 [17] on semantic interoperability.

The M2M standardization should also go toward concrete implementation, which would imply the elaboration of testing plans as a basis to prepare certification of M2M standard compliant systems. Interoperability test events are becoming a must in this multi-actor activity to guarantee the industry the value of such a horizontal standardized framework.

References

[1] ETSI M2M Workshop. All presentations available from: http://www.etsi.org/index.php/news-events/events/542-2nd-etsi-tc-m2m-workshop-from-standards-to-implementation for M2M Workshop 2011 (presentations of the subsequent ETSI M2M Workshops are also available from: http://www.etsi.org/news-events/news/609-2012-m2mworkshop for M2M Workshop 2012; and http://www.etsi.org/news-events/news/10-news-events/events/666-2013-m2mworkshop for M2M Workshop 2013).

[2] ETSI Technical Specification 102 689 Release2 M2M Service Requirements, 2013.

[3] ETSI M2M Standard Approved Technical Specifications (and Approved Technical Reports). Available from: http://www.etsi.org/technologies-clusters/technologies/m2m – Standards folder.

[4] ETSI Technical Specification 102 690 Release2 M2M Functional, Architecture, 2013.

[5] ETSI Technical Specification 102 921 Release2 mIa, dIa and mId, interfaces, 2013.

[6] BBF Broadband Forum webpage, Accessible from: http://www.broadband-forum.org/.

[7] Continua Health Alliance webpage, Accessible from: http://www.continuaalliance.org/.

[8] OMA webpage, Accessible from: http://openmobilealliance.org/.

[9] R.Th. Fielding, Architectural styles and the design of network-based software architectures, University of California, Dissertation for the Degree of Doctor of Philosophy. Available from: https://www.ics.uci.edu/~fielding/pubs/dissertation/fielding_dissertation.pdf, 2000 (accessed 31.03.14).

[10] HGI webpage, Accessible from: http://www.homegatewayinitiative.org/.

[11] HGI RD-039 Requirements for Smart Home Services Wireless Home Area Networks (WHANs), 2014. Available from: http://www.homegatewayinitiative.org/documents/Current_HGI_Publications.asp.

[12] ICT4SP Event organized by IDEAS Project and EC DG-CONNECT (10th September 2013), 4th Workshop on EEBuildings Data Models, Nice. Presentations available from: http://www.resilient-project.eu/en/web/guest/ict4sp-presentations.

[13] TNO Study on "Available semantics assets for the interoperability of smart appliances. Mapping into a common ontology as a M2M application layer semantics," requested by the European Commission DG Communications Networks, Content and Technology SMART 2013/0077 2013. Available from: https://www.tno.nl/downloads/Extended_Summary_Smart_Appliances_EU_2013-0077.pdf.

[14] OSGi Alliance webpage, Accessible from: http://www.osgi.org/Main/HomePage.

[15] ETSI Technical Report 102 966 machine-to-machine communications (M2M); Interworking between the M2M Architecture and M2M Area Network Technologies, 2014.

[16] OASIS oBIX webpage, Available from: http://www.obix.org/.

[17] IERC – Internet of Things European Research Cluster www.internet-of-things-research.eu.

Overview of 3GPP machine-type communication standardization

P. Bhat[1], M. Dohler[2,3]
[1]Vodafone Group Services Ltd, Newbury, UK; [2]King's College London (KCL), London, UK; [3]Worldsensing, Barcelona, Spain

3.1 Introduction

The emerging Internet of Things (IoT) is envisaged to connect billions of devices over the next decade [1]. A solution that is of low cost, high coverage, and low power would be fundamental as a solution targeted for applications for IoT for majority of use cases. Strong industry alignment on technology standards and protocols will drive greater economies of scale and ease end-to-end integration and interoperability.

Machine-to-machine (M2M) is a huge segment of the overall IoT ecosystem. Notably, it enables the connectivity between machines in an autonomous, reliable, and cost-efficient manner. Only a ubiquitous connectivity of objects, similar to the connectivity of computers in the Internet, would enable the Internet paradigm in the IoT. Traditionally, the said connectivity has been achieved by means of wires, mainly in the context of industrial automation and control. Wireless alternatives have emerged over the past decade, which offer notable business benefits, while also suffering shortcomings.

The choice of wired versus wireless solutions, however, is heavily debated in the M2M industry. Wireless solutions are attractive in comparison to wired solutions for their relative ease of deployment and robustness against single point of failures, while wired solutions can ensure coverage in difficult to reach locations for radio. The main challenge for wireless solutions for M2M applications in comparison to wired solutions is power consumption and cost of the radio. Different M2M services have different requirements, and a single standardized solution for all the requirement for mobility, power consumption, range, ease of roll out, etc., is key to successful adoption ensuring cost target for mass market M2M solution if is to be achieved. Solutions for isolated clustered deployment are based on wireless personal area networks (WPANs), for example, Bluetooth, ZigBee, and 6LoWPAN. Solutions based on cellular technologies such as those developed by 3GPP are more suitable for M2M services requiring mobility and a communication wider range.

A further challenge is to cater for the critical link between the M2M network and the core enterprise systems through an API, a standards-based interface that allows real-time data exchange. Its standardization will pave the way for real-time interfaces that are lightweight, high scale, and continuously available. The industry is currently developing standards for harmonized APIs to enable the value chain to be realized.

Machine-to-machine (M2M) Communications. http://dx.doi.org/10.1016/B978-1-78242-102-3.00003-4

All of the above characteristics are fundamental for delivering an optimal solution for M2M communications. Cellular MTC (the working term for M2M in 3GPP) solutions are really well positioned to meet these characteristics. It is thus to no surprise that 3GPP, through its industry partners, started studying and standardizing specific technical machine-type communications (MTC) embodiments.

This chapter details these developments in great depth by first outlining the advantages, challenges, and suitability of cellular 3GPP-enabled MTC and then discussing specific 3GPP developments in the subsequent section. The chapter is concluded by discussing possible future avenues and developments for 5G and beyond.

3.2 Pros and cons of M2M over cellular

This section outlines in detail the trade-offs one encounters when utilizing a cellular M2M connectivity solution. Notably, we discuss the advantages of using cellular, the challenges that need to be addressed, and how suitable prior and current generations of cellular are to meet these challenges.

3.2.1 Advantages of using cellular M2M

Using cellular systems to provide M2M connectivity enjoys numerous advantages. Notably, ubiquitous coverage, roaming support, guaranteed interoperability, quality of service (QoS) guarantees, ability to offer service-level agreements (SLAs), and the availability of operational service platforms make cellular a technically and commercially competitive M2M solution. A more detailed discussion on each item follows.

One of the biggest worries of an M2M solution provider, such as company the Worldsensing, is to guarantee reliable and robust connectivity, even for static deployments. Compared to ZigBee, Wi-Fi, or low-power wide area networks, cellular enjoys virtually ubiquitous coverage globally, even in the most remote spots one can think of. Therefore, there is no need to install additional gateways or repeaters, which significantly lowers the deployment barrier of entry and the running costs. Overall, we found that the total cost of ownership, despite the higher modem costs and data plans, comes more cost-efficient than competing connectivity solutions.

Another enormous advantage of cellular M2M is the ability to support mobility and roaming. This immediately expands the market to, for example, automotive and tracking applications. There is currently no other connectivity technology available that yields such a coverage and mobility support as cellular. This gives enormous peace of mind to M2M solution providers.

From a business point of view, another big advantage is the availability of SLAs in the licensed cellular spectrum bands. Regulations currently limit the ability to offer SLAs in license-exempt ISM bands, which significantly limits business-to-business M2M uptake. To exemplify this, imagine a critical infrastructure is instrumented and a serious accident happens, the buildup to which is not reported because the wireless connectivity was in outage. Using ISM-band technologies leaves all parties in a legal limbo since the M2M solution provider has little influence over connectivity but is not able to leverage the fault on to the modem provider. This is very different in licensed cellular bands where SLAs can be undersigned and enforced, thus providing additional peace of mind to M2M solution providers.

Yet another strong business aspect is the availability of horizontal service platforms, such as those offered by telco giants like Vodafone. Notably, these platforms lower the barrier of entry into markets where such platforms are already used in a different context. To exemplify this, imagine a city hall having used Vodafone's service platform for a long time to provide IT connectivity and services in the city; the barrier of entry to provide additional services such as M2M/IoT applications is much lower if the same service platform can be used/reused.

3.2.2 Challenges to facilitate cellular M2M

However, several challenges remain to date (Q3 2014), which would guarantee a true uptake of M2M cellular solutions. Notable challenges pertain to the ability to uniquely identify an M2M device, optimize the system for small and infrequent data transmissions, be able to provide additional coverage in very challenging propagation environments, handle congestion in access and core networks, lower complexity and energy consumption, and facilitate viable service exposure. These challenges in device, network, and service provisioning design are discussed now in some more details.

Challenges related to the cellular *M2M device* design pertain to (i) energy consumption, (ii) modem complexity, and (iii) modem cost. All these three are intertwined where, for example, a reduction in complexity typically comes along with a decrease in cost. Processing energy can be significantly reduced if the device complexity is reduced. Furthermore, given that transmission powers of cellular devices are very high compared to the low-power ZigBee/Wi-Fi devices, energy consumption related to the actual wireless transmission can only be reduced if the system allows a significant reduction in the time it takes to deliver the data over the air interface.

As for the M2M *network design*, the issues to be addressed pertain to communication protocols able to handle communications efficiently and above all address congestion in the access; other important issues are the ones addressing and provisioning coverage and mobility. Indeed, communication protocols are highly inefficient today when it comes to the support of M2M traffic since connection setup takes too long to facilitate energy-efficient operations. Also, current physical random-access channel protocols are not able to cope with a large amount of devices accessing the network simultaneously. Furthermore, there is a requirement to optimize the system for small and infrequent data transmission: a device that has only a small amount of data, if it can quickly transmit or receive its data and then become dormant, would reduce interference to other devices and would conserve its battery. As for addressing, the device will need to be identified and addressed; existing addressing schemes may not be adequate for supporting a large number of devices.

3.2.3   Suitability of current cellular solutions

Cellular networks have evolved from Total Access Communication System (TACS) of the early 1980s with voice as primary service to GSM/EGPRS of the early 1990s with voice and SMS as primary service to UMTS/HSPA of mid-2000 for voice and data. Early 2010s saw LTE being deployed with some deployments of its evolution in subsequent years by early adopter operators. Evolution from each of the generations from the first-generation TACS to the fourth-generation (4G) LTE can be approximated at 10 years, and deployment of fifth-generation cellular systems is envisaged for the 2020s.

A key constituent for said cellular system is spectrum, considered to be a valuable resource, and therefore, one of the key considerations for technology evolution is spectrum efficiency. Spectrum efficiency of 4G has improved 20-fold compared to that achieved with first-generation systems. 4G LTE system can be deployed in six different bandwidth configurations ranging from 1.4 to 20 MHz. This flexibility is key to better utilization of spectrum where the spectrum is fragmented.

Latency and user throughput are important attributes for user experience. 4G LTE systems have been optimized for mobile broadband delivery with end-to-end latency of the order of few tens of milliseconds and average user data rate of few ten's of Mbps. High performance of 4G system for broadband data comes at increased device complexity, increased power consumption, a not so mature ecosystem, and a roaming challenge due to limitation of device support for fragmented deployment with a large number of bands globally.

In terms of suitability of technology to support the stringent requirements of M2M, the 2G technology family, that is, GSM and GPRS/EGPRS, is ideal as power consumption and cost are low, coverage is global, and the ecosystem is developed; however, from an economic point of view, it is perceived to be much more viable to refarm the bands for next-generation systems. The 3G family, that is, UMTS and HSPA, is a less desirable technology because power efficiency is inferior and the cost of the modem too; overall, it is a technology, which is an "overkill" in terms of design since it provides much more than needed. 4G technologies, that is, LTE and LTE-A, are interesting again since the air interface, OFDM(A), allows the scaling of the bandwidth according to needs; modem cost and global coverage are an issue however.

In summary, the economics for the majority of MTC applications rely on lower cost by leveraging on economies of scale, and current 2G systems are capable of addressing the needs. The argument for use of 4G LTE system for MTC application is to benefit from improved spectral efficiency and the bandwidth flexibility offered by 4G systems and longevity of the technology as a future cellular system. As said above, high data rate and low latency are however not key attributes for the majority of MTC applications. The challenge is in adapting a 4G LTE system that was specifically designed for efficient broadband communication to also deliver and support MTC applications.

Recent efforts in 3GPP and other industry forums have been to address these and other challenges for a system optimal for MTC applications. These are now discussed in the subsequent section.

3.3 MTC standardization in 3GPP

In 3GPP, M2M is referred to as machine-type communication (MTC); and we will be using the terms M2M and MTC interchangeably. Subsequently, we will outline the technical requirements—from a 3GPP point of view—and then discuss 3GPP releases on MTC along with the proposed solutions.

3.3.1 Technical requirements

3GPP radio technologies for packet data are widely deployed and can be broadly categorized based on physical layer characteristics as GPRS using TDMA, HSPA using W-CDMA, and LTE system using OFDMA. Enhancements to each of these technologies are staged in different releases in 3GPP. An important aspect contributing to the success of 3GPP technologies for cellular use is maintaining backward compatibility with legacy releases and tight interworking between technologies and efficient roaming support, which is also key to M2M applications that require support for mobility.

As an established technology that is widely deployed, 3GPP technologies have a cost advantage in providing new services with reduced rollout costs with reuse of existing cellular sites, radio equipment, and spectrum. However, 3GPP cellular systems were not specifically designed for machines; they were primarily designed for human interactions and are not optimized for all M2M applications specifically for greenfield deployments. Ubiquitous coverage for short messages, terminal cost, address space limitations, IPR costs, possibilities of phasing out of technologies, and terminal power consumption are some of the challenges for M2M facing operators. The need for maintaining backward compatibility also presents challenges in optimizing the system for low-end M2M applications. Unlike mobile handsets, M2M devices have a longer replacement timeline and may pose challenges to technology migration.

Identifying a specific MTC device, enablement of mass adoption of M2M communication with increased coverage and lower cost, and service enablement are some of the key aspects addressed by 3GPP for standardization of MTC communication.

3GPP specification work is broadly grouped into RAN, SA, GERAN, and CT, and these groups are responsible for defining functions, requirements, and interfaces of 3GPP systems. RAN is specifically responsible for radio access part of 3G and its evolution to 4G and beyond, while GERAN is responsible for radio access part of 2G and its evolution. SA is responsible for the overall architecture and service capability, while CT is responsible for specification of terminal interfaces and capabilities and the core network part of 3GPP systems. 3GPP features are phased into releases and the work may be preceded by a study. Typically, all new features are preceded by a study phase.

3.3.1.1 The need for MTC user identification

There are two important aspects to an MTC module, the SIM and the MTC device. For the majority of the solutions worked by 3GPP, it is essential to identify the MTC user. The majority of low-cost MTC subscriptions are envisaged to have the SIM integrated during surface mounting on PCB or have soldered SIM chip; there are also MTC devices where the SIM is not integrated with the device. For the latter case, it is essential for operators to be able to individually regulate access using SIM profile or the device capability. It may also be of interest to detect a change of the device IMEI in case of SIM card misappropriation.

Rate policing, where usage restriction is applied when cap on usage is reached, would also require identification of MTC users. Throttling of data speed is already used by operators as a rate control policy for normal cellular operation, and this can be seen as a variation of the same. One challenge this may present is due to the wide range of MTC applications, for example, usage restriction for low data rate, but a large volume cannot utilize volume-based restriction of services. Further, the SIM (when not integrated with the device) could be swapped with a non-MTC-specific subscription.

The SIM card contains the IMSI of the subscriber with association in the HLR for the subscriber profile, which includes details about subscribed services and feature profile. Operators are already able to support customized MTC services based on the subscription profile, such as optimal data packet size and optimal routing with dedicated Access Point Name for MTC services. The IMSI, specific charging policy for MTC subscription, is provisioned by the operator, and the operator has complete control over the subscriber that is allowed in the network.

3GPP-based cellular devices are identified by IMEI. For LTE, the LTE user equipment (UE) category information is provided to the base station. The eNodeB is then able to determine the performance of the UE and communicate with it accordingly. User experience w.r.t. peak data rate and system performance w.r.t. spectrum efficiency are related also to device category. Devices, not requiring to support high data rate and/or low latency, will require to be specified with a new MTC-specific LTE device class/ category. It is envisaged that 3GPP will be defining one or more new LTE UE categories for MTC. This will be one of the means to identify and isolate MTC devices if it is impacting performance of the network and be able to restrict access for MTC devices. One of the concerns operators share is restricting access to roaming devices. Notably, the operator should be able to identify such roaming MTC devices from MTC-specific UE category and be able to restrict access to the devices if the operator do not wish to service those devices.

3.3.1.2   The need for coverage improvement

There are numerous MTC applications. Some of the MTC applications require extensive coverage. An MTC user who moves around is unlikely to be out of coverage for long. On the other hand, there are many M2M applications where terminals will be fixed but have no access to a fixed line. MTC applications such as smart metering and parking meter sensors not only can have challenging deployments but also are not mobile. There is a substantial market for a service that can provide much more nearly ubiquitous connectivity. The most obvious application today is utility metering. Even if almost all premises have outdoor coverage, many of them do not have coverage where the electricity meter is located, for example, in basements. Increasing the number of base stations is great for coverage and capacity improvement but with each additional base station are the challenges of backhaul, site acquisition, rental, power supply, etc. The goal of 100% coverage will never be achieved, but there is a need to reach the small number of special cases without adding significantly to the total cost of the complete solution. Hence, it is important to ensure system cost is not increased in improving coverage for MTC devices. 3GPP is specifying low complexity and improved coverage M2M modules to facilitate mass deployment of M2M modules.

A 3GPP study identified several features that are not required for MTC devices and could reduce device complexity significantly. 3GPP identified limiting device capability to single receive RF, restricting supported peak data rate and reducing supported data bandwidth and support of half-duplex operation as key to reducing device complexity. This simplification while reducing device complexity does introduce additional specification impact. Maintaining system performance with normal LTE devices with additional scheduler restrictions to serve these low complexity devices is a fine balance.

Coverage improvement is achieved primarily by repetition of information. 3GPP Release 12 stage 3 functional freeze is planned June 2014. Considering the timeline, 3GPP has identified some of the complexity reductions for specification in ongoing Release 12, while the remaining are for consideration in Release 13.

3.3.1.3 Service exposure and enablement support

3GPP is standardizing support for third parties to interact with the 3GPP system to use 3GPP functions in order to provide third-party services to their customers. Standardization work related to M2M service enablement is ongoing in standardization organizations outside 3GPP (e.g., ETSI TC M2M and the oneM2M Global Initiative; see the previous chapter) under the assumption that M2M service enablement can be offered by a network operator to third parties with business agreements with operators. 3GPP support for service exposure and enablement allows for the use of 3GPP capabilities beyond pure IP-based data transmission that can already be offered by 3GPP networks. Provisioning support for service enablement is being facilitated by provisioning additional information (e.g., transmission scheduling information or indications for small data and device triggering) and defining new interfaces between the 3GPP Core Network and application platforms. To ensure privacy, exposed network information ensures delinking of private user information, that is, no link with UE identity.

3.3.2 3GPP MTC-related releases

3GPP has adopted notion of releases to allow for a stable platform for implementation while facilitating introduction of new features. Tables 3.1 and 3.2 provide a list of high-level features and the release targeted.

Standardization work started as early as 2005 when 3GPP TSG SA1 (group defining services) started with a feasibility study to conclude with a report during 2007 as captured in 3GPP Technical Report (TR) 22.868. 3GPP Release 10 Technical Specification (TS) 22.368 specifies the machine-to-machine communication requirements.

Table 3.1 3GPP MTC-specific feature enhancements (core network and service architecture aspects)

Name	Release	Groups responsible
Service exposure and enablement support	Rel-13	S1
Service requirements maintenance for MTC	Rel-13	S1
Study on enhancements for infrastructure-based data communication between services	Rel-13	S1
Machine-type and other mobile data application communication enhancements	Rel-12	S2, S1, S3, C1, C3, C4, C6, R2, R3
Study on alternatives to E.164 for MTC	Rel-12	S1
Study on enhancements for MTC	Rel-12	S1
System improvements to MTC	Rel-11	S1, S2, S5, C1, C3, C4, C6, R2, R3, R5
Network improvements for MTC	Rel-10	S1, S2, S5, C1, C4, C6, G2, R2, R3, R5, G3new
Study on security aspects of remote provisioning and change of subscription for M2M equipment	Rel-9	S3, C6
Study on facilitating M2M communication in GSM and UMTS	Rel-8	S1

Table 3.2 3GPP MTC-specific feature enhancements (radio access network aspects)

Name	Release	Groups responsible
Low-cost and enhanced coverage MTC UE for LTE	Rel-12	R1, R2, R4
Study on provision of low-cost MTC UEs based on LTE	Rel-12	R1, R2, R4
Study on RAN aspects of machine-type and other mobile data application communication enhancements	Rel-12	R2
Study on GERAN improvements for MTC	Rel-12	G1, G2
Study on RAN improvements for MTC	Rel-11	R2, R1, R3, R4

Refinement of requirements and logical analysis were captured in 3GPP Technical Report 23.888, and protocol implementation was staged for Release 11 and captured in respective TS of responsible working groups. Charging requirements were addressed by SA5 to reuse existing 3GPP functions (e.g., session initiation and control) to possible extent. 3GPP architecture work on MTC has started in Rel-10, and in Rel-12, SA2 worked on efficient transmission of small data transmissions and low power consumption UEs.

3GPP efforts have initially focussed on identification of machine-type devices to allow operators to selectively handle such devices in overload situations. Subsequent efforts have been on radio-level enhancements for complexity reduction, coverage improvement, reduction of UE power consumption, and optimization for handling small data.

3.3.3 MTC feature enhancements

Some of the key MTC feature enhancements are elaborated below.

3.3.3.1 Overload and congestion control at core network and RAN

3GPP identified possible issues with network overload when M2M devices have been deployed in a large number. M2M devices may cause signaling overload as they simultaneously recover triggering registration and other procedures causing increased signaling and core network overload. Specifically on network failure, there could be a sudden surge in signaling as MTC UE associated with the network reselects to alternate network (possibly roaming) simultaneously to the not yet failed network.

The selected network could be a roaming network, which is not dimensioned to serve all the roaming MTC users let alone simultaneous access. Even normal signaling from a large number of UEs may cause network overload irrespective if used for MTC or not. To perform overload and congestion control, 3GPP specified specific core network and RAN overload and congestion control features to protect the networks.

Rel-10 study item on network improvements for machine-type communications studied core network aspects of overload and congestion control. Rel-11 study item on RAN improvements for MTC was started Q4, 2009. A Rel-11 umbrella work item on system improvements to machine-type communication started in May 2010. This WI also addresses RAN overload control for UTRAN and E-UTRAN.

A mechanism to prevent and control such scenarios is performed by configuring MTC UE "low access priority" indicator identifying the devices as delay-tolerant devices, and this can be used by the network to control various procedures. In UTRAN or E-UTRAN, the RAN (RNC for UTRAN and eNB for E-UTRAN) would reject the connection request from the UE with an extended wait time of up to 30 min when overload is indicated by the core network to the RAN. Such UEs can be then be identified by the network to signal enable "extended wait timers" after rejecting access request from the UE when the UE will be subjected to longer back-off timers at overload. Access can also be barred for those UEs that indicate low access priority. Rejection of RRC connection establishments can be rejected by the MME in the E-UTRAN for UEs that access the network with low access priority. Longer tracking area update can also be allocated to reduce congestion.

In GERAN, the BSS enables implicit reject configurable for up to 200 s, and devices configured with "low access priority" will start an implicit reject timer preventing any further access for the duration of the timer allowing network to control the abnormal congestion. GERAN also supports RAN overload control by extending legacy mechanism of access control barring by excluding devices configured for extended access barring only.

3.3.3.2 Low-cost and enhanced coverage MTC UE for LTE

A study on provisioning of low-cost MTC UE based on LTE targeting complexity reduction of LTE modem for M2M applications was started in September 2011. The study identified various techniques for RF and baseband complexity reduction. The study of complexity reduction is aimed at MTC devices not requiring high data rates and/or low latency. With techniques for complexity reduction, some penalty with coverage reduction was observed, and the study was extended to study coverage improvement for low-cost MTC UEs.

As a conclusion of the study, 3GPP started specification work to specify a new LTE UE category with reduced complexity and improved coverage compared to normal LTE user equipment. Details of techniques are captured in 3GPP Technical Report 36.888. Options considered for complexity reduction for delay-tolerant MTC are mainly aimed at reducing the RF and baseband cost of the device. Simplifications considered for complexity reduction included reducing the maximum bandwidth for data to a narrower bandwidth of 3 MHz or 5 MHz, reducing the number of receive antenna to 1 in the device, which also restricts support of some MIMO modes and support of half-duplex operation, and reducing the peak data rate (to $\sim$1 Mbps) supported by the device. Coverage improvement for delay-tolerant MTC devices is achieved by with repetition and bundling of information.

3.3.3.3 Other enhancements

3GPP Release 12 is expected to be completed between end of 2014 and early 2015. MTC enhancements in 3GPP Release 12 among others will focus on below improvements that were not addressed in earlier releases.

To address optimizations for cluster of MTC devices, MTC group addressing is being considered when a network is able to broadcast messages to a group of MTC devices within a particular geographic area. Access control restriction or QoS policy would then be possible on the MTC group to set limits on data rate and to set a time limit for data transfer for each MTC device of the group until the restriction expires.

An enhancement to steering of roaming MTC devices is being specified and is performed primarily by using the low access priority indication provided by the UE. eNodeB steers UEs configured for low access priority to specific MMEs.

An enhancement to small data and triggering is primarily to enable efficient transmission of a small amount of data. For a normal LTE UE in RRC IDLE state, service request procedure would be used for transfer of data. For small data from MTC devices, this may cause significant overhead. Pre-established NAS security if used to transfer the IP packet as NAS signaling instead of establishing RRC security for UE in RRC IDLE state would result in a more optimal system for transferring a small amount of data. This reduces signaling overhead, enables efficient use of network resources, and reduces delay for reallocation.

Event monitoring is an important aspect of an M2M system. Configuration of monitoring events of MTC UEs requires activation of monitoring of all or a subset of events, detection of events, and reporting of events. Monitoring enhancements enable the detection of events reporting to the services capability server or application server.

3GPP acknowledges the need to address the UE power consumption, specifically for devices powered by battery. Solutions for UE power consumption optimization envisaged by 3GPP by extending DRX in idle mode and configuring long DRX cycles in connected mode are being studied.

3GPP Release 13 and future releases are expected to enable further support for efficient delivery of M2M services and mass deployments of M2M devices.

3.4 Concluding remarks

This chapter introduced some of the challenges faced by the cellular industry for mass adoption of M2M and the rationale for standardized solutions for M2M. We also provided insights into solutions that are currently being specified for efficient support of M2M communications. This chapter specifically explored important aspects of M2M-specific standardization activities in 3GPP for delivering standardized solutions optimized for cellular usage.

As for the future, M2M standards will continue to evolve and continue to challenge industries in the near term with multiple standards with each standard possibly optimal for a specific attribute of M2M communications (e.g., cost, coverage, mobility, and security). On a longer term, longevity of M2M devices and demand for reduction in cost for mass deployment will drive the industry in developing unified standards that will also require coexisting with the current ecosystem. Longevity of M2M devices will specifically demand a future communication network that is likely to be maintained for foreseeable future and will not be obsolete in the near term. A single radio access solution that is highly adaptable but optimized and standardized specifically for M2M application providing low complexity and low power utilization with very low infrastructure cost would need to be developed as a single standardized solution. For cellular operators, this should essentially also support good coexistence with their current communication infrastructure.

An interesting shift to observe is if the cellular MTC solution will qualify for industry control application; this would require the network to exhibit virtually zero outage and also support guaranteed end-to-end delays of a few milliseconds rather than hundreds of milliseconds. Related to the latter, this requires fundamental shifts in the access procedures to make sure that congestion is handled appropriately as well as the core network to make sure that no additional delays are incurred due to the particularity of the very hierarchical design in form of serving gateway (S-GW) and packet gateway (P-GW). To this end, the authors of reference [2] proposed an amended architecture as shown in Figure 3.1, which advocates for the use of femto-cells, that is, Home eNBs (HeNBs), to lower congestion, delay, and energy consumption w.r.t. to a pure macrocell deployment. The mandatory usage of a HeNB-GW in the backhaul ensures that data and signaling congestion does not occur in the core since data and control traffic can be bundled and throttled. Furthermore, it provides direct injection points of other M2M technologies, such as low-power Wi-Fi, without the need to go via the P-GW, which often resides on the other end of the country.

The benefits offered by using this example architecture can be summarized as follows [2]:

- *One Stream Control Transmission Protocol (SCTP) association between the HeNB-GW and MME.* In LTE-A, a SCTP association is created between two entities exchanging control plane traffic. Through the use of the HeNB-GW, the presence of a large number of HeNBs does not imply congestion at the MME. Indeed, only the HeNB-GW transmits SCTP *heartbeat* messages to the MME instead of each single HeNB. Furthermore, the number of SCTP association establishments and releases due to HeNBs switching on/off is minimized.
- *S-gateway scalability.* The number of GPRS Tunneling Protocol (GTP), UDP, and IP connections between HeNB-GW and S-GW is drastically reduced compared to a direct HeNB connection. In this way, the number of HeNBs may increase without an increase in the number of UDP/IP paths and GTP echo messages managed by the S-GW.
- *Paging optimization.* Optimized paging mechanisms for downlink data transmission to the managed HeNBs can be implemented within the HeNB-GW to reduce latency.

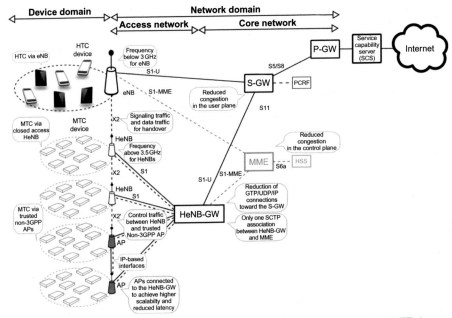

Figure 3.1 Enhanced network architecture for ultradense MTC access to the 3GPP LTE-A core via HeNBs/HeNB-GWs and trusted non-3GPP APs.

In the future, 3GPP may take up some suggestions put forward from the academic and industrial community, such as reference [2] discussed here or the large set of contributions available to date [3–56]. Overall, the market will show in the coming years if cellular MTC is a valid value proposition to the end users, i.e., heavy industries and the automotive sector, among others.

References

[1] L.M. Ericsson, More than 50 billion connected devices, Ericsson Review (2011) http://www.ericsson.com/res/docs/whitepapers/wp-50-billions.pdf.

[2] M. Condoluci 1, M. Dohler, G. Araniti, A. Molinaro, K. Zheng, Towards 5G DenseNets: architectural advances for effective machine-type communications over femtocells, Centre for Telecommunications Research (CTR) seminars, May 2014, http://www.ctr. kcl.ac.uk/seminars/slides/MASSIMO_CONDOLUCI.pdf.

[3] D. Astely, E. Dahlman, G. Fodor, S. Parkvall, J. Sachs, LTE release 12 and beyond, IEEE Commun. Mag. 51 (7 (July)) (2013), 154–160.

[4] V. Gonclves, P. Dobbelaere, Business Scenarios for Machine-to-Machine Mobile Applications, International Conference on Mobile Business and Ninth Global Mobility Roundtable (ICMB-GMR), June 2010, 2010, pp. 394–401.

[5] J. Alonso, M. Dohler, Machine-to-machine technologies & markets – shift of industries, In: Tutorial, IEEE WCNC, 6 April 2014, Istanbul, Turkey 2014.

[6] K. Zheng, F. Hu, W. Wang, W. Xiang, M. Dohler, Radio resource allocation in LTE-advanced cellular networks with M2M communications, IEEE Commun. Mag. 50 (7 (July)) (2012), 184–192.

[7] 3GPP Technical Specification 22.368, Service Requirements for Machine-Type Communications (MTC), Rel. 12, Dec. 2013. Available from: www.3gpp.org.

[8] 3GPP Technical Report 37.869, Study on Enhancements to Machine-Type Communications (MTC) and Other Mobile Data Applications; Radio Access Network (RAN) Aspects, Rel. 12, Sep. 2013. Available from: www.3gpp.org.

[9] ETSI Technical Specification 123 682, Architecture Enhancements to Facilitate Communications with Packet Data Networks and Applications, Sep. 2013.

[10] J.G. Andrews, H. Claussen, M. Dohler, S. Rangan, M.C. Reed, Femtocells: past, present, and future. IEEE J. Select. Areas Commun. 30 (3 (April)) (2012), 497–508.

[11] A. Laya, L. Alonso, J. Alonso-Zarate, Is the random access channel of LTE and LTE-A suitable for M2M communications? A survey of alternatives, IEEE Commun. Surv. Tut. 16 (1 (December)) (2013), 4–16.

[12] 3GPP Technical Report 38.868, RAN Improvements for Machine-type Communications, Rel. 11, Oct. 2011. Available from: www.3gpp.org.

[13] G. Wu, et al., M2M: from mobile to embedded internet, IEEE Commun. Mag. 49, (2011), 36–43.

[14] S. Andreev, et al., Efficient small data access for machine-type communications in LTE, in: Proceedings of the IEEE ICC 2013.

[15] S.-Y. Lien, et al., Cooperative access class barring for machine-to-machine communications, IEEE Trans. Wireless Commun. 11 (2012), 27–32.

[16] A. Gotsis, et al., M2M scheduling over LTE: challenges and new perspectives, IEEE Vehicular Tech. Mag. 7, (2012), 34–39.

[17] O. Dementev, et al., Analyzing the overload of 3GPP LTE system by diverse classes of connected-mode MTC devices, in: 2014 IEEE World Forum on Internet of Things (WF-IoT), 6–8 March 2014, 2014, pp. 309–312.

[18] X. Jian, et al., Beta/M/1 model for machine type communication, IEEE Commun. Lett. 17 (2013), 584–587.

[19] M.-Y. Cheng, et al., Overload control for machine-type communications in LTE-advanced system, IEEE Commun. Mag. 50 (2012), 38–45.

[20] M. Hasan, et al., Random access for machine-to-machine communication in LTE-advanced networks: issues and approaches, IEEE Commun. Mag. 51 (2013), 86–93.

[21] M. Gerasimenko, et al., Impact of machine-type communications on energy and delay performance of random access channel in LTE advanced, Trans. Emerging Tel. Tech. 24 (2013), 366–377.

[22] P. Jain, et al., Machine type communications in 3GPP systems, IEEE Commun. Mag. 50 (2012), 28–35.

[23] A. Ksentini, et al., Cellular-based machine-to-machine: overload control, IEEE Network 26 (2012), 54–60.

[24] T. Tirronen, et al., Machine-to-machine communication with LTE with reduced device energy consumption, Trans. Emerging Tel. Tech. 24 (4) (2013), 413–426.

[25] C. Antón-Haro, et al., Machine-to-machine: an emerging communication paradigm, Trans. Emerging Tel. Tech. 24 (4) (2013), 353–354.

[26] H. Thomsen, et al., Code-expanded radio access protocol for machine-to-machine communications, Trans. Emerging Tel. Tech. 24 (4) (2013), 355–365.

[27] M. Gerasimenko, et al., Impact of machine-type communications on energy and delay performance of random access channel in LTE-advanced, Trans. Emerging Tel. Tech. 24 (4) (2013), 366–377.

[28] A. Gotsis, et al., Analytical modelling and performance evaluation of realistic time-controlled M2M scheduling over LTE cellular networks, Trans. Emerging Tel. Tech. 24 (4) (2013), 378–388.

[29] C. Bockelmann, et al., Compressive sensing based multi-user detection for machine-to-machine communication, Trans. Emerging Tel. Tech. 24 (4) (2013), 389–400.

[30] A. Bartoli, et al., Energy-efficient physical layer packet authenticator for machine-to-machine networks, Trans. Emerging Tel. Tech. 24 (4) (2013), 401–412.

[31] S. Plass, et al., Machine-to-machine communications via airliners, Trans. Emerging Tel. Tech. 24 (4) (2013), 427–440.

[32] F. Cabral Pinto, et al., The business of things architecture, Trans. Emerging Tel. Tech. 24 (4) (2013), 441–452.

[33] L. Karim, et al., Fault tolerant, energy efficient and secure clustering scheme for mobile machine-to-machine communications. Trans. Emerging Tel. Tech. (2014). http://dx.doi.org/10.1002/ett.2801.

[34] B. Yang, et al., M2M access performance in LTE-A system. Trans. Emerging Tel. Tech. (2014). http://dx.doi.org/10.1002/ett.2746.

[35] M. Popović, et al., Evaluation of the UTRAN (HSPA) performance in different configurations in the presence of M2M and online gaming traffic. Trans. Emerging Tel. Tech. (2013). http://dx.doi.org/10.1002/ett.2738.

[36] C. Lai, et al., A novel group access authentication and key agreement protocol for machine-type communication. Trans. Emerging Tel. Tech. (2013). http://dx.doi.org/10.1002/ett.2635.

[37] A. Cimmino, et al., The role of small cell technology in future Smart City applications, Trans. Emerging Tel. Tech. 25 (1) (2014), 11–20.

[38] F. Bader, et al., Cognitive radio in emerging communications systems—small cells, machine-to-machine communications, TV white spaces and green radios, Trans. Emerging Tel. Tech. 24 (7–8) (2013), 633–635.

[39] C. Perera, et al., Sensing as a service model for smart cities supported by Internet of Things, Trans. Emerging Tel. Tech. 25 (1) (2014), 81–93.

[40] M. Dohler, et al., Feature issue: smart cities – trends & technologies, Trans. Emerging Tel. Tech. 25 (1) (2014), 1–2.

[41] L. Lei, S. Shen, M. Dohler, C. Lin, Z. Zhong, Queuing models with applications to mode selection in device-to-device communications underlaying cellular networks. IEEE Trans. Wireless Commun. PP (99) (2014. http://dx.doi.org/10.1109/TWC.2014. 2335734.

[42] K. Zheng, S. Ou, J. Alonso-Zarate, M. Dohler, F. Liu, H. Zhu, Challenges of massive access in highly dense LTE-advanced networks with machine-to-machine communications, IEEE Wireless Commun. 21 (3 (June)) (2014), 12–18.

[43] M. Dohler, C. Ratti, J. Paraszczak, G. Falconer, Smart cities, (Guest Editorial). IEEE Commun. Mag. 51 (6 (June))(2013), pp. 70, 71.

[44] P. Blasco, D. Gunduz, M. Dohler, A learning theoretic approach to energy harvesting communication system optimization, IEEE Trans. Wireless Commun. 12 (4 (April)) (2013), 1872–1882.

[45] M.R. Palattella, N. Accettura, L.A. Grieco, G. Boggia, M. Dohler, T. Engel, On optimal scheduling in duty-cycled industrial IoT applications using IEEE802.15.4e TSCH, IEEE Sens. J. 13 (10 (October)) (2013), 3655–3666.

[46] P. Blasco, D. Gunduz, M. Dohler, Low-complexity scheduling policies for energy harvesting communication networks, in: 2013 IEEE International Symposium on Information Theory Proceedings (ISIT), 7–12 July 2013, pp. 1601,1605.

[47] K. Wang, J. Alonso-Zarate, M. Dohler, Energy-efficiency of LTE for small data machine-to-machine communications, in: 2013 IEEE International Conference on Communications (ICC), 9–13 June 2013, pp. 4120, 4124.

[48] C. Wang, M. Daneshmand, M. Dohler, X. Mao, R.Q. Hu, H. Wang, Special issue on internet of things (IoT): architecture, protocols and services, [Guest Editorial]IEEE Sens. J. 13 (10 (October)) (2013), 3505–3510.

[49] N. Accettura, M.R. Palattella, G. Boggia, L.A. Grieco, M. Dohler, Decentralized traffic aware scheduling for multi-hop low power lossy networks in the internet of things, in: 2013 IEEE 14th International Symposium and Workshops on a World of Wireless, Mobile and Multimedia Networks (WoWMoM), 4–7 June 2013, pp. 1, 6.

[50] S.A. Meybodi, M. Dohler, A.N. Askarpour, J. Bendtsen, J.D. Nielsen, The feasibility of communication among pumps in a district heating system, IEEE Antennas Propagat. Mag. 55 (3 (June)) (2013), 118–134.

[51] M.R. Palattella, N. Accettura, X. Vilajosana, T. Watteyne, L.A. Grieco, G. Boggia, M. Dohler, Standardized protocol stack for the internet of (important) things, IEEE Commun. Surv. Tut. 15 (3 (Third Quarter)) (2013), 1389–1406.

[52] T. Predojev, J. Alonso-Zarate, M. Dohler, Energy efficiency of cooperative ARQ strategies in low power networks, in: 2012 IEEE Conference on Computer Communications Workshops (INFOCOM WKSHPS), 25–30 March 2012, pp. 139, 144.

[53] N. Accettura, M.R. Palattella, M. Dohler, L.A. Grieco, G. Boggia, Standardized power-efficient & internet-enabled communication stack for capillary M2M networks, in: 2012 IEEE Wireless Communications and Networking Conference Workshops (WCNCW), 1 April 2012, pp. 226, 231.

[54] T. Predojev, J. Alonso-Zarate, M. Dohler, Energy analysis of cooperative and duty-cycled systems in shadowed environments, in: 2012 IEEE 17th International Workshop on Computer Aided Modeling and Design of Communication Links and Networks (CAMAD), 17–19 September 2012, pp. 226, 230.

[55] M.R. Palattella, N. Accettura, M. Dohler, L.A. Grieco, G. Boggia, Traffic aware scheduling algorithm for reliable low-power multi-hop IEEE 802.15.4e networks, in: 2012 IEEE 23rd International Symposium on Personal Indoor and Mobile Radio Communications (PIMRC), 9–12 September 2012, pp. 327, 332.

[56] A. Bartoli, J. Hernandez-Serrano, M. Soriano, M. Dohler, A. Kountouris, D. Barthel, Optimizing energy-efficiency of PHY-Layer authentication in machine-to-machine networks, in: 2012 IEEE Globecom Workshops (GC Wkshps) 2012, pp. 1663–1668.

Lower-power wireless mesh networks for machine-to-machine communications using the IEEE802.15.4 standard

4

T. Watteyne
Dust Networks Product Group, Linear Technology, Union City, CA, USA

4.1 Introduction

Low-power wireless mesh networks are often described as the "fingers of the Internet." These large networks of tiny devices sit at the edge of the Internet and serve as the extension of the IT world into the physical world. Wireless "motes" are often equipped with sensing devices and feed information gathered in the physical world back into monitoring systems living in the Internet. Similarly, those systems can send back commands to motes connected to actuation devices, thereby influencing the physical world. Low-power wireless mesh networks are an essential component of a machine-to-machine (M2M) solution to allow for the reliable and secure transportation of sensor data and actuation commands into physical systems.

This chapter gives an overview of low-power mesh technology. We will start by a historical perspective of the decade or so of the development of these networks and the emergence of the IEEE802.15.4 standard. This standard is the foundation of virtually all low-power mesh solutions and the common denominator of the technology presented in this chapter. We will highlight the challenges a low-power mesh network faces: reliability and power efficiency. In the second part of this chapter, we will look at the different solutions that have emerged, from single-channel star-network solutions to fully fledged multihop mesh networks. This chapter ends with a discussion of the challenges ahead of us and the different ongoing standardization efforts to tackle them.

4.2 The origins

4.2.1 Early low-power technologies

Low-power mesh technology started as a wild academic idea, which was subsequently transformed into a commercially viable technology. The concept of wireless sensor networks dates back to a 1980 report by the Lincoln Labs at MIT, which dealt with distributed sensor networks. The field was really born through the "Smart Dust"

Defense Advanced Research Projects Agency (DARPA) project, lead by Prof. Kris Pister from the University of California, Berkeley.

The Smart Dust concept started from the desire to make microrobots using microelectromechanical systems (MEMS) technology. In 1992, this technology led to a workshop on future technology-driven revolutions in military conflicts. What came out is the clear sense that three different technologies were following exponential decreasing cost curves: sensing (driven by the MEMS revolution), computation (following Moore's law), and communication. Similarly, it was clear at the time that the size and the power of such devices would follow trends similar to cost: Everything you needed to build a wireless sensor node was decreasing in size, power, and cost. That was the seed of the Smart Dust idea.

The concept of Smart Dust resonated with several communities of people. In 1997, Smart Dust[1] was funded by the US DARPA. Its main goal was to put a wireless node into a 1 mm^3, including MEMS sensors, computation, communication, and power supply. Following the DARPA proposal, Smart Dust went down the miniaturization path and ended up building a 5-mm^3 mote, about the size of a grain of rice, with an accelerometer, a light sensor, analog-to-digital converter, a digital microcontroller, optical communication, and an array of solar cells.

The miniaturization path clearly pointed out the issues related to ultra low-power design and systems integration. At the same time, as the miniaturization path was followed, researchers started looking at building wireless sensors with off-the-shelve hardware. Radios started to replace optical devices for communication. The "Berkeley mote" was born, initially using 900 MHz radios, with it the TinyOS operating system, small enough to fit on their tiny 8-bit microcontrollers. With the arrival of RF communication came also the challenges related to the unreliable nature of wireless communication.

4.2.2 First demonstrations

At the Intel's development forum in 2001, 800 Berkeley motes where placed at night in the main auditorium, one under each seat. During the keynote session the next morning, participants were asked to take out the motes and switch them on. The motes formed a multihop communication infrastructure, showing up in real time on the projector screen. Self-healing was demonstrated by pulling out the batteries of randomly chosen motes and seeing the network reorganizing around the remaining motes. This was the birth of multihop, self-organizing, and self-healing low-power wireless networking.

During the 29-palm demonstration,[2] eight wireless sensors were placed under the wings of a 2-m span unmanned aerial vehicle. The plane was programmed to drop sensors along a road. Once deployed, the motes, equipped with magnetometers, recorded the time of passage of large military vehicles. The plane, which flew back

---

[1]http://robotics.eecs.berkeley.edu/~pister/SmartDust/.

[2]http://robotics.eecs.berkeley.edu/~pister/29Palms0103/.

and forth along the road, queried each sensor on its passage and reported vehicle passage times back to a base station.

Another demonstration was done on the iconic San Francisco Golden Gate Bridge during the summer of 2006 [1]. Forty-six nodes equipped with accelerometers were attached to the 1280 m structure. More than providing deformation information to the bridge engineers, this project posed some interesting problems when having a 46-hop deep linear network.

4.2.3 IEEE802.15.4, the foundation

The IEEE802.15.4 [2] standard was developed around the same time of these early demonstrations, which has contributed to making it the standard adopted by virtually all current low-power mesh technology. IEEE802.15.4 is a double standard: it defines both the physical layer (i.e., the modulation, communication rate, and other considerations when designing a radio chip) and the medium access control (MAC) layer (i.e., when a device should talk and at what frequency).

The success of this standard can be attributed to the fact that its physical layer provides a healthy trade-off between range (typically 10–100 m), baud rate (typically 250 kbps), maximum frame size (127 bytes), and power consumption (5–20 mA at 3.6 V, depending on the vendor). While there have been three revisions since the first version of the IEEE802.15.4 standard in 2003, the physical layer at the most popular 2.4 GHz frequency band has remained backward-compatible.

At the 2.4 GHz band, an IEEE802.15.4-compliant radio is capable of transmitting frames up to 127 bytes long, on 1 of 16 orthogonal frequency channels. These channels, each 2 MHz wide, are equally spaced by 5 MHz and span the frequency band from 2.405 to 2.480 GHz.

The upper half of the IEEE802.15.4 standard is the MAC layer, that is, the rules to allow for multiple devices to talk over a wireless medium, which is broadcast and shared in nature. As depicted in Figure 4.1, an IEEE802.15.4 frame starts with a physical synchronization header, followed by a physical payload that can be up to 127 bytes. The physical header contains a synchronization header,[3] allowing the receiver to "lock" onto the signal it receives, followed by a 1-byte length field containing the

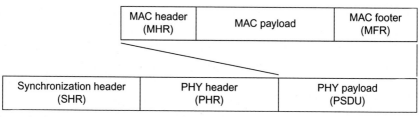

Figure 4.1 Format of an IEEE802.15.4 frame.

---

[3]Some MAC protocols are qualified as "preamble sampling." This is a MAC layer technique, which is different from the physical preamble defined here.

number of bytes in the physical payload. Nothing in the design of an IEEE802.15.4 radio enforces any format of that physical payload; the MAC part of the IEEE802.15.4 defines the format depicted in Figure 4.1.

Each IEEE802.15.4 radio is typically equipped to have a unique 64-bit identifier known as EUI64 (Extended Unique Identifier, 64 bits). Most vendors burn that identifier in the radio and ensure global uniqueness. This is very similar to the use of "MAC addresses" in Ethernet network cards. When a packet is exchanged between two neighboring nodes, the MAC header contains the source and destination MAC addresses of the devices.

4.3   Challenges of low-power mesh networking

Most major chip manufacturers have low-power radio in their catalog. Some of these radios use proprietary technology (i.e., they do not interoperate with radio chips from different vendors), yet most of them now comply with the IEEE802.15.4 standard, which has become ubiquitous in low-power mesh technology. Some manufacturers couple those radio chips with a microcontroller in a single package (e.g., Ref. [3]); others sell a stand-alone chip to which an external microcontroller connects over some digital bus.

When choosing a low-power system, it is important to understand the challenges such a system faces. This section starts by highlighting the two main challenges: reliability and power consumption. It then highlights what communication protocols do to take those into account.

4.3.1   The unreliable nature of wireless

The wireless medium is unreliable in nature. It doesn't matter what frequency channel radios communicate on, their transmission power settings, or the orientation of their antennas, some packets sent by the transmitter will not be received. These networks are therefore sometimes referred to as "lossy" network. Packet delivery ratio (PDR), the ratio between the number of packets received and the number of packets sent, is usually used to qualify how "lossy" a link between two nodes is.

When a packet is not received correctly, it has to be retransmitted. To know when to retransmit, the transmitter has to know that its packet was not received. This is especially important in ultra low-power networks, where every retransmission consumes precious energy. Link-layer acknowledgments are therefore typically used. It is important to take into account the additional traffic caused by retransmissions when provisioning a network or predicting its goodput.

In the context of low-power mesh networking, the two main phenomena to impact reliability are external interference and multipath fading.

Interference is an intuitive phenomenon, where different technologies (or different networks using the same technology) compete to use the same frequency band. This causes packets to be lost.

The second phenomenon, multipath fading, is both less intuitive and far more destructive. Multipath fading happens when the signal emitted by a transmitter "bounces" off objects in the environment, forming multiple "echoes." What the antenna of the destination receives is the superposition of the line-of-sight signal, with the echoes arriving a number of *ns* later. If the receiver is positioned exactly at the *wrong* location, those different copies of the same signal destructively interfere. That is, even if the destination node is located only 5 m from the transmitter, and the transmitter is transmitting at 0 dBm—a transmission power high enough for the signal to cover the distance—multipath fading causes the receiver not to receive anything.

Reference [4] shows experimental data highlighting the destructive nature of multipath fading on low-power wireless networking solutions. In the experiment, two motes are separated by 5 m, and the transmitter sends 1000 packets to the receiver. The packets contain a counter, allowing the receiver to determine what fraction of packets it has received. After those 1000 packets are sent (which takes a handful of seconds), a robotic arms moves the receiver by 1 cm, and the measurement repeats.

Figure 4.2 shows the PDR between the transmitter and receiver nodes, with the receiver positioned on 735 locations, on a 20 by 35 cm area (roughly the area of a sheet of printer paper). In most of the locations, the PDR is high (i.e., above 85%), but at some locations, the PDR drops down to zero, meaning the link between the nodes "disappears." The topography depicted in Figure 4.2 is due to the environment, that is, it changes when an object in the environment is moved or a door is opened. This extreme

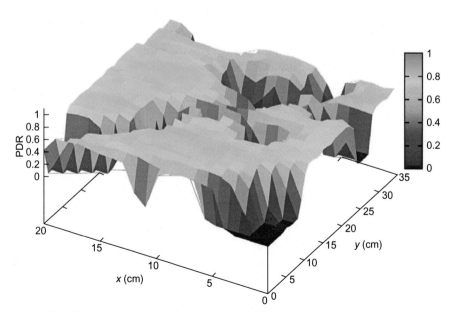

Figure 4.2 Visualizing the destructive nature of multipath fading.

volatility of the wireless links affects single-channel networks, that is, networks in which all the modes operate at the same frequency channel, all the time.

Reference [4] shows that, by simply changing the communication frequency (e.g., going from channel 11 (2.405 GHz) to channel 16 (2.430 GHz)), the topography is different. This means that, when communication fails, it is better to retransmit on a different frequency than on the same frequency. Reference [4] quantifies the coherence bandwidth to be less than 5 MHz (the spacing between consecutive IEEE802.15.4 channels) in the general case.

4.3.2   Low-power operation

In a typical mote, the radio chip is the component that accounts for most of the energy budget. Depending on the vendor, an IEEE802.15.4 radio consumes between 5 and 20 mA at 3.6 V. Since the transmission power is low (0 dBm, or 1 mW, is typical), the circuitry modulating and demodulating the signal consumes more than the power amplifier. This means that an IEEE802.15.4 radio draws approximately the same amount of current when transmitting or receiving. The only way to reduce its contribution to the energy budget is to turn it off as often as possible.

It takes a radio around 4 ms to transmit or receive a maximum-sized 127-byte packet. If the radio were switched on all the time (corresponding to 100% radio duty cycle), it would deplete a typical pair of AA batteries (holding 2200 mAh of charge) in around a week. Turning the radio duty cycle down to 25% extends the lifetime to about a month. Turning it further down to 1% yields years of lifetime.

4.3.3   Protocol considerations

The previous sections highlight the challenges on reliability and power consumption faced by low-power mesh networks. This section details the best-practice rules one can infer from those.

Channel hopping is a technique well known to combat multipath fading. It is used in several wireless technologies, such as IEEE802.15.1 (Bluetooth). In the 2.4 GHz frequency hand, an IEEE802.15.4 radio can operate on 16 orthogonal frequency channels. Channel hopping is obtained by scheduling the communication between two neighboring nodes and having them exchange consecutive packets on different frequencies. By using link-layer acknowledgment, the transmitter can know when the receiver has not received the data packet correctly. Retransmission happens on a different frequency.

Scheduling communication is needed when using channel hopping, to ensure that the receiver is listening on the same frequency the transmitter is sending on. This implies synchronizing the nodes. A beneficial side effect of time synchronization is that the transmitter and receiver nodes only need to switch their radio on when communicating. The radio duty cycle is therefore a direct function of the amount of data exchanged in the network and can be brought well below 1%, possibly yielding a decade of lifetime.

4.4 The past

4.4.1 Proprietary solutions

Proprietary solutions are a necessary step, while standards-based solutions are being finalized. We call "proprietary" a solution, which is not interoperable with other vendors. With proprietary technology, to enable a sensor of vendor A to send commands to an actuator of vendor B, one has to design an application-layer translator, which understands both A's and B's solutions and serves as an interface between them. Building application-level translators for each pair of solutions is not a scalable strategy.

The push for standardized solutions comes mainly from end users, who require standards compliance and hence the ability to interoperate between different vendors. In the context of low-power wireless mesh technology, interoperation can happen at several layers, from the physical layer (i.e., different devices are part of the same physical network) to the networking layer (i.e., devices can talk to each other over the Internet without needing an application-level translator). The latter is made possible by the adoption of IPv6 as a common networking denominator, as will be detailed later.

4.4.2 ZigBee

The ZigBee Alliance was arguably the first to finalize a complete protocol stack for low-power mesh networks. The ZigBee protocol stack is rooted entirely in the legacy IEEE802.15.4-2003 and IEEE802.15.4-2006 MAC protocol. It differentiates between full-function devices (FFD) and reduced-function devices (RFD), where the former are typically used as routing nodes, the latter as nonrouting leaf nodes. A ZigBee network consists of a number of FFDs collecting data from RFDs located around it.

A ZigBee network is not time-synchronized. Routing nodes therefore need to leave their radio on all the time to be ready to receive data packets from an RFD at any time. In a practical setting, this often means having the router nodes be mains-powered. As a result, most ZigBee networks have been used in a "star" topology, where RFDs send data to single collector FFDs a single hop away.

A ZigBee network operates on a single frequency channel. While it is possible to reconfigure the complete network to communicate at a different frequency, the network does not channel hop and is therefore prone to the effects of external interference and multipath fading highlighted above.

4.4.3 Time synchronized mesh protocol

The first solution to bring time synchronization and channel hopping to low-power mesh networking was a proprietary solution by Dust Networks (a product group within Linear Technology) called Time Synchronized Mesh Protocol (TSMP) [5]. It was designed with two goals in mind: wirelike reliability and long battery lifetimes.

The core idea of a TSMP network is for the motes to communicate following a schedule. All nodes in a TSMP network are synchronized. Time is sliced up into time slots, where time slots are grouped into slotframes, which continuously repeat over time. A schedule indicates to each mote what to do in each timeslot: transmit, receive, or sleep. Additionally, for each transmit or receive slot, the schedule indicates the specific neighbor to communicate with, as well as a channel offset to communicate on.

TSMP nodes communicate using IEEE802.15.4 radios. There are 16 channel offsets, corresponding to the 16 frequencies available to an IEEE802.15.4-compliant radio communicating at 2.4 GHz. The channel offset serves as a parameter in the equation used for calculating the frequency to communicate on. Each time a mote has a transmit or receive slot, it calculates the frequency to communicate on on the fly, based on the associated channel offset. The schedule is built in such as way that both the transmitter and receiver use the same channel offset. This ensures that the receiver listens on the frequency the transmitter is sending on.

We call "slot" a time slot, and "cell" the tuple *(<time slot>, <channel offset>)*. At each iteration of the slotframe, two neighboring nodes communicate on the same cell(s). The equation used to translate the channel offset of that cell into the frequency is done in such a way that a different frequency is obtained at each slotframe iteration. That is, channel hopping behavior is obtained even with a static schedule.

Typically, a TSMP network uses a collision-free schedule. Cells in the schedule are dedicated for a unicast transmission between two well-identified neighbors, and the same cell is not scheduled for two different pairs of motes in the network.

The combination of time synchronization, channel hopping, and collision-free communication enables a TSMP network to exhibit a performance vastly superior to single-channel unsynchronized networks. Reference [6] presents results of a 2006 real-world deployment. Forty-five nodes were deployed in a printing facility and were configured to report per-frequency PDR measurements for each link in the network. Reference [6] shows that this network has produced, over the course of 26 days, 3.6 million packets traveling multiple hops and of which only 17 were lost. This yields an end-to-end reliability of over 99.999%.

The performance of TSMP, in particular the fact that it allows for wirelike reliability and for battery-powered device more than a decade of lifetime, made it particularly applicable to industrial applications, including large-scale monitoring of industrial processes and peel-and-stick deployments. TSMP served as the basis for WirelessHART [7]. WirelessHART is the wireless extension of HART, the *de facto* wired standard for remote monitoring of industrial equipment. The idea of time-synchronized channel hopping developed in TSMP was later on also the basis for the IEEE802.15.4e [8] standard, as will be detailed later on.

4.5 The present

4.5.1 Reliability, reliability, reliability

Reliability will always be one of the most important conditions for adopting low-power mesh networking technology. In most real-world applications, a wireless mesh network complements or replaces what otherwise would have been a wired infrastructure. End users expect the wireless system to behave similarly to the wired network;

it is therefore not acceptable for the wireless system to exhibit widely different end-to-end reliability figures.

Reliability and low-power are not independent. At the MAC layer, unreliability translates into more local retransmissions; at the application layer, unreliability translates into end-to-end retries. Both translate into more packets being exchanged, hence a large energy consumption.

Besides reliability and low-power, security is the other aspect of wireless networks that cannot be compromised. Luckily, low-power mesh networking technology, and in particular the IEEE802.15.4 standard, was developed at the time when IEEE802.11 Wi-Fi networks were attacked and their security weakness (especially the WEP (Wired Equivalent Privacy) mode) publicly demonstrated. The result is that, since the first revision of the IEEE802.15.4 standard, strong security has been present. IEEE802.15.4 uses the CCM* combined authentication and encryption scheme [2], which builds upon an AES 128-bit cipher. As a result, most IEEE802.15.4 radio chips come with an AES 128-bit hardware component, which enables secure operation of IEEE802.15.4-based networks, provided a secure key distribution mechanism is in place.

Time-synchronized channel hopping-based low-power mesh networks, such as TSMP, WirelessHART, and IEEE802.15.4e, are able to answer the reliability, low-power, and security requirements.

4.5.2 Industrial-grade standards

Standardization bodies have played a major role in the success of low-power wireless mesh technologies, in particular the IEEE and the Internet Engineering Task Force (IETF). The IEEE has traditionally focused on lower layers (physical layer (layer 1) and the data link layer (layer 2), which includes the MAC layer). The IETF standardizes the protocols "between the wire and the application," which includes the networking, transport, and application layers. Standards by the IEEE and the IETF are therefore complementary, and a fully standards-based protocol stack typically consists of standards from both.

As detailed earlier, IEEE802.15.4 [2] is the standard with the longest-standing influence of low-power mesh technology. The physical layer has been virtually unchanged since its first revision in 2003, and since it presents a healthy trade-off between range, packet size, reliability, and power consumption, it has become the foundation of most low-power wireless solutions.

The shortcomings of the legacy IEEE802.15.4 MAC layer (single-channel operation and the fact that router nodes need to be mains-powered in any practical deployment) became apparent in early ZigBee deployment. Recognizing this, the IEEE formed the IEEE802.15.4e task group, with a goal to "enhance and add functionality to the 802.15.4-2006 MAC to better support the industrial markets." The IEEE802.15.4e standard [8], published in April 2012, uses the concept of time synchronization and channel hopping (similar to TSMP) in its Time Slotted Channel Hopping (TSCH) mode. This allows an IEEE802.15.4e network to exhibit a performance similar to WirelessHART, with the added advantage that IEEE802.15.4e limits itself to the MAC protocol. This allows the IETF upper layer to be used in conjunction.

Strictly speaking, IEEE802.15.4e is a "MAC amendment," that is, it replaces the (now legacy) IEEE802.15.4 MAC layer while not modifying the physical layer. This means that IEEE802.15.4e can be considered a "software update" and can function on any IEEE802.15.4-compliant radio chip. Similarly, IEEE802.15.4e was designed to be flexible and run on top of different physical layers, including IEEE802.15.4g.

IEEE802.15.4g [9] is a recently finalized physical layer amendment to the IEEE802.15.4-2011 standard. Much like IEEE802.15.4e, it targets only one layer of the IEEE802.15.4 standard; unlike IEEE802.15.4e, it targets the physical (PHY) layer. It is targeted specifically at low-data-rate, wireless, smart metering utility networks. IEEE802.15.4e being a MAC amendment, it is possible to use IEEE802.15.4e with IEEE802.15.4g-compliant radios.

An additional amendment to the physical layer of IEEE802.15.4 is IEEE802.15.4k [10], created by the Low Energy Critical Infrastructure Task Group, which defines two additional physical layers.

The IETF is the standardization body that is behind most of the protocols used in today's Internet, including Internet Protocol (IP), TCP, UDP, and HTTP. It is therefore natural for the IETF to have taken a leading role in standardizing the Internet of Things. Devices on the Internet today have widely varying capabilities, resources, and usages, yet all agree on one common denominator: the IP. "Things" are no exception, so integrating this new class of devices into the Internet means having them speak IP and in particular its latest version, IPv6. The 6LoWPAN group[4] was created in 2005 to design a scheme allowing (long) IPv6 headers and packets to be carried in (short) IEEE802.15.4 frames. 6LoWPAN allows low-power nodes to form a star topology and connect to the Internet through a 6LoWPAN gateway.

The IETF Routing Over Low power and Lossy networks (ROLL) working group[5] was created in 2008 to design a routing protocol for low-power mesh networks. The result of this work is the IPv6 Routing Protocol for Low-Power and Lossy Networks (RPL), standardized in 2012, a distance vector routing protocol. An RPL network is IPv6-compliant (thanks to 6LoWPAN), but in which the nodes form a multihop mesh, rather than a star, enabling a much wider variety of applications.

Devices on the Internet (often called "hosts") often interact with each other using a client/server model. This is the case for HTTP traffic, in which a browser (the client) consults resources hosted on a Web server. The IETF Constrained RESTful Environments (CoRE) working group[6] was created in 2009 to enable the same style of interaction when using low-power nodes. The results of that work in the Constrained Application Protocol (CoAP) protocol, published in 2013.

Both a host on the Internet and a low-power wireless sensor node can implement CoAP. A CoAP-enabled device plays the role of both host (similar to a Web browser) and server. This enables Weblike interactions. A low-power wireless node attached to a gardening sprinkler can contact a CoAP-enabled weather server to decide whether it

[4]http://datatracker.ietf.org/wg/6lowpan/charter/.
[5]http://datatracker.ietf.org/wg/roll/charter/.
[6]http://datatracker.ietf.org/wg/core/charter/.

needs to switch on. CoAP-enabled smartphones can be used to remotely switch on the heating system in a smart house.

Data traveling over the Internet transit through equipment managed by different entities. It is therefore important to have an end-to-end secure session in place between the CoAP server and CoAP host (akin to "https://"). The new DTLS In Constrained Environments (DICE) working group develops a security mechanism similar to HTTPS (including public/private keying and certificate-based authentication) while being targeted at highly constrained wireless devices.

Combining the wirelike reliability and ultra low-power nature of IEEE802.15.4 with the ease of integration of an IPv6-enabled IETF "upper stack" is an open problem. This is, however, changing with a new standardization activity inside the IETF, known as "IPv6 over the TSCH mode of IEEE 802.15.4e" (6TiSCH). This activity defines the missing communication protocols to allow the TSCH schedule to be managed by a scheduling entity, as detailed later on.

All the standards presented in the previous section form a fully standards-based protocol stack designed for low-power wireless mesh.

4.5.3 Internet integration

The goal of this section is to illustrate the use of the protocol stack described above. We describe what happens when a mote in a low-power mesh network sends data to a server on the Internet, a typical interaction model for these types of networks. This is illustrated in Figure 4.3.

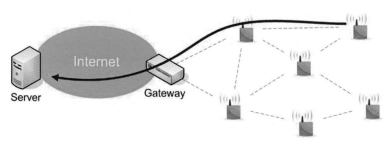

Low-power wireless mesh network

Figure 4.3 End-to-end communication between a mote and a server on the Internet.

The mote's application is programmed to periodically read a value from a sensor. The goal of the application is to send this value to a CoAP resource located on the Internet host.

Inside the mote's protocol stack, the CoAP layer prepends the CoAP header. The IPv6/6LoWPAN layer then prepends the (compressed) IPv6 header. The forwarding engine is invoked and looks into the routing table maintained by the RPL protocol to find the address of the next hop. The IEEE802.15.4e layer prepends the IEEE802.15.4 MAC header, which includes this next hop's MAC address. The frame is stored in the IEEE802.15.4e transmit queue, ready for transmission. When the IEEE802.15.4e schedule indicates the appropriate transmit cell, the packet is transmitted.

The packet is relayed hop by hop inside the low-power mesh network. When the packet reaches the gateway of the network, it inflates the 6LoWPAN header into a fully fledged (but equivalent) IPv6 header. The packet traveling over the network is IPv6-compliant and is routed through the Internet until it reaches the server.

At the server, the CoAP layer identifies the resource targeted by the initiator mote and consumes the payload. A typical use case is to write the received data into a database.

4.6   The future

4.6.1   Today's challenges

Low-power mesh networking technology is a good example of fruitful interaction between academia, industry, and major standardization bodies. A fully standards-based protocol stack is being finalized. Commercial products are available [3], which, based on this protocol stack, enable wirelike reliability, more than a decade of mote lifetime, and secure communication. These commercial solutions make low-power wireless networking a reality *today*.

A challenge that still remains partially open is the building and maintenance of the IEEE802.15.4e TSCH schedule, in a standards-compliant way. While IEEE802.15.4e details what a mote has to do given a schedule, it does not detail how to build that schedule, maintain it, and match it to the communication needs of the network. In the absence of a standard defining the scheduling layer, today's commercial solutions necessarily rely on some proprietary techniques.

Scheduling an IEEE802.15.4e TSCH network is essential to its efficient operation. The TSCH schedule allows a clean trade-off between throughput, latency, redundancy, and low-power consumption. By adding more transmit/receive cells to the schedule, more data can be generated by the network, but since the motes' radios are "on" more often, this also reduces the lifetime of the network. The scheduler needs to take into account the lifetime, throughput, and latency requirements from the network and compute the schedule that satisfies those while remaining as low power as possible. While the network is running, the scheduler also needs to remain aware of possible topological changes and continuously maintain the schedule.

4.6.2   IETF 6TiSCH: combining IPv6 connectivity
with industrial performance

A new working group was created inside the IETF, called "IPv6 over the TSCH mode of IEEE 802.15.4e" (6TiSCH) to address this issue. The 6TiSCH group[7] is considering two modes of operation. In the centralized mode, a path computation element is responsible for centrally building and maintaining the schedule. A second, distributed, mode is considered. In this case, the motes in the low-power mesh network use a reservation

---

[7]https://datatracker.ietf.org/wg/6tisch/charter/.

protocol to reserve TSCH cells hop by hop between the source and destination nodes. The choice between the centralized and distributed modes needs to take into account parameters such as architectural and schedule optimality requirements.

Initial work in the 6TiSCH working group has focused on the envisioned architecture [11] and the definition of the 6top protocol layer to serve as management entity for the 6TiSCH schedule [12]. The overarching goal of this standardization effort is to glue together the IPv6-based IETF upper stack (with standards such as 6LoWPAN, RPL, and CoAP) and the IEEE lower stack (IEEE802.15.4e TSCH mode and IEEE802.15.4 physical layer).

4.6.3 Toward hybrid systems

Designing a complete M2M solution is a complex task. Given the variety of deployment and traffic requirements, different solutions will be combined in different ways. IEEE802.15.4-based technologies, as presented in this chapter, are designed to interconnect a large number of wireless motes deployed in the same geographic region and with tight constraints on energy and computational power. As described throughout this chapter, commercial solutions exist that enable wirelike reliability, a decade of lifetime for battery-powered devices, and secure communication [3].

For example, the Dust Networks SmartMesh Power and Performance Estimator [13] allows one to predict the performance of a SmartMesh IP network, a product by Linear Technology. For example, let's take the example of a 100-mote network, with motes deployed in such a way that they form a multihop network, which is 4 hops deep and in which each mote reports 80 bytes of payload every 30 s. [13] allows us to predict that, in this case, the average current draw of a mote is between 17.7 μA and 34.2 μA at 3.6 V, depending on the location in the topology. If this node is powered using a typical 2200 mAh battery, this would yield a battery lifetime between 7 and 14 years.

IEEE802.15.4-based technology does not answer all M2M requirements. A large M2M system typically spans beyond the use cases for low-power mesh networks. IEEE802.15.4-based technology is a complement to the different other M2M technologies developed in this book.

For example, in a smart parking application, one can combine low-power mesh and cellular technology. Car presence sensors are mounted on the ground and form a mesh network together with sensor mounted on the parking meters. The parking data collected by this low-power mesh network are reported to a cellular-enabled gateway device, which enables the data to flow the management system located in the cloud. Such a hybrid solution allows the operator of this network to combine the strengths of cellular and low-power mesh technology.

4.7 Conclusion

Low-power mesh networks are a reality today. Solutions exist that are developed by major standardization bodies such as the IEEE and the IETF and that reflect the lessons learned during almost a decade of work by academic and industrial research.

Commercial products exist that allow for low-power wireless nodes to securely operate in a multihop mesh network for less than 50 μA on average, yielding a decade of battery lifetime, and exhibit an end-to-end reliability of over 99.999%.

The protocol stack that is becoming the *de facto* standard for low-power wireless mesh networks includes the IEEE802.15.4e Time TSCH mode, which uses time synchronization to aggressively reduce the radio duty cycle and channel hopping to effectively combat external interference and multipath fading. Standardization efforts are being finalized, in particular in the IETF 6TiSCH group, to efficiently manage the IEEE802.15.4e TSCH.

Low-power mesh networking technology is part of the M2M answer. Together with the other M2M technologies developed in this book, it enables the secure operation of large-scale M2M networks and systems.

Acknowledgments

This chapter is based on the tutorial entitled "Standardizing the Internet of (Important) Things," which was given at the Summer School on Cyber-Physical Systems on 11 July 2013 in Grenoble, France. The author would like to thank the attendees and their valuable feedback, which helped shape the contents of this chapter.

References

[1] S. Kim, S. Pakzad, D. Culler, J. Demmel, G. Fenves, S. Glaser, M. Turon, Health monitoring of civil infrastructures using wireless sensor networks, in: In the Proceedings of the 6th International Conference on Information Processing in Sensor Networks (IPSN'07), April 2007, ACM Press, Cambridge, MA, 2007, pp. 254–263.

[2] 802.15.4-2011: IEEE Standard for Information technology, Local and metropolitan area networks, Wireless Medium Access Control (MAC) and Physical Layer (PHY) Specifications for Low Rate Wireless Personal Area Networks (WPANs), 2011.

[3] LTC5800-IPM—SmartMesh IP Mote-on-Chip, available online http://www.linear.com/docs/41870.

[4] T. Watteyne, S. Lanzisera, A. Mehta, K.S.J. Pister, Mitigating multipath fading through channel hopping in wireless sensor networks, in: IEEE International Conference on Communications (ICC), 2010.

[5] K.S.J. Pister, L. Doherty, TSMP: Time Synchronized Mesh Protocol, in: IASTED International Symposium on Distributed Sensor Networks (DSN), Orlando, Florida, USA, 2008.

[6] L. Doherty, W. Lindsay, J. Simon, Channel-specific wireless sensor network path data, in: IEEE International Conference on Computer Communications and Networks (ICCCN), 2007.

[7] WirelessHART Specification 75: TDMA Data-Link Layer. HCF SPEC-75, 2008.

[8] 802.15.4e-2012: IEEE Standard for Local and metropolitan area networks–Part 15.4: Low-Rate Wireless Personal Area Networks (LR-WPANs) Amendment 1: MAC sublayer, 16 April 2012.

[9] 802.15.4g-2012: IEEE Standard for Local and metropolitan area networks–Part 15.4: Low-Rate Wireless Personal Area Networks (LR-WPANs) Amendment 3: Physical Layer (PHY) Specifications for Low-Data-Rate, Wireless, Smart Metering Utility Networks, 27 April 2012.

[10] 802.15.4k-2012: IEEE Standard for Local and metropolitan area networks–Part 15.4: Low-Rate Wireless Personal Area Networks (LR-WPANs) Amendment 5: Physical Layer Specifications for Low Energy, Critical Infrastructure Monitoring Networks, 2013.

[11] P. Thubert, T. Wattteyne, R. Assimiti, An Architecture for IPv6 over the TSCH mode of IEEE 802.15.4e. draft-ietf-6tisch-architecture-03, July 2014 (work-in-progress).

[12] Q. Wang, X. Vilajosana, T. Watteyne 6TiSCH Operation Sublayer (6top). draft-wang-6tisch-6top-sublayer-01, July 2014 (work-in-progress).

[13] Dust Networks SmartMesh Power and Performance Estimator, available online http://www.linear.com/docs/42452.

M2M interworking technologies and underlying market considerations

B. Cendón
TST, Cantabria, Spain

5.1 Interworking technologies for M2M communication networks: introduction

This chapter takes a bird's-eye view of the heterogeneous protocols used in the machine-to-machine (M2M) ecosystem. Furthermore, we analyze the mechanisms to match capillary with IP-based networks or, in other words, transpose the information in an effective way from a wireless sensor network to the Internet. The third part of the chapter discusses possible ways to effectively interconnect the previous with the cloud and the different strategies to do so, whether by lowering down to end devices the use of legacy technologies or creating new effective standards with the M2M constraints in mind. Then, the fourth part analyzes the implications of the M2M interoperability in the market from different points of view; that is, the hardware providers, the software developers, the service integrators, and the telecom operators, both traditional and new ones. Finally, future directions in the M2M ecosystem are discussed.

5.2 A panorama of heterogeneous technologies

Though various standardization bodies have attempted to establish what an M2M architecture shall be, the many different applications to be implemented and all the possible technical approaches that could be considered leave a lot of room for interpretation.

Nevertheless, after analyzing how current M2M deployments are being planned and integrated, it can be stated that in most of the cases, three different approaches are taken when presenting the internetworking architecture:

- Three-level architecture with non-IP end devices
- Two-level architecture with IP-enabled end devices
- Two-level architecture with non-IP end devices

In all three cases, there is a common layer, the bottom one. At this level, the presence of a physical device is always understood. It could be something simple, any small end device, capable of interacting with sensors or actuating on relays, couplers, or switches, or a more complex item, with more advanced processing capability,

Machine-to-machine (M2M) Communications. http://dx.doi.org/10.1016/B978-1-78242-102-3.00005-8

embedded applications, and intelligence. The device shall have connectivity capabilities and shall not be an isolated item.

It is from this level when the implementation of the interworking communications might differ, and it will depend on the application, cost, and network coverage capabilities. In some cases, however, it will depend also on the regulations or convenience of the device manufacturer, system integrator, or telecom operator. Do they want to sell SIM cards or just create the most effective deployment? Shall the Wi-Fi network of the city be used at any cost? Do they have an expensive but unused fiber-optic network? These are questions that will also be decisive at the moment of choosing one of the approaches presented in this chapter, and eventually, it might not be the most optimal one.

5.2.1 Three-level architecture with non-IP devices

This architecture conception is the one used in deployments that require an important number of low-cost end devices, in some cases things as simple as battery-powered meters. In this case, the IP-enabled technologies are more expensive or just more power consuming, so a capillary network is deployed, relaying the IP connectivity through one or more gateways (Figure 5.1).

The first layer of connectivity in this architecture is usually brought by two different approaches:

- A point-to-point connectivity, non-IP, toward a gateway. For example, a connection over IEEE 802.15.4 or wireless M-Bus
- A mesh or routed connectivity, non-IP, toward a neighbor relay device or a gateway, for example, a connection over ZigBee, Z-Wave, WirelessHART™, 6LowPAN, or similar

Over this initial level of connectivity, the objective is to be able to route the data through these elements by using a second layer that provides an IP-enabled

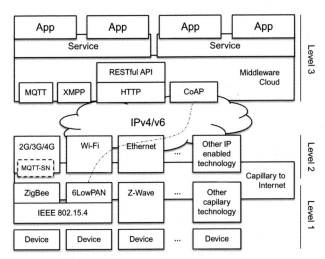

Figure 5.1 Simplified representation of a three-level architecture with non-IP devices.

connection. This layer is usually implemented as a gateway and will provide routing, data aggregation, and, in some cases, network management capabilities.

The gateway connectivity will be offered by means of an IP-enabled interface over popular technologies such as Wi-Fi (IEEE 802.11), Ethernet (IEEE 802.3), and cellular (GPRS, EDGE, UMTS, HSxPA, and LTE in 3GPP networks). Other technologies like WiMax (IEEE 802.16), PLC (power-line communications), fiber optics (e.g., FTTx and HFC), and xDSL might be used, but these are less popular in emerging M2M deployments.

Once this IP connectivity is provided, connections over UDP or TCP are established. Most common architectures will relay on proprietary implementations, RESTful APIs over HTTP, with JSON, XML, or custom format data exchange or specific M2M protocols like MQTT and CoAP. The chapter will elaborate more on those in later sections.

The third level of this architecture will interconnect with a remote entity, in most of the cases a service provider able to collect the data, route the commands, and perform the tasks related to device management. Over this layer, a set of services might be offered in order to build applications that enable the interaction with the implemented system.

5.2.2 Two-level architecture with IP-enabled devices

This architecture interpretation will represent those deployments where the end devices are equipped with an IP-based connectivity, e.g., a connection over Wi-Fi, Ethernet, or a cellular modem. This is the case where the presence of gateways is not necessary, because of different possible reasons. It might present a low-end device/gateway ratio that does not justify such deployment, or the required parameters of device cost, power consumption, and network coverage are met (Figure 5.2).

These IP-enabled devices will be able to autonomously interact with the next level, in this case equivalent to the third layer presented in the previous section. As the protocols required are usually more complex or just have higher throughput rates, an extra degree of intelligence and memory footprint is expected in the hardware of these

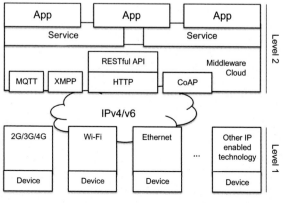

Figure 5.2 Simplified representation of a two-level architecture with IP-enabled devices.

devices as they shall provide extra capabilities on top of the IP stack to implement the required protocol (e.g., HTTP and MQTT).

5.2.3 Two-level architecture with non-IP devices

As an evolution of the existing network communications protocols, new technical solutions are being introduced in M2M in order to simplify the deployments and gain wider network coverage without compromising cost or power consumption requirements. These technologies might present an IP-based backhaul, but as this is going to be transparent to the system integrator, we can consider them as only two levels, as we can do with a cellular-based implementation (Figure 5.3).

New M2M protocols and operators like Weightless and SIGFOX have implemented such approach and present deployments where the end devices are provided with direct connectivity to a base station without compromising the low-power-consumption requirements.

5.3 From capillary to IP networks

5.3.1 The gap between IP and capillary networks

The IP stack of protocols was originally created to interconnect big data centers in a non-reliable network environment. The Ethernet (IEEE 802.3) standard was created to provide a reliable but simple wire connection in a computer local network. The Wi-Fi (IEEE 802.11) standard came with the objective of creating a wireless Ethernet. All the cellular data protocols, ranging from GPRS to LTE, have been created to enable data communications in mobile phones, that is, battery-powered personal devices that are often recharged.

Though more examples about the conception of the different protocols used in IP-based communications could be provided, the previous ones might be enough to state the message that none of them were created having in mind the requirements that many of the end devices in M2M deployments might have today, that is, low data volumes, low power consumption, low-cost devices, and low processing CPUs. These limitations, combined with the fact that the hardware costs of these technologies have

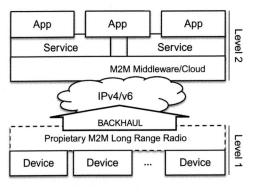

Figure 5.3 Simplified representation of a two-level architecture with non-IP devices.

been traditionally high, though this is less and less the case nowadays, have caused the emergence of new low-power communication technologies able to provide point-to-point connectivity at a small price and, in some cases, mesh network capabilities.

It is under the scope of these constraints that protocols like IEEE 802.15.4, wireless M-Bus, ZigBee, 6LowPAN, Z-Wave, WirelessHART™, TinyMesh™, and Digi-Mesh™ were created. Some of them have become standards and others are just proprietary implementations. Not all of them operate at the same level in the OSI stack of protocols, as some are link protocols and others have networking capabilities, but in all cases, all of them need a gateway device in order to aggregate the data and provide global IP connectivity. We gather all these technologies M2M implementations into what it is called capillary networks.

5.3.2 The big IP funnel

The protocols commonly used in capillary network deployments are, in most cases, not ready to directly map the data communications with an IP network, that is, the Internet. Though standardized solutions exist for easing this connectivity, in most of the current systems, these are not used, and in order to implement an optimal solution, the key differences and the limiting factors between the capillary and the IP networks shall be taken into account:

- *Data throughput*: Compared with the capabilities of current IP networks, the capillary ones usually provide limited throughputs as their underlying technology has been created to meet low-cost and low-power requirements, i.e., have the cost of material and energy per bit of data transmitted reduced to the minimum. Theoretically speaking, the maximum throughput in communications over IEEE 802.15.4, including the network protocols on top of it, is not greater than 250 Kbps, and when measured, this throughput can go down to 150 Kbps, even in ideal conditions, operating in the ISM 2.4 GHz band, with no network collision and in pure point-to-point communications [1]. The use of mesh technologies or different operating frequencies provides significant smaller throughput.
- *Addressing*: Another point in which the communication between the IP and the capillary networks becomes affected is the different addressing mechanisms to be used. This issue has been solved by protocols like 6LowPAN and ZigBee IP, with the intent to provide seamless connectivity till the end point by means of a standardized stack implementation, but most of the existing deployments deal in a nonstandard way with the data aggregation and addressing between the IP world and the capillary world.
- *Payload size*: There is also a mismatch in the size of the packets and payloads between the IP and the capillary networks. Take as an example IEEE 802.15.4 protocol, its maximum message size is 133 bytes; taking into account the physical, MAC, and security headers, the maximum data payload might vary between 127 and 102 bytes, and the minimum can be as smaller as 2 bytes [2]. A minimum IP datagram size is 576 bytes [3]. Addressing in IP is completely different than in IEEE 802.15.4, the first uses a 32-bit address for IPv4 and a 128-bit size for IPv6, while the latter has 16-bit reduced implementation and a wider IEEE standard 64-bit one [4]. Both are incompatible, and a hardcoded address translation needs typically be done.

It is obvious that the protocols used in capillary networks were not designed with the compatibility with the Internet in mind. It can be easily explained by the fact that they were created in a moment when the penetration of the Internet was smaller, the IP stack was not optimized for such applications, and the cost of the hardware was significantly higher. Standard solutions like 6LowPAN/CoAP and MQTT-SN are available to allow this integration and shall be privileged to custom or proprietary ones. In any case, it is important to be aware of these limitations when planning the internetworking of data in new M2M deployments.

5.4   Going up to the M2M cloud

Once the gap between the capillary and the IP networks has been filled, whether with a gateway able to work on IP networks or with an IP-enabled end device, a protocol shall be established to deal with the exchange of information toward a middleware in the cloud. This middleware will be used as the base of the services and applications to be provided by a M2M vertical.

Traditionally, the web services have been responsible for filling this gap up, but more recent systems are favoring the use of custom M2M standards like CoAP and MQTT, simpler and created to communicate all types of IP-enabled devices, even for the most limited ones.

5.4.1   Enabling M2M data exchange with web services

The exchange of information on IP networks is currently flooded by the use of the HTTP application protocol. Created in the mid-1990s as the mechanism to exchange hypermedia in the World Wide Web, it is nowadays commonly used for the exchange of any kind of media over the Internet regardless of their use as web content. And under the scope of M2M communications, one of the main methods that HTTP has enabled is the so-called web services that establish a way of providing communication between machines by the way of changing structured data information where commands or information can be included.

A typical web service will offer a set of services over a service provider. In order to use them, a service requester will get the information put on a data file, e.g., a XML or a JSON file; send it over an HTTP connection to the provider; and get an answer in the same file format. Depending on the behavior of the web service, it will be defined as compliant to RESTful rules or not. Most of the available M2M cloud/middleware solutions available on the market offer an API based on RESTful web services, with data representation in XML, JSON, or just CSV, that is, plain text separated by commas.

5.4.2   RESTful web services

RESTful web services are most commonly used in M2M communications than as they offer a stateless service, which can be quite valuable in unreliable communications that might not guarantee the stability of the data connection. They also avoid the use of extra memory on end devices to manage any kind of session-driven connection or manage connection states.

A RESTful web service will define unique resources through identifiers encoded in a URI that hierarchically will allow operating into a more detailed level. Many device vendors have adopted RESTful APIs to operate their solutions. A good example is the Philips Hue™ lighting system [5]. They provide an API that allows light operation, group management, and scheduling lights and configures all the system. To do so, the API defines a generic resource called "/api." From this resource, a "username" has to be created and then four third-level resources: "/lights," "/groups," "/schedule," and "/config," With these resources, it will be possible to fully operate the lights, for example:

- Accessing through an HTTP GET to <URL>/api/<username>/lights, a JSON listing all the available lights will be returned, and doing so through an HTTP POST, a search for new lights will start. A confirmation JSON will be returned.
- Sending a configuration JSON file through an HTTP PUT to <URL>/api/<username>/lights/<id>/state will allow to switch on/off the lights and set brightness, hue, color temperature, and several other light features.

Web services, RESTful compliant in special, are a great solution for managing M2M communications, but when it comes to optimizing the data consumption, battery performance, or CPU power, their conception is not the best choice for constrained M2M devices as HTTP headers are big and the data represented with JSON or XML are big too. To avoid these issues, several proposals of custom protocols have been introduced. Between those, the chapter will focus on MQTT, a new simple protocol over TCP/IP; CoAP, a RESTful protocol that intends to provide a low-complexity substitute for HTTP; and XMPP.

5.4.3 MQTT

As it has been stated, IP communications and the application protocols traditionally used over it are not the most optimal ways of getting low-cost and low-energy-consumption M2M solutions. In a time when M2M was only reserved for computer-to-computer communications, they completely make sense, but as the M2M has evolved toward a more reduced world of connected things and the hardware to enable IP connectivity has become more affordable, custom protocols to ease this communications are gaining momentum. Between those, MQTT [6] is gaining worldwide recognition as something simple, powerful, and secure enough to become a *de facto* standard in M2M communications.

Created in 1999 by Dr. Andy Stanford-Clark of IBM and Arlen Nipper of Arcom (now Eurotech), it is a very simple protocol over TCP/IP based on a publish-and-subscribe mechanism toward an MQTT broker, so only MQTT clients will use end devices, offering different QoS levels providing three degrees of assurance delivery. It also includes a "ping" mechanism to check the status of the client connection. The communication goes through a well-known port, the 1883, an arbitrary one, or a websocket. It also provides three levels of security: none, user and password, or exchange of TLS/SSL certificates. When it comes to the protocol itself, it defines a set of messages with very basic headers and variable payloads, all of that in a standard document that makes less than 50 pages, so the complexity to implement it is quite reduced.

In order to access and perform a subscription on an MQTT resource, a topic shall be defined. The topics are hierarchical representation of resources and are created the moment a part publishes something related to that topic or gets subscribed to it. There is no need to perform additional operations. An example of that use is a sensor device publishing over a "test/temperature" topic and a remote entity subscribing to that one. Once the device publishes the data, the remote entity is immediately notified. If the remote entity is in a sleep period, the MQTT broker has caching capabilities, so the message can be delivered in the moment the entity wakes up and connects to the topic.

Another advantage of MQTT is the availability of multiple programming languages and open-source implementations, being the most known Mosquitto [7] and Paho [8].

A variation of MQTT, the MQTT-SN [9], has been introduced for sensor networks, and it is intended to be used over WSN network protocols like ZigBee. In this case, it moves up from a two-level architecture with MQTT clients and an MQTT broker toward a three-level architecture with MQTT-SN clients; MQTT-SN gateways, which could be aggregating or transparent; and an MQTT broker, given the non-IP nature of ZigBee and similar protocols.

5.4.4 CoAP

CoAP, Constrained Application Protocol, is a RESTful application protocol running over UDP that is used for resource-constrained, low-power devices in lossy networks, especially optimized for deployments with a high number of end devices within the network. Already released as a suite of IETF RFCs, it intends to provide an M2M optimized alternative to HTTP and yet provide further advantages.

Defined by the IETF CoRE group [10] under the scope of the task force that rules the Internet protocol definition, it is the ideal companion to the 6LowPAN protocol, the IPv6 over low-power wireless personal area networks [11], and it provides connectivity from just between devices in the same constrained network to end-to-end connectivity over the Internet. It can be also used to create CoAP proxies that map more complex HTTP RESTful APIs. Those proxies will be able to provide cached information when an HTTP request arrives and the end device is in a sleep period.

CoAP is able to create and manage resources on devices, publish and subscribe data, manage multicast of data, provide device description when requested, and give mechanisms to tell if a device is powered or not.

5.4.5 XMPP

XMPP [12], Extensible Messaging and Presence Protocol, is an application protocol originally conceived as a communications protocol for open messaging applications; it was later adopted by the IETF and became a set of RFCs. Widely used in popular messaging applications as Google Talk, it uses XML as the data representation language and defines publish-and-subscribe mechanism. Given these capabilities, the protocol has made their way out of the messaging world and is also used as signaling protocol for rich media communication systems, gaming, file transfer, social networks, communication with connected consumer goods, smart grids, and even M2M/IoT applications.

As in MQTT, XMPP uses hierarchical topics, but it addresses initial unique identifier by using a "Jabber" e-mail address following a format like an e-mail address, that is, "user@domain." The original XMPP implementation requires a constant TCP link to be established and lacks an efficient binary encoding, so it is not the more suitable solution for low-power M2M networks, but recent works (2014) are improving the standard with M2M constraints on mind by the following extensions, all of them yet in experimental state: XEP-0322 (Efficient XML), XEP-0323 (Sensor Data), XEP-0324 (Provisioning), XEP-0325 (Control), and XEP-0326 (Concentrators) [13].

5.5 M2M market as internetworking enabler

By just providing an overview of the existing internetworking issues, the chapter has introduced to the reader into the enormous fragmentation existing in current M2M deployments. Until now, only technical considerations have been taken into account, but the technical analysis is incomplete if M2M market actors and drivers are not scrutinized in more details.

5.5.1 Market actors

The M2M market has been fragmented into several roles depending on the services that they were able to provide. A noncomprehensive list of those M2M roles includes the following ones:

- *Hardware manufacturers:* Including device developers, communication module providers, and IC manufacturers. The outcome of these will be to provide M2M-ready hardware able to sense, act, or manage something on one side and connect over a capillary network or the Internet on the other. This category also includes the developers of the different M2M gateways available on the market.
- *Telecom operators:* Initially focused on just one objective, selling as much SIM cards as possible, they are reinventing their roles lately by offering new disruptive operator services, point-to-point dedicated M2M connection, or pure Wi-Fi offers or becoming a full vertical solution provider.
- *Cloud and middleware service providers:* Evolved from traditional ISPs or cloud provider services in some cases or conceived over traditional ones, the objective of these cloud and middleware offers for M2M is to fill the nonstandardized gap that traditionally has existed between the M2M gateways or IP-enabled end devices and the service providers. Presenting additional added value features, these services still have to find their precise role in the M2M world by integrating new standards (MQTT, CoAP, etc.) and creating alliances with application developers, hardware manufacturers, and telecom operators to be able to provide full validated verticals.
- *User application developers:* Focused on M2M applications, they will develop applications over custom implementations or a given middleware.
- *System integrators and vertical solution providers:* This last group is the one able to gather all the previous actors and provide a full vertical M2M solution. In some cases, it just might be one of them who takes this responsibility, though telecom operators and big telecom companies are taking the lead on this role.

5.5.2 Preliminary CAPEX and OPEX considerations

When it comes to the real world and assuming the costs and the risk of starting the deployment of a new M2M technology or any given service, a detailed study of the impact of CAPEX and OPEX is needed. A procedure ought to be set up in order to have a clear view on the financial impact, the time of amortization, the expected incomes, and the costs associated with the future operation of the deployment expects a positive outcome.

Many could be the questions to ask in the early stages of planning a new deployment of M2M technology; all those related to the convenience of the technology, maturity, reliability, cost of equipment, cost of installation, evolution, diversity of providers, compatibility with standards, and existing ecosystem of devices and developers shall be consistently asked in order to recreate the best picture of what might be the result, its affordability, and a coherent financial plan.

Given this initial picture, the technical approach might be different depending on which actor is taking the leading role in the deployment, but the outcomes from this phase shall always be taking into account regardless of the company or government interests.

5.5.3 The role of traditional operators

Traditional operators are "giants with feet of clay." They have overwhelming resources, the best network capabilities, and the necessary decision power to become influencers in the evolution and standardization in the market. But they are slow to react toward new trends in the market. And in the special case of the M2M technologies, having such heterogeneity of technologies makes them to move even slower. Until now, the core of the M2M business for these market players has been mostly focused on the volume of SIM cards and data service contracts signed. There are no innovation, no device development, and no middleware or cloud support. It is only in the latest times when some of them have made significant moves to become full vertical and horizontal providers by creating the necessary ecosystem of knowledge and companies and by introducing new cloud/middleware systems and custom applications.

Being strong in cellular and wired connectivity, they suffer the lack of standardized M2M radio technologies, and the commercial constraints limit the radius of action of their deployments in some cases. That is why new operators and alternatives are appearing in the market.

5.5.4 New M2M operators

New operators have been emerging in recent years, and their gaining momentum might be crucial in the communications role for M2M deployments. They are not always independent from the traditional operators as in some cases they might just be new business divisions, but they can be considered apart from the existing xDSL, cellular, or cable ISPs.

In this area, two groups of technologies are leading the way as the basis of new full M2M operators:

- *Wi-Fi-only operators*: Wi-Fi is undeniably one of the most popular wireless access technology today (2014), and though it is generally not as reliable as cellular data communications solution (e.g., there are no handover procedures and no quality of service guarantees), its enormous market penetration in the home markets has created areas with great coverage. This fact, combined with the deployment of high-capacity access points by some operators and the lower cost of the enabling hardware, is making Wi-Fi a technology to be considered as a potential bearer for M2M communications.

- *Long-range M2M communications operators*: As it has been stated before in the chapter, it is a fact that there is a lack of custom M2M protocol and radio technologies, especially when it comes to optimizing the coverage and minimizing the energy consumption and cost of the devices. It is under these premises that new technological proposals have been introduced to fill this technological gap. As an example, there are two names that can be highlighted as disruptive proposals:

  - The French company SIGFOX™, who has developed a very low-cost technology over the 868 MHz (Europe) or 902 MHz (the United States) ISM band and has led the deployment of base stations throughout Europe in association with local partners. With simple base stations covering a few kilometers, and with a very limited spectrum usage, just a maximum of 140 messages uploaded per day with a maximum payload of 12 bytes [14], the usages in M2M are mostly suited for applications in telemetry and metering. The yearly cost of the service depends on the size of the deployment and the data transmitted, but prices start at 1$ per year, so that is a real competitor for traditional operators.

  - In the United Kingdom, there is a proprietary but open technology called Weightless [15], which intends to use the spectrum liberated by TV stations, the so-called TV white spaces, for M2M data communication. The premises of the technology are a network coverage of around 5 km, an IC cost of 2$, low yearly fees starting at 1$, and 10-year battery duration. Compared to SIGFOX, Weightless is not an individual company and it is managed by the Weightless SIG, or Special Interest Group, and no commercial deployments are yet available, though several trials have been executed and the technology is backed by big industry names such as ARM, BT, Accenture or Huawei, who has recently acquired Neul, the company behind the initial deployment of Weighless in the UK. This is a must-follow technology with an audacious proposal that combines low-cost hardware architectures with cognitive radio techniques.

5.5.5 M2M hardware vendors

The hardware vendor market is evolving very quickly, thus meriting a separate section. Instead of dealing with specific devices and vendors, the focus will however be put on the key issues that M2M has brought to the hardware vendors and how the market is evolving fast to solve them. These can be summarized in the following topics:

- *Cost and capabilities of processors and memory:* Both in the configurations of end devices or gateways, the features and cost of current processor solutions have boosted the M2M market enabling lower-cost but high-capacity features. In this area, it has been the ARM Cortex™ family that has created a revolution in the market by enabling cheap access to 32-bit technology with higher memory footprints that significantly improve the legacy solutions, dominated during years by very limited 8-bit solutions. As an example, we can see a clear evolution in commercial development solutions during the last years. The company TST, which the author of the chapter is the CTO, provides (as of 2014) development solutions of under 100€ based on a 32-bit STM32F1x microcontroller based on an ARM Cortex™-M3 core, running at 72 MHz, with a memory footprint of 96 kbytes of RAM and 1 Mbyte of flash. Only a couple of years before, the company Libelium released their Waspmote, based on an (only) 8-bit ATMEGA 1281, capable of running at a maximum speed of 16 MHz, with memory footprint of 8 kbytes of RAM and 128 kbytes of flash. Comparing similar processor solutions, STM32F103RB vs. ATMEGA2560, by the measurements done by the Embedded Microprocessor Benchmark Consortium [16], it can be pointed that the performance of the processor is improved by factors of 25 times, going down to 3 when both processors run at the same clock frequency. The conclusion to this is that these new solutions enable the inclusion of operating systems, multitasking, network protocols, concurrent applications, and enhanced security that was not possible to have with old legacy solutions.

- *Cost and commercial offer of communication modules:* The cost of including autonomous technologies, such as full-stack Wi-Fi (IEEE 802.11b/g), Ethernet (IEEE 802.3, 100 Mbytes), GPRS, and selected mesh or point-to-point radio solution, has moved below the barrier of the 10$. Though some others like 3G, 4G, and Wi-Fi based on 802.11n are usually over this threshold, this breakthrough enables nowadays the inclusion of the first ones in conjunction with limited capability processors and creates connected M2M devices at a very reasonable cost.

- *Cost of fabrication and going to the market:* Producing and mounting PCBs were never easier than today. Complex free or low-cost software CAD solutions are available on the market (e.g., CadSoft EAGLE PCB), and the Internet has created a big market of PCB and mounters quite price-competitive. Though it is true that for optimized and large deployments there should be a deep study and negotiation of the prices and periods, the cost of prototyping and creating small series of electronics (tens or hundreds) is completely affordable, even for start-ups and SMEs.

- *The impact of open hardware in M2M:* The launch of open hardware solutions like Arduino and Raspberry Pi has created a huge ecosystem of "makers" interested in electronics as hobbyist, but that has enabled a very easy access to real hardware for prototyping at a very low cost and using completely free software solutions. Nowadays, protocols like MQTT and XMPP for M2M can be easily tested on a 25$ Raspberry Pi or any high-featured Arduino clone board like Intel Galileo [17] or STM32 Nucleo [18]. Just a few years ago, only the cost of development platforms and development IDEs, including JTAGs, needed the investment of thousands of dollars to have similar results.

5.5.6 M2M cloud/middleware and software application providers

Another important sector in the M2M landscape that has seen an important growth is the one dedicated to offer a cloud or middleware service for exchange data and control information between the deployed devices and the final user applications. The more strict requirements in areas such as reliability, size of the network, scalability, security, and standards compilation have made the "in-house" solutions used till now in M2M as less and less valid choices. The market has interpreted this need by the apparition of different proposals that will cover these requirements and in several cases will also have extra features like scripting capabilities, multiple choices of access protocols, and ready-to-use APIs for web, mobile, and desktop development.

Solutions like Xively™ (previously Cosm/Pachube), theThings.IO, ThingWorx™, EVRYTHNG™, Carriots™, Cumulocity™, Axeda™, Telefonica's SBC™, IBM's IOC™, and Lhings™ will provide APIs in different protocols like the ones presented before on the chapter (RESTful, MQTT, XMPP, etc.) or ready to use for different platform and programming languages, a price plan based on connected devices, data transferred, and number of connected devices and an ecosystem of compatible devices, that is, gateways, end devices, and repeaters.

In fact, many solutions are available on the market today, and big cloud players like Google and Amazon have not entered yet. Given that this is a sector that is going to evolve very quickly, it needs to be built around standards and needs to offer added value and integrated M2M verticals to provide an attractive product.

5.6 Future trends

M2M has been around for a while and currently provides the underlying connectivity paradigm of the emerging Internet of Things (IoT). The fact is that IoT deployments by nature base connectivity on IP compliant solutions if true scalability is sought. Issues and problems around internetworking will thus likely reduce over the coming years. Though there are different efforts to regulate and find a solution to these issues and problems through standardization, it is the market that is setting the trends by providing simple and working solutions.

The keys for a successful future in the world of M2M and IoT will be the following:

- A future 5G definition shall succeed in what current 4G/LTE was not able to and define the necessary mechanisms for enabling a real and optimized M2M communications. The market needs low-cost, low-complexity, long-battery-life, and SIM-less devices with worldwide connectivity.
- In the meantime, solutions like Weightless and SIGFOX will evolve their deployments, open their standards, make optimal usage of the frequency bands by means of really performing a fair use of white spaces in the spectrum, and manage different qualities of service, always assuring reliability and having best-in-class cost and power consumption figures.
- The current scope of heterogeneous internetworking standards that solve only part of the problem is causing the creation of different consortiums to provide solutions to this problem. Under this scope, the activity of projects like Qualcomm's AllJoyn [19] or Internet standardization bodies like the IETF, working on CoAP and 6LowPAN, and OASIS [20], working on MQTT and MQTT-SN, shall be closely followed.
- New players should enter the M2M and IoT world. New innovative networks based on disruptive technologies like open air optics, low orbit satellites, and stratospheric drones or balloons will have their role to play in the future. It is significant that big companies like Google [21] and Facebook [22] are taking positions in those areas.
- The market shall also tend to have a more optimal usage of existing communication networks, by means of having communication modules with lower cost or enhancing legacy protocols with low-power versions (e.g., Bluetooth low energy or low-power Wi-Fi).

In conclusion, the future of a successful M2M and IoT internetworking will rely on the fast adoption of industry-accepted standards able to provide seamless connectivity between all the elements in the network and create the mechanisms to provide all the required associated services. The work is ongoing, but it is at a stage where many scattered pieces still need to be put together.

References

[1] B. Latré, P. De Mil, I. Moerman, B. Dhoedt, P. Demeester, Throughput and delay analysis of unslotted IEEE 802.15.4. Available from: http://www.academypublisher.com/jnw/vol01/no01/jnw01012028.pdf, 2006 [Accessed 30 March 2014].

[2] D. Chen, M. Nixon, A. Mok, WirelessHART™: Real-Time Mesh Network for Industrial Automation, Springer, 2010.

[3] A.S. Tanenbaum, D.J. Wetherall, Computer Networks, fifth ed., Pearson, USA, 2011.

[4] IEEE 802.15 WPAN™ Task Group 4 (TG4), Available from: http://www.ieee802.org/15/pub/TG4.html, 2014 [Accessed 30 March 2014].

[5] Philips hue API, 201, Available from: http://developers.meethue.com/ [Accessed 30 March 2014].

[6] MQ Telemetry Transport, 2014, Available from: http://mqtt.org/ [Accessed 30 March 2014].

[7] Mosquitto, an Open Source MQTT v3.1/v3.1.1 Broker, Available from: http://mosquitto.org/, 2014 [Accessed 30 March 2014].

[8] Paho project, Available from: http://www.eclipse.org/paho/, 2014 [Accessed 30 March 2014].

[9] A. Stanford-Clark, H.L. Truong, MQTT For Sensor Networks (MQTT-SN) Protocol Specification, Available from: http://mqtt.org/new/wp-content/uploads/2009/06/MQTT-SN_spec_v1.2.pdf, 2013 [Accessed 30 March 2014].

[10] Constrained RESTful Environments, CoRE working group, Available from: https://datatracker.ietf.org/wg/core/charter/, 2014 [Accessed 30 March 2014].

[11] RFC 4919, IPv6 over Low-Power Wireless Personal Area Networks (6LoWPANs): Overview, Assumptions, Problem Statement, and Goals, Available from: https://datatracker.ietf.org/doc/rfc4919/, 2013 [Accessed 30 March 2014].

[12] XMPP Standards Foundation, Available from: http://xmpp.org/, 2014 [Accessed 30 March 2014].

[13] XMPP Extensions, Available from: http://xmpp.org/xmpp-protocols/xmpp-extensions/, 2014 [Accessed 30 March 2014].

[14] SIGFOX technology, Available from: http://www.sigfox.com/en/#!/technology, 2014 [Accessed 30 March 2014].

[15] Weightless SIG, Available from: http://www.weightless.org/, 2014 [Accessed 30 March 2014].

[16] The Embedded Microprocessor Benchmark Consortium (EEMBC), Available from: http://www.eembc.org/, 2014 [Accessed 30 March 2014].

[17] Intel Galileo, Available from: http://arduino.cc/en/ArduinoCertified/IntelGalileo, 2014 [Accessed 30 March 2014].

[18] STM32 MCU Nucleo, Available from: http://www.st.com/web/catalog/tools/FM116/SC959/SS1532/LN1847, 2014 [Accessed 30 March 2014].

[19] About AllJoyn, Available from: https://www.alljoyn.org/about, 2014 [Accessed 30 March 2014].

[20] OASIS, Advancing Open Standards for the Information Society, Accessed from: https://www.oasis-open.org/, 2014 [Accessed 30 March 2014].

[21] Project Loon by Google, Available from: http://www.google.com/loon/, 2014 [Accessed 30 March 2014].

[22] Internet.org Project, Available from: http://internet.org/, 2014 [Accessed 30 March 2014].

Weightless machine-to-machine (M2M) wireless technology using TV white space: developing a standard

W. Webb
Weightless SIG, Cambridge, UK

6.1 Why a new standard is needed

Forecasts for the global machine-to-machine (M2M) market are of 50 billion connected devices by 2020. However, the market for machine communications to date has been weak with perhaps a few tens or hundreds of millions of devices. There are some cars with embedded cellular modems and some relatively high-value items such as vending machines are equipped with cellular packet-data modems. But the market today is only a tiny fraction of the size it has long been predicted to grow to. Our judgment is that this is because the existing wireless solutions do not meet the needs of the majority of the M2M marketplace. These needs include the following:

- Low cost, of both the hardware and the service. Many machines are individually of relatively low value—imagine, for example, a temperature sensor. Wireless modem costs need to be in the region $1–$2 and annual service charges less than $10, in some cases less than $1, to make it worth embedding wireless technology in such devices.
- Excellent coverage. To make applications such as smart metering viable, there needs to be coverage of near 100% of all devices. With many terminals deep within the home or even in basements, this implies better coverage than achieved with today's cellular networks.
- Long battery life. Many machines are not connected to the mains and so have to operate on batteries. Having to change the battery is at best an annoyance and at worst a significant expense. Battery life of 10 years or more is essential for many applications such as smart dustbins or gas and water meters.
- M2M functionality. M2M systems have some requirements that are unique. These include guaranteed message delivery, multicast transmission of messages to many devices, and the ability to handle widespread "alert" messages.

There is no current wireless system that comes close to meeting all of these requirements.

Cellular technologies do provide sufficiently good coverage for some applications, but the hardware costs can be $20 or more depending on the generation of cellular used, and the subscription costs are often closer to $10 per month than $10 per year. Battery life cannot be extended much beyond a month. Cellular networks are often ill-suited to the short message sizes in machine communications resulting in massive

Machine-to-machine (M2M) Communications. http://dx.doi.org/10.1016/B978-1-78242-102-3.00006-X

overheads associated with the signaling needed in order to move terminals from passive to active states, report on status, and more. And the most suitable cellular systems for M2M tend to be 2G, but these are now starting to be switched off—for example, AT&T recently announced the closure of its GSM network in 2017. So while cellular can capture a small percentage of the market that can tolerate the high costs and where devices have external power, it will not be able to meet the requirements of the 50 billion device market. Indeed, if it could, it would have done so already and there would be no further debate about the need for new standards.

There are many short-range technologies that come closer to the price points. These include Wi-Fi, Bluetooth, ZigBee, and others. However, being short range, these cannot provide the coverage needed for applications such as automotive, sensors, asset tracking, healthcare, and many more. Instead, they are restricted to machines connected within the home or office environments. Even in these environments, there may be good reasons why a wide-area solution is preferred. For example, an electricity supply company is unlikely to accept that their meter is only connected via, e.g., ZigBee, into a home network, which in turn connects to the home broadband. Were the homeowner to turn this network off fail to renew their broadband subscription or even just change the password on their home router, then connectivity could be lost. Restoring it might require a visit from a technician with associated cost. Maintaining security across such a network might also be very difficult.

Finally, it is critical that the technology is an open global standard, rather than a proprietary technology. With a wide range of applications, there will need to be a vibrant ecosystem delivering chips, terminals, base stations, applications, and more. The manufacturer of a device such as a temperature sensor will need to be able to procure chips from multiple sources and to be sure that any of them will interoperate with any wireless network across the globe.

Without a wide-area machine communications network that meets all of the sector requirements, it is unsurprising that forecasts for connected machines have consistently been optimistic.

While the needs of the machine sector have long been understood, the key problem to date has been a lack of insight as to how they could be met. Ubiquitous coverage requires the deployment of a nationwide network, and the conventional wisdom has been that such networks are extremely expensive. For example, a UK-wide cellular network can readily cost $2 billion with costs of spectrum adding another $1–2 billion. With the machine market unproven, such investments were not justifiable and would result in an overall network cost that would not allow the sub-$10/year subscription fees needed to meet requirements.

6.2 The need for spectrum

If existing technologies cannot address the market need, then the obvious answer is to develop a new one. However, before this can be done, radio spectrum for it to operate in must be found. Until this is done, the system design cannot be tailored for the spectrum available, and there will be little credibility that the technology will be able to be successful. The characteristics of spectrum bands vary. In the case of M2M, there is a need for a band that is as follows:

- Low cost
- Plentiful in terms of being some tens or hundreds of MHz wide
- Globally harmonized
- Low frequency, ideally below around 1 GHz where long-range propagation can be achieved

It needs to be free, or at least very low cost, to keep the investment cost low. An emerging technology cannot afford to pay billions of dollars in each major market around the world. It needs to be plentiful to provide the capacity to service billions of devices without needing a very dense network of base stations that would be expensive to construct in order to deliver sufficient capacity. It needs to be globally harmonized in order to allow devices to roam across countries and to enable the economies of scale needed to deliver <$2 chipsets. Finally, it needs to be low frequency to enable good range from each base station and therefore a relatively small number of base stations to provide ubiquitous coverage. All of these effectively relate to the cost of building large-scale networks, which in turn will impact the annual subscription levels that operators need to change M2M users. If these costs cannot be kept to below $10, then many applications will not be viable.

Unfortunately, low-frequency spectrum is in very high demand and so rarely becomes available in sizable quantity, is almost never globally harmonized, and where even a few of these attributes hold true is extremely expensive.

The lack of spectrum that meets all these requirements has meant that—up until now—the only option for wide-area machine communications has been to make use of existing networks, predominantly cellular.

6.3   TV white space as a solution

Since around 2010, companies and regulators have been actively investigating a new way of accessing spectrum by sharing it with the existing license holder rather than owning it outright [1,2]. This use of spectrum is often called white space access.

To understand white space access, recall that all wireless communications needs some radio spectrum. For cellular communications, this is dedicated spectrum often sold at auction. For applications such as Wi-Fi, this is unlicensed spectrum, set aside by the regulator to allow all to make use of it on an unprotected basis. As evidenced by the high prices paid at auction, spectrum is in scarce supply and finding bands that can be repurposed for new uses is becoming ever harder. This stimulated a search to find ways to make more intensive use of the existing bands. Observations of utilization of spectrum that was licensed showed that often, there appeared to only be about 20% in use in any given time and place, prompting the concept of shared use whereby others could opportunistically access the unused parts of the spectrum as long as they did not cause interference to the license holders.

Shared access could occur in any frequency band, but proponents chose to focus on the band used for terrestrial TV broadcasting. This tends to span the frequency band from around 400 to 800 MHz, varying somewhat from country to country. This band is advantageous because the relatively low frequencies allow radio signals to propagate further, because it is harmonized around the world and because TV transmitters are static and well characterized and hence relatively easy to share with. TV transmitters typically do not use the same frequencies in neighboring areas to avoid interference. The net result of this is that in any given area, there may only be around 25% of the frequencies actually being used, in principle leaving the remainder for white space devices.

Initial thinking focused on the white space device sensing whether the spectrum was in use in an area by looking for existing transmissions, but experiment and theory showed that it was too difficult to make a device able to be sufficiently sensitive to ensure it received transmissions in all circumstances. So subsequent work has moved to a geolocation database approach whereby devices determine their location, e.g., using GPS,[1] and report this to a database using pre-existing communications channels (i.e., not white space) and then, the database determines which frequencies are available for use for the device. The database does this through having as input the known location of licensed transmitters in the band, the anticipated location of receivers, and a set of rules to derive availability for white space usage. These rules include understanding how sensitive TV receivers are to interference and predicting how far interfering signals might travel.

However, white space is not without its issues. These are broadly regulation and interference.

Regulation for white space is still developing in many countries [3–5]. However, it is clear that white space access will require devices that have the following characteristics:

- Relatively low output power. The FCC has specified 4W EIRP for base stations and 100 mW EIRP for terminals. Ofcom is currently suggesting power levels even lower than this. These are an order of magnitude lower than cellular technologies.
- Stringent adjacent channel emissions. White space devices must not interfere with existing users of the spectrum, predominantly TVs. Hence, the energy that they transmit must remain almost entirely within the channels they are allowed to use. The FCC has specified that for higher power devices, the adjacent channel emissions at the edge of the band between the wanted and the adjacent channels need to be 55 dB lower than in-band emission. This specification is much tighter than most of today's wireless technologies.
- The need to frequently consult a database to gain channel allocation. Devices may need to rapidly vacate a channel if it is needed by a licensed user. They must consult a database to be informed as to the channels they can use and must quickly move off these channels as required.

Interference from licensed users can be problematic in white space. Many channels have residual signals from TV transmissions. Either these can be in-band emissions from

---

[1]Strictly, the master device needs to determine its location, and then, it can control "slave" devices that do not need to know their location but can be assumed to be in the coverage area of the master. This is how the low-cost Weightless system is configured with base stations reporting location but devices acting as slaves that do not need to be location-aware.

distant TV masts that are too weak for useful TV reception but still significantly above the noise floor, or, alternatively, they can be adjacent channel emissions from nearby TV transmitters, some of which are transmitting in excess of 100 kW. In addition, since the band is unlicensed, other shared users might deploy equipment and transmit on the same channels as the machine network, causing local interference problems.

These are not insurmountable issues. But no current technology has been designed to operate in such an environment and so would be suboptimal at best. For example, we have shown that in the United Kingdom, an optimized technology could access around 90 MHz of white space after all the interference issues are taken into account, whereas an existing technology such as Wi-Fi or WiMAX could only access around 20 MHz [6].

White space meets all of the requirements for M2M communications. It is unlicensed and so access to it is free. It is plentiful with estimates of around 150 MHz of spectrum available in most locations—more than the entire 3G cellular frequency band. It has the potential to be globally harmonized since the same band is used for TV transmissions around the world. Finally, it is in the perfect low-frequency band that enables excellent propagation without needing inconveniently large antenna in the devices. Access to this shared spectrum provides the key input needed to make the deployment of a wide-area machine network economically feasible. Hence, white space spectrum provides the key to unlock the machine network problem. But it comes at the cost of needing to design a new standard optimized for the particular access requirements and interference issues of white space.

6.4   Designing a new technology to fit M2M and white space

It would be possible to take an existing technology and to try to use it, or adapt it, for machine communications. However, almost all existing technologies are designed for personal communications[2]—their transmissions are at the direct instigation of things people want to do such as make a call or browse a web page. Machines are very different from people. Typically, their requirements vary in the following manner:

- Much shorter message size than most human communications (with the exception of SMS text messages). Most machines only send a few bytes of information, whereas a person may download many megabytes of information.
- More tolerant of delay. Most machine communications is relatively unaffected by a few seconds of delay, whereas people quickly find this frustrating.
- Generally predictable communication patterns. Machines often send data at regular intervals and so can be "pooled" on these occasions. People's communication needs are typically unpredictable and so contended access for resources is needed.
- Generally massively lower "willingness to pay" in that people will tolerate monthly bills of $50 or thereabouts, whereas for many machine applications, monthly bills must be $1 or less.

---

[2]There are efforts to adapt LTE to become a better fit for "machine-type communications," but these are moving slowly and finding it difficult to re-engineer LTE to a low-enough cost point.

Taking advantages of these differences allows the design of a system that is much more efficient, providing greater capacity than would otherwise be the case and hence having low cost. The predictability of most communications allows a very high level of scheduled communications as opposed to unscheduled, or contended, communications. The difference is akin to prebooking passengers on flights so that each flight is full, but not overcrowded, rather than just letting passengers turn up, as with most trains, and suffer the crowding problems that occur. By telling terminals when their next communication is scheduled, future frames of information can be packed very efficiently and terminals can be sent to sleep for extended periods prolonging battery life.

Scheduling brings many other advantages. The first is efficiency. Contended access schemes can only operate up to about 35% channel usage—above this level, the probability of access messages clashing becomes so high that very little information gets through. By comparison, scheduled access can achieve close to 100% efficiency. Scheduling can be enhanced by complex algorithms in the network that prevent terminals close together in neighboring cells transmitting simultaneously that ensure terminals suffering local interference are scheduled on frequency transmissions where interference is minimized and much more.

Another attribute for M2M is that coverage is typically more important than data rate. For example, it is more critical that all smart meters can be read than what the data rate of transmission is—as long as it is sufficient to transfer data regularly. In fact, most machine communications can be measured in bits/s rather than kbits/s or Mbits/s. As an example, a smart meter will typically send around 20–40 bytes of information perhaps once every 30 min. This equates to an average of 240 bits per 30 min or 8 bits/min. There are applications that will require higher data rates, but speed is rarely critical. Hence, a good M2M system design will trade off data rate against range. This can be achieved by spreading the data to be transmitted, a technique used by GPS satellites so that the weak signal they transmit can be reliably received thousands of miles away on the surface of the earth. Spreading involves multiplying the data by a predefined codeword such that one bit of transmitted data becomes multiple bits of codeword. The receiver can then use correlation to recover the codeword at lower signal levels than would otherwise be possible. Codewords are selected to have particular correlation properties and typically have length 2^n (e.g., 16, 32, and 64). So, for example, multiplying the transmitted data by a codeword of 64 results in an improvement in link budget of some 18 dB ($64 = 2^6$, and for each factor of 2, there is a 3 dB gain) but reduces the data rate by a factor of 64. Most buildings have a penetration loss for signals entering them of around 15 dB, so spreading by this factor would provide indoor coverage to machines where only outdoor coverage previously existed. Spreading can be extended indefinitely, and some M2M solutions have spreading factors extending as far as 8192, providing great range but very low capacity. Large spreading factors do add complexity to the system design since they extend the time duration of important system control messages that all devices must hear, which in turn requires long frame durations. Such spreading would rarely be used in cellular communications. Here, data rate is often as important, if not more so, than coverage. A reduction in data rate of 64-fold would make many links very frustrating for users. Here is a clear area where designing for machines rather than people results in very different design choices.

Another important factor, at least at this early stage of the market, is flexibility. It is far from clear what M2M applications will emerge. Even the balance between uplink and downlink is unclear—for example, smart meters will likely generate predominantly uplink traffic, while software updates, perhaps for car engine management systems, will have large downlink messages. This suggests that systems should be time division duplex (TDD) in order that the balance between downlink and uplink can be changed dynamically.

M2M systems should make the terminal as simple as possible, keeping complexity within the network. This is contrary to the trend in cellular communications where handsets have been becoming ever more powerful and complex and there is a general design goal to move intelligence to the edge of the network as far as possible. There are two key reasons to keep M2M terminals simple. The first is to keep the cost as low as possible—as mentioned earlier, many applications require chips with costs of the order $1–$2. The second is to minimize power consumption for terminals that are expected to run off batteries for $10+$ years. This means that, for example, multi-antenna solutions should be avoided and that terminals should not be expected to make complex calculations to decode their messages. Even an apparently simple decision, such as requiring a terminal to respond on the uplink of a frame where it receives a message on the downlink could require it to process the downlink message much more rapidly, needed a more powerful processor. Careful design throughout is needed to achieve minimal terminal complexity. As an example, cellular communications tend to adopt turbo-coding error correction systems. These increase the link budget by 1–2 dB typically over simpler error correction systems such as convolutional coding but massively increase the complexity of the receiver. With cellular handsets having multicore processors, this is of little consequence, but the impact on a simple low-complexity ARM core in a machine would be severe.

Finally, there is likely to be an imbalance within an M2M network where the base station has much more power and processing at its disposal and so can have a greater range than the terminals. This is of no value since the terminals need to be able to signal back and so the link budget must be balanced. With base stations transmitting often around 4 W (36 dBm) but battery-powered terminals restricted to around 40 mW (16 dBm), there is a 20 dB difference. This can be balanced by a combination of using narrower bandwidths in the uplink and using greater spreading factors.

Designing M2M solutions does not require any technological breakthrough. But it does require great care in understanding the implications of each decision, and it needs a system design that is radically different from a cellular network, with design decisions often appearing contrary to the conventional wisdom of the day.

As well as optimizing for machines, the design of a new system needs to be optimized for shared white space access. Shared spectrum is different. Shared access to the TV band is the first time when unlicensed users are allowed to mix with licensed users as long as they do not cause any interference to those users [7,8]. Any system operating in TV white space should adhere to the following design rules.

First, it needs very low levels of out-of-band emissions. This minimizes interference caused to licensed users and so maximizes spectrum availability. Achieving such low emission levels can be simplified by selecting modulation formats that have excellent out-of-band characteristics. These tend not to be the modulation systems in use for high-data-rate communications such as OFDM, which are optimized for throughput and overcoming multipath reflections rather than adjacent channel emissions but instead the simpler single-carrier approaches used in technologies like GSM.

Next, the system needs to tolerate interference caused by other unlicensed users that can be random and sporadic. This can be achieved using frequency hopping to rapidly move off compromised channels. However, hopping in a network works much better with central planning to avoid neighboring cells using the same frequencies. Optimal planning, where different frequencies may be available in different cells and the sequence may need to dynamically adapt to interference, is complex and requires new algorithms.

Where interference cannot be avoided, the system needs to be able to continue to operate. Spreading, as discussed above, can also be useful to work in channels with interference, again trading off range (or tolerance to interference) against data rate. Base stations can often experience significant interference from distant TV transmissions and require mechanisms such as interference cancelation to reduce its impact. This can be achieved by the base station steering a null toward the TV transmitter. The downside of such a null is that any terminals located in this direction will not be able to communicate with the base station. However, typically, a base station will hop over a number of frequencies, and the interfering TV transmitter will differ for some of these and so be in a different direction. This allows communications with terminals in a null on one of the hops by scheduling their communications on a different hop.

Finally, where there are few white space channels available, it can often be possible to increase availability by transmitting with lower power and hence causing less interference. Power control is therefore critical, again coupled with spreading where needed to regain the range lost from the lower power.

Either of these issues—communicating with machines and making use of white space spectrum—would be sufficient to justify the design of a new standard. Taken together, they make a compelling case for designing a new technology optimized for machine communications in TV white space.

6.5 Weightless: the standard designed for M2M in shared spectrum

Weightless is the name given to a new technology designed to meet the requirements of the M2M sector. An overview of Weightless is shown in Figure 6.1. The design of Weightless can be segmented into the classical layers of communications systems—the physical layer or PHY, the medium access layer or MAC, the security features, and the applications interface. Broadly, the PHY is designed to handle white space issues and the key needs of machines, while the MAC provides for the features and facilities needed. Here, we discuss each of the layers, starting from the top down. This is necessarily a high-level overview of Weightless, and much more detail can be found in Webb [9].

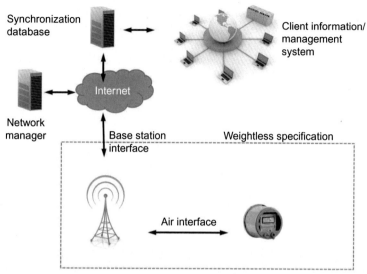

Figure 6.1 Overview of Weightless.

6.5.1 Applications

There are likely to be thousands of different M2M applications. It did not seem sensible to define a separate "profile" for each within Weightless, and instead, a general applications interface was standardized. This was kept as simple as possible to maximize flexibility. Broadly, it allows a terminal to register applications with the network and then to send and receive data from these applications. A terminal can register multiple applications, and indeed, most terminals are expected to have at least two, one being for device maintenance such as software upgrade and application registration and the other being the data channel for the main application.

As well as simple sending and receiving of data, machines have requirements for multicast downlink calls and alert uplink messages. Multicast calls are when the same data are sent to multiple machines—for example, a change of electricity tariff or a device software update. Alert messages are when the machine needs to send an urgent, unscheduled message such as a notification of a power failure. Within Weightless, it was decided to treat each of these as a separate application. This then allows, for example, selective cancelation by the network of alert messages without canceling other messages from the same device. A good example of why this is important is when a power outage occurs in a town. All the electricity meters might try to simultaneously send an alert message, potentially overwhelming the network. Once the network has received a few such messages, it does not need any more and can cancel all others. It can send a message to all the "alert" applications in all electricity meters asking to cancel all transmissions until further notice without fear of impacting other traffic such as scheduled meter reading. A terminal can register in as many multicast groups and with as many alert applications as it wishes, subject to verification by the network that it has appropriate rights to do so.

6.5.2 Security

Security is important in any wireless communications system. For machines, there are slightly different requirements to people. Often, authentication is more critical than encryption. For example, a smart meter needs to make sure it is sending data to the correct network, and equally, the network needs to be sure it is getting readings from the right meter. However, the readings themselves may not be overly confidential. Nevertheless, Weightless decided on a security solution as robust as that used for cellular or Wi-Fi, with 128-bit encryption. Security in Weightless is based on a "shared secret"—the same approach as used in cellular. Each chipset has an individual secret key known as K-Master that is securely embedded within it before activation. This key can never be read directly from the chip. A master security database securely stores the K-Master for each device against its unique device identity. The master key is then used to derive multiple different operational keys for authentication, for encrypting user data and for encrypting control information. The key derivation process, the authentication process, and the encryption system are all based on open industry standards, mostly published by NIST. The advantage of using such standards is that they have been subject to rigorous tests and attacks and have been proven to be secure.

When a terminal is first activated, it completes a mutual authentication process with its service provider. The service provider then encrypts any data prior to passing it to the network operator. This ensures that data cannot be read or modified by the network operator allowing the terminal to use multiple networks if it wishes.

Multicast keys are sent by the network to the terminal when it registers for particular multicast applications. These keys are shared among all the terminals within an application and so are inherently less secure than for one-to-one communications. However, they can be frequently changed and groups disbanded and reformed as needed.

6.5.3 MAC

The MAC layer is responsible for dividing the radio resource among the various terminals. To do so, it defines a frame structure. This starts with the base station sending synchronization information to allow terminals to lock onto transmission. It then sends information about the base station such as the base station identity, the frequency hopping pattern in use, and the frame number. This allows a terminal to understand which base station it has attached to and how to stay attached to that base station. Next, it sends a resource map setting out how the rest of the frame is structured. This shows how much of the frame is dedicated to downlink transmission from the base station to the terminal, how much is scheduled to uplink where resource is given to specific terminals, and how much is contended to uplink, which any terminal can use to attempt to contact the base station for unscheduled transmission such as attaching to the network or sending an alert. For the downlink and scheduled uplink, each piece of resource is allocated to a particular terminal, with the terminal being informed as to the start time and duration of the resource. It then either listens or transmits during this period as appropriate.

M2M traffic is often characterized by very short messages, for example, a 50 byte smart meter reading. The MAC protocol is designed to add minimal signaling overhead to such messages to avoid highly inefficient transmission. This is done through flexible small packets with highly optimized header information. Every effort is made to minimize per terminal signaling—for example, the 128-bit terminal address is not used after registration but replaced with a 32-bit temporary address so that the resource map can transmit less information.

In order to maximize battery life, terminals do not need to listen to every frame. However, they should listen to the first frame in each superframe. These occur about every 17 min and contain information such as whether the hopping sequence will change, any other changes to cell parameters, and potentially unscheduled paging messages for terminals. The system has been designed such that terminals with limited information to transmit can have a battery life of up to 10 years.

6.5.4 PHY

A key starting point in defining the PHY is the conflict between excellent coverage requirements and yet low-power constraints due to both white space regulation and the need for long battery life in terminals. The only way to achieve long range with low power is to spread the transmitted signal. Hence, variable spreading factors from 1- (no spreading) to 1024-fold are a core part of the Weightless specification. Spreading is essentially a mechanism to trade range against throughput, so using high spreading factors can achieve significant range extension but at the cost of lower data rates. Happily, there is sufficient bandwidth in the white space frequencies, and M2M data rates are sufficiently low that more than adequate capacity and throughput can still be achieved even with high levels of spreading.

Use of the white space spectrum does not provide guaranteed uplink and downlink pairing, making time division duplex (TDD) operation essential. This in turn leads to a frame structure with a downlink part and then an uplink part that repeats periodically. The maximum spreading factor informs what this repetition should be since the header information at the start of the frame needs to be spread by the maximum spreading factor in use in the network in order that all terminals in the cell can decode it. If this header takes up more than around 10% of the frame length, then the system starts to become inefficient as signaling becomes a significant percentage of the total traffic. Simple calculations that show frame lengths of around 2 s are optimal. This would be overly long for person-to-person communications, with such a delay being highly annoying, but is not an issue for M2M communications.

The need for stringent adjacent channel emission levels suggests the use of single-carrier modulation (SCM). Because the terminals are often very low power (e.g., 40 mW) compared to base stations (which can be up to 4 W), the link budget needs to be balanced. This is achieved with a narrower band uplink such that the noise floor is lower. Using around 24 uplink channels per downlink has the effect of balancing the link budget.

Operation in white space requires good interference tolerance. This is achieved primarily using frequency hopping at the frame rate (2 s) so that the impact of any interference is restricted to a single hop rather than degrading the entire transmission. Frequencies with persistent interference can be removed from the cell hopping sequence. Other mechanisms to remove interference include the base station directing antenna nulls toward strong sources of interference, careful scheduling of transmissions to terminals to avoid the frequencies where they perceive the strongest interference, and the use of spreading to make the signal more resistant to interference when all these other techniques are insufficient.

The transmit part of the radio chain is shown in Figure 6.2.

The key features are as follows:

- Forward error correction (FEC) encoding uses convolutional coding to add extra redundant bits to the MAC level message.
- Whitening randomizes the bit stream by multiplying it by a known random sequence to make it approximate to white noise. This overcomes problems that can be caused if the data contain long strings of 1s or 0s, which might confuse synchronization systems or result in unwanted spurious emissions.
- Phase shift keying (PSK) or quadrature amplitude modulation (QAM) mapping encodes the data onto "symbols" depending on the signal-to-noise level available.
- Spreading multiplies the data by a codeword trading off extra range against a reduced data rate.
- The cyclic prefix insertion adds a repetition of the end of the frame to the start of the frame. This allows the received frame to be readily converted into the frequency domain uncontaminated by multipath from previous transmissions.
- The synchronization insertion adds known patterns of bits that can be used by the receiver.
- Interleaving is not used as most bursts of data are too short for it to bring benefit.

One of the key features of the PHY layer is that it accommodates a very wide range of path loss values corresponding to terminals close to the base station to some distance away—perhaps 10 km. Table 6.1 shows how the various combinations of modulation scheme, coding rate, and spreading can be used to span an extremely wide range of path loss and data rate values.

6.6   Establishing a standards body

A global standards body—the Weightless SIG—has been established to start the process of developing the Weightless standard. As it matures, the intent is to move it into ETSI in order to gain all the benefits that a well-respected recognized standards body can bring.

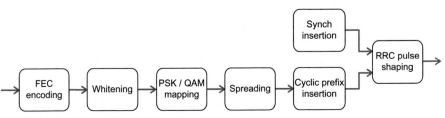

Figure 6.2 The transmit chain in Weightless.

Table 6.1 Modulation, coding, and spreading rates used

Modulation scheme	Coding rate	Spreading factor	Downlink PHY data rate (Mbps)	Required signal level (dBm)
16-QAM	1	1	16.0	−82.5
16-QAM	3/4	1	12.0	−86.5
16-QAM	1/2	1	8.0	−90.0
QPSK	3/4	1	6.0	−93.5
QPSK	1/2	1	4.0	−97.0
BPSK	1/2	1	2.0	−100.0
BPSK	1/2	4	0.5	−106.0
BPSK	1/2	16	0.125	−112.0
BPSK	1/2	63	0.040	−118.0
BPSK	1/2	255	0.010	−124.0
DBPSK	1/2	1023	0.0025	−128.0

A key element of a successful wireless technology is to be an open global standard. History tells us that the only wireless technologies that are widely successful are open standards—for example, DECT, Wi-Fi, Bluetooth, 3G, LTE, and GPS. There are many obvious reasons for this including global economies of scale, the benefits of competition, and the value of multisourcing and interoperability of multiple different devices. It was clear from the start that Weightless would need to be an open standard but not clear as to which path it should take to achieve this.

There are broadly three different kinds of standards entities:

1. Formally recognized standards-making organizations such as ETSI. These are bodies that have powers conferred upon them to develop standards that are automatically formally recognized often worldwide.
2. *De facto* standards organizations such as the IEEE. These are bodies that have no formal standards-making powers but have been widely recognized throughout the world as delivering standards.
3. Bespoke bodies established to deliver a particular standard such as the Bluetooth SIG. These are bodies set up for one specific purpose.

Each has advantages and disadvantages. In the 1980s and 1990s, the formal bodies and *de facto* entities were particularly effective. ETSI delivered DECT and GSM, while the IEEE delivered the 802.11 Wi-Fi standard. However, by the 2000s, their effectiveness appeared to be lessening as a result of bureaucracy and tactical behavior by member companies. This often resulted in long delays to standards that were sub-optimal because of the need to include a range of different technologies to appease members. As a result, few if any of these bodies have delivered any new standards that have been successful since 2000, although they continue to effectively deliver enhancements to existing standards such as 802.11. More recently, the bespoke approach has demonstrated success with Bluetooth and can be broadened to include certification, type approval, and even marketing and publicity. However, it has many disadvantages including the need to establish a new body with costs, legal complications, and so on.

It also needs to gain sufficient credibility that the documents it produces are widely recognized as *de facto* standards even if they have no formal recognition. Alternatively, they can seek formal recognition through ratification by a formal standards body, for example, the white space standard called Cognea was ratified by the standards body ECMA.

Weightless made the decision to take the third option—the establishment of a bespoke body. This was because its forming members felt that delivering a standard quickly and effectively was more important than a route that automatically conferred formal recognition. For that reason, the Weightless Special Interest Group (SIG) was formed in 2011 with work on the standard starting in earnest in the first quarter of that year and the first complete standard—version 1.0—being published in April 2012.

For a standard to be successful, it must both be technically excellent and have widespread support from an ecosystem comprising industry, regulators, application providers, users, and others. History has shown that the former does not automatically result in the latter—sometimes, standards that are less advanced technically win out because of their greater commercial acceptance. The technical design of Weightless has been covered in the earlier sections; here, we provide some information on how widespread support for the technology was achieved.

Support comes from awareness, credibility, and fit with business needs.

First, key players need to become aware of the new standard. This is a marketing function, using a range of tools such as conferences, publications, and direct approaches to inform key players. The target set of companies is relatively small, so widespread publicity as used in consumer advertising is not appropriate, instead targeted conference, publications, analysts reports (via briefing analysts), and so on can be used effectively. Early engagement of a number of key opinion formers in the industry can be very helpful as these in turn can brief a large number of key players. This stage can consume large amounts of effort, for example, in attending conferences around the world. Weightless representatives have been attending conferences around the world, seeding articles and conducting a wide range of programs to stimulate awareness.

The second stage is credibility. Once a company has become aware of the standard, they will seek to learn more through resources such as websites and emerging standards documents. Often, a first test that they will apply is to rate the likely chances of ultimate success based upon the companies and individuals involved. In the case of Bluetooth, the size and international reach of founder companies such as Intel, Toshiba, and Ericsson conferred immediate credibility (although in practice, none of these companies subsequently profited or drove the Bluetooth market forward). Weightless has established credibility through a number of mechanisms. One, like Bluetooth, is the use of major companies as "promoters" or board members. In the case of Weightless, this includes ARM, Cable & Wireless Worldwide (now owned by Vodafone), and CSR. Another is the personal credibility of the team driving the standard forward including key members of the SIG and of the companies writing key parts of the standards. Credibility, though, is very nebulous and also tends to exhibit a "tipping point" when enough large companies join that most others perceive they need to join as well. Standards entities need to work hard to reach this point.

The final stage is for prospective members to be able to see a business case to join or use the standard. This may be because they can sell products, services, training, and type approval; their core offering is improved by the product; and so on. In the case of Weightless, the breadth of its usage is both a potential help and a hindrance. It is helpful because it increases the potential pool of interested companies. It can be a hindrance because it can make it difficult to define clearly what the benefit might be. Business cases might include the following:

- Direct sales of chips for a semiconductor company
- Improvements to water-measurement systems for a company selling flow meters and similar into the water industry
- Provision of testing services for a direct fee for a test house

Many more are likely to emerge over time.

6.7 Conclusions

The value in machines having wireless communications has long been understood and a large market predicted for many years. That this has not transpired has been because of the difficulty of meeting all the requirements within the constraints of the available radio spectrum. These constraints changed significantly with the advent of shared spectrum that provides near-perfect spectrum with free access. However, the combination of the unique and unusual nature of that access and the very different characteristics of machine traffic compared to human traffic means that using any existing standard is far from optimal, hence the need for a standard designed specifically for machine communications within white space. The Weightless standard was completed in early 2013 and work is now under way on delivering products.

References

[1] D. Gurney, G. Buchwald, L. Ecklund, S. Kuffner, J. Grosspietsch, Geo-location database techniques for incumbent protection in the TV white space, in: DYSPAN 2008—3rd IEEE Symposium on New Frontiers in Dynamic Spectrum Access Networks, vol. 3, no. 1, October, 2008, pp. 232–240.

[2] M. Nekovee, Cognitive radio access to TV white spaces: spectrum opportunities, commercial applications and remaining technology challenges, in: DYSPAN 2010—4th IEEE Symposium on New Frontiers in Dynamic Spectrum Access Networks, vol. 4, no. 4, April, 2010.

[3] See the Ofcom consultation published at http://stakeholders.ofcom.org.uk/consultations/geolocation/.

[4] See the US Notice of Proposed Rulemaking published at http://hraunfoss.fcc.gov/edocs_public/attachmatch/DOC-301650A1.doc.

[5] The "Blue Book" available from http://www.dvb.org/technology/standards/.

[6] W. Webb, On using white space spectrum, IEEE Commun. Mag. 50 (8) (2012) 145–151.

[7] J. Wang, M. Ghosh, K. Challapali, Emerging cognitive radio applications: a survey, IEEE Commun. Mag. 49 (3) (2011) 74–81.

[8] M. Fitch, M. Nekovee, S. Kawade, K. Briggs, R. MacKenzie, Wireless service provision in TV white space with cognitive radio technology: a telecom operator's perspective and experience, IEEE Commun. Mag. 49 (3) (2011) 64–73.

[9] W. Webb, Understanding Weightless, Cambridge University Press, Cambridge, UK, 2012.

Supporting machine-to-machine communications in long-term evolution networks

7

A. Laya[1,2], K. Wang[3], L. Alonso[1], J. Alonso-Zarate[3], J. Markendahl[2]
[1]Universitat Politècnica de Catalunya (UPC), Barcelona, Spain; [2]KTH Royal Institute of Technology (KTH), Stockholm, Sweden; [3]Centre Tecnològic de Telecomunicacions de Catalunya (CTTC), Barcelona, Spain

7.1 Introduction to M2M in LTE

The main objective of this chapter is to discuss key challenges related to the support of machine-to-machine (M2M) and machine-type communications (MTC) in LTE. This technology was designed to solve mobile broadband demand for human beings and not for the particular requirements of MTC, which are quite different from human-type communications (HTC).

LTE is very flexible and provides capacity on the move to transmit real-time data for bandwidth-hungry applications. Unfortunately, such flexibility and capacity come at the cost of high hardware expenditures, high energy consumption, and, overall, a system that is too powerful for what MTCs demand for the majority of envisioned applications (not all of them). To make things more complex, 4G coverage is still not widely available, in contrast to 2G technologies that can be considered ubiquitous and universal [1].

Energy efficiency is a must; in many industrial verticals, communication network deployments are expected to have long device lifetimes, with many devices estimated to remain operational for decades [2]. A side effect of such long lifetime is the need to appropriately select the communication technology to be put in place. Based on this criterion, LTE can be regarded as the safest choice for the future. According to the OECD [1], 2G networks are planned to be decommissioned within the next 5–15 years, and therefore, acquiring GSM modules for MTC solutions might result in a loss of connectivity services in the case of spectrum refarming or if an operator decides to shut down a legacy network [2].

This chapter presents and describes the imminent technology and business-related challenges that must be faced in order to efficiently support MTC in LTE networks of the future.

Machine-to-machine (M2M) Communications. http://dx.doi.org/10.1016/B978-1-78242-102-3.00007-1

7.2 Main technical challenges and existing solutions

A nonexhaustive list of open challenges, which are the main focus of this chapter, is the following:

1. *Handling large numbers of devices*, orders of magnitude above what current networks can manage.
2. *Attaining low-energy consumption requirements*, measuring lifetime of deployments in decades.
3. *Ensuring low cost of LTE MTC devices*, to facilitate market penetration of solutions built upon LTE.
4. *Extending coverage*, to ensure single-hop transmissions between transmitter and receiver, thus facilitating simple communications.

These challenges are discussed in further detail in the following subsections.

7.2.1 Handling a very large number of devices

A huge number of devices are expected to connect to the communication networks in the foreseeable future, with some forecasts predicting figures of 50 billion devices connected by the year 2020 [3]. This figure does not even consider the coexistence with the current (and future) human users. Such increasing MTC traffic load may result in radio network and signaling congestion, which will inevitably increase delay and packet losses. Additionally, conventional HTC services may be affected by the arrival of MTC devices, and even total service outage may occur. Therefore, supporting a large number of devices per cell and securing the network availability are fundamental challenges that LTE needs to face and solve.

Access control is an effective means to prevent the overload of networks. The random access channel (RACH) of LTE is used for unconnected users to get connection to the network and thus be able to exchange data. The operation of the RACH is essentially based on Frame-Slotted ALOHA, which suffers from congestion when the number of contending devices is high.

To improve upon this, Release 9 of 3GPP specified the access class barring (ACB) mechanism to perform access control [4], as presented in Figure 7.1. With such an approach, all users are randomly divided into 10 different access classes (AC). The LTE network broadcasts a barring rate and a barring timer (mean duration of access control) to guide devices to run the random access procedure in the case of network overload. When user equipment (UE) tries to access the network, it first draws a random number and compares the number with the barring rate. If it is below the barring rate, it starts the random access procedure. Otherwise, it waits for a random period of time before reattempting access to the network. By varying the barring rate, the network can have some degree of control over the access contention. For example, different ACs can be allocated to high-priority users, thus allowing barring a particular AC to realize a prioritized access control if necessary.

3GPP Release 11 enhanced the ACB with a mechanism referred to as Extended Access Barring (EAB) for MTC. This mechanism aims at highly delay-tolerant

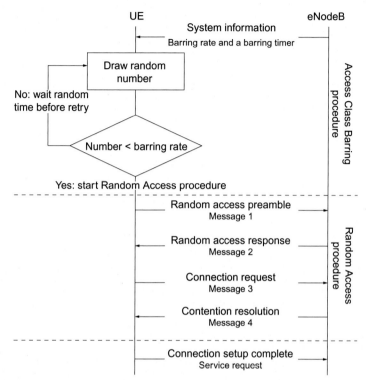

Figure 7.1 Access Class Barring procedure and Random Access procedure.

application, which can be controlled individually through specific control information called System Information Block (SIB). In a nutshell, the idea is that some devices can be blocked from accessing the system during periods of high contention.

In addition to the EAB, other methods have been proposed to cope with the RACH congestion [5], such as the following:

- Separate RACH resources for MTC and HTC devices coexisting in the same cell, thus ensuring that congestion caused by MTC load will not affect conventional HTC traffic.
- Dynamically allocate additional RACH resources for MTC traffic in the case that the network can predict in advance an overload caused by an MTC load increase, e.g., time-controlled MTC applications.
- Smartly assign MTC back off times to reschedule random access attempts in order to minimize congestion in a particular time.
- Allocate MTC devices in different access time slots. In a sense, this is similar to separating MTC from HTC, but adding further granularity to the classification of MTC traffic.
- Completely remove random access from MTC devices by using pull-based models where devices do not attempt to get access to the network, but it is the network that is the one polling devices when it is either known or predicted that they will have data to transmit in the uplink.

The RACH has been identified as a key area for innovation where improvements are needed to support a higher number of contending devices. However, it is not the only one. For example, avoiding congestion in the core network is necessary to avoid congestion of both data and control transmissions. For instance, an MTC application moving along cells will cause a lot of control information to be exchanged among collocated sites. Such kind of use cases requires particular attention, and thus, even though the problem starts at the radio interface with the RACH, it propagates along the LTE architecture. In this sense, the simplicity of the LTE architecture becomes appealing for MTC traffic. However, redesign of the control information for limited MTC applications must be explored as well.

7.2.2 Low-energy-consumption solutions for MTC

Low-energy consumption is always a desirable requirement for battery-operated devices. It is even more crucial for MTC devices when deployed in a large number and in scattered areas where battery replacement may be very costly, hard, or even impossible. Therefore, attaining high-energy-efficiency wireless communications is a key challenge for LTE. Different approaches can be considered here:

1. Use of the Discontinuous Reception (DRX) scheme, by which UE's radio circuitry is switched to sleep mode when there is no data exchange. This simple approach helps in reducing energy consumption during idle periods [6]. However, it inevitably increases latency and delay for bursty traffic, since there is no continuous connection established. The DRX scheme can be very appropriate for delay-tolerant applications, thus reducing energy consumption [7].

2. Use of Adaptive Modulation and Coding (AMC) and Uplink Power Control (UPC) mechanisms to adapt transmission rate and reliability. These two techniques must be exploited to improve energy efficiency considering the particular requirements of MTC applications, such as the case of small data transmissions [8].

3. Signaling reduction. Signaling can also be optimized to reduce energy consumption associated with MTC applications. For example, some MTC applications feature low mobility, having static MTC devices or simply moving along pre-established routes. Reduction, or even suppression, of control data related to mobility can be fostered.

4. Exploitation of short-range communications through cooperative schemes [9] and Device-to-Device (D2D) communications. LTE Release 12 studies the D2D scheme for the proximity and public safety services [10]. In traditional wireless networks, devices communicate with each other by always using the network infrastructures, as shown in the left side of Figure 7.2. D2D communications allow devices to communicate directly with each other without having to route traffic through the eNodeB, as shown in the right side of Figure 7.2.

7.2.3 Supporting low-cost MTC UEs

Legacy cellular technologies, such as 2G, 2.5G, and 3G, are capable of providing the requested service to a great majority of MTC applications at lower cost than LTE. For that reason, users are likely to choose GPRS instead of LTE modems if they are able to do so [11,12]. However, in the long term, operators are not willing to maintain two, or even three, network infrastructures. Therefore, reducing the cost of LTE devices is critical. Having noted this need, the 3GPP has presented a technical report on the provision of low-cost MTC devices based on LTE [11]. This report studies the Radio

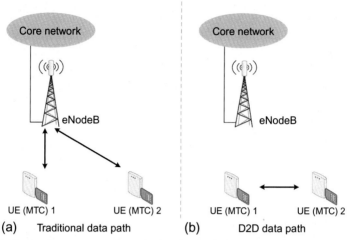

Figure 7.2 Two paths to communicate devices in proximity. (a) Traditional data path using the network infrastructure and (b) Device-to-Device (D2D) data path.

Access Network specifications and assesses the feasibility of defining a new type of UE, which is suitable for the low-end MTC market in terms of the price and performance. The report lists the requirements for the solutions devised to provide low-cost MTC, which can be grouped into three aspects:

1. The price of low-end MTC devices based on LTE should be comparable to that of the devices based on the GPRS.
2. The performance of low-end MTC devices based on LTE should be higher than that of the devices based on the GPRS in terms of data rate, spectrum efficiency, and power consumption.
3. The low-end LTE devices should hold backward compatibility and have no impacts on other conventional devices and the existing network architecture.

The intuitive solution to decrease the device cost is to reduce the UE complexity and simplify its implementation with controlled performance degradation. Therefore, the 3GPP report describes the current cost structure of a reference LTE UE and identifies the complexity by simplifying methods that may bring significant UE cost savings [11]. Such methods include the following:

- *Reduction of maximum bandwidth:* Reducing the maximum bandwidth of 20 MHz can significantly reduce the UE cost [11]. This can simplify the baseband processing, which may utilize lower complexity Fast Fourier Transforms (FFT)/Inverse Fast Fourier Transforms (IFFT).
- *Single RF receiver chain:* A regular UE has two antennas and two RF receiver chains to attain diversity in reception. Using only one RF receiver chain can reduce the receive filtering cost and the baseband processing functional blocks in terms of FFT, the channel estimator, and data buffer memory. However, using one RF receiver chain reduces the receive sensitivity, thus leading to a reduction of the link budget (coverage).
- *Reduction of peak rate:* To support the highest data rate, a UE needs to support large Transport Block Sizes (TBS) for DL and UL, a maximum number of simultaneously assigned

Physical Resource Blocks (PRB), and highest modulation orders. If the peak data rate is reduced, the UE cost can be reduced by means of simplifying the UL processing, turbo coding, and Hybrid Automatic Repeat Request (HARQ) buffering.

- *Reduction of transmit power:* The maximum transmission power of a regular UE is 23 dBm. This is typically achieved through a power amplifier circuit. Therefore, reducing the output power or even removing the power amplifier stage completely could contribute to a reduction of the UE cost. On the downside, a UE with reduced transmit power will decrease its link budget, thus shortening coverage. In order to maintain the UL coverage, the UE may need to use the low-level modulation coding and low-speed data rates to perform transmissions, which also lead to a lower spectral efficiency.

- *Half-duplex operation:* Duplexing in LTE is typically done in frequency—Frequency Division Duplexing (FDD). Therefore, a regular UE can transmit and receive data simultaneously using a duplexer. Alternatively, using Time Division Duplexing (TDD) would be desirable for MTC traffic, thus contributing to a reduction of the overall cost. In addition, a UE operating in TDD does not need to provide the processing power and memory for both downlink and uplink operations, thus reducing complexity of memory requirements. It is worth mentioning that such approach increased the complexity on the eNodeB schedulers, since they would still have to operate in full-duplex mode for regular users while in TDD mode for simplified MTC devices.

- *Reduction of supported downlink transmission modes:* For the LTE DL, a Release 10 UE supports one layer of spatial multiplexing and up to nine kinds of transmission modes. If the supported transmission modes are reduced to the two basic kinds, the Demodulation Reference Signal (DMRS)-based channel estimation and the Pre-coding Matrix Indicator (PMI) computation can be removed. Moreover, the Multiple Input and Multiple output (MIMO) detection/equalization algorithms can be simplified. The main drawback is that removing the precoding scheme may affect the DL performance due to a lack of precoding gain.

7.2.4 Providing extended coverage for MTC devices

As it has been exposed in the previous subsection, some methods that achieve complexity reduction do so at the cost of decreasing coverage in both the DL and UL. Unfortunately, some MTC networks are envisioned to be deployed in extreme coverage circumstances, such as the basements of buildings (e.g., structure monitoring) or beneath the ground (e.g., smart parking detectors), where radio signals suffer from severe attenuation. Therefore, coverage extension in such deployments is a key challenge for LTE. The study item in LTE Release 12 [11] also investigates the possible methods to overcome this challenge.

In particular, the focus has been so far on use cases where devices have fixed locations and is also assumed that UEs have low data rates and can tolerate large delays (non-time-critical applications). According to such characteristics, some coverage improvement methods that have been proposed are the following:

- *Hybrid-ARQ (HARQ) retransmission schemes*: When a receiver cannot decode a received message correctly, it does not discard the message but requests the sender to retransmit the message again through an HARQ mechanism. Each retransmission carries the same data information but with different redundancy bits (different redundant versions). Therefore, the receiver can combine the different copies to perform incremental decoding. This approach improves the decoding probability even when the attenuation of the channel is high.

- *TTI bundling*: The Transmission Time Interval (TTI) is the time unit for the eNodeB to schedule UL and DL data transmissions. TTI bundling functionality allows a sender to transmit the same data through consecutive TTIs; each TTI carries redundant versions of the same message. Therefore, the receiver can improve the decoding probability and thus extend coverage. Compared to the use of reactive HARQ, proactive TTI bundling can avoid the control channel overhead used for acknowledgments.
- *Repetition*: HARQ retransmission and TTI bundling can be generalized to the concept of *repetition*. Similarly, the control information can also use repetition to improve the coverage, such as the functionality of HARQ–ACK repetition.
- *Code spreading*: Through code spreading, an original code with small amounts of bits can be spread to a larger code with more bits, which is then transmitted using a greater number of TTIs.
- *Low rate coding*: Using more redundancy information increases the decoding probability in the receiver side and can be used to extend coverage.
- *Radio Link Control (RLC) segmentation*: Large data packets can be split into smaller fragments. Each small packet can be coded with low rate or low modulation order.
- *New decoding techniques*: New decoding techniques, such as correlation or reduced search space decoding, can be used to improve the decoding probability.
- *Power boosting/(Power Spectral Density) PSD boosting*: Power boosting technique can be used by the eNodeB to improve the power on the DL transmissions. Furthermore, for the eNodeB or UEs with limited power level, the boosting technique can be used to aggregate all available power together on a smaller bandwidth, which effectively increases the transmit power density on such bandwidth, thus enlarging the link budget.
- *Relaxing timing requirements*: Control channel performance requirements for the normal UEs are sometimes too high for the UEs in extreme deployments. For example, after sending a Physical Random Access Channel (PRACH) preamble, a UE needs to receive the random access response (RAR) in a time window. A UE implementing repetition coding would need more time to accumulate multiple copies of the RAR message to decode it correctly. Therefore, it would be desirable to relax the duration of waiting window.
- *Small cells*: Small cells can be deployed to improve coverage and increase the network capacity. At the cost of more complex network planning, the use of smaller cells has demonstrated to be an efficient solution to reduce interference, reduce transmission powers, increase network capacity, and boost transmission rates. Even though the study of small cells has received lots of attention, future work must be aimed at dynamically creating these small cells, exploiting cooperation among devices, and using the capability to exchange data among devices in close proximity (Device-to-Device Communications).

7.3 Integrating MTC traffic into a human-centric system: a techno-economic perspective

The variety of applications exploiting MTCs is astonishingly wide. Therefore, MTC traffic is very heterogeneous, turning into a challenge its classification in different levels. For this reason, it is widely accepted that there will be no single solution that fits for all. Indeed, different business strategies and technologies must be considered depending on the service requirements and scenarios of different applications based on MTCs. The focus in this chapter is to discuss the techno-economic view of the problem, aiming at identifying ways to solve the open challenges in the future.

Considering the ICT historical evolution from the operators' perspectives and their revenue models, there was an early pricing strategy for voice communications based on charging per usage. With the introduction of mobile data services, operators restructured their billing strategy for charging flat rates for data, and later on, they adopted data block models. MTC might be the next inflection point forcing a new pricing structure; especially if we consider that most MTC devices will only transmit very low amount of data but applications can generate higher profits.

In the following section, the most relevant aspects from a techno-economic point of view on MTC are discussed which correspond to the following:

1. *The impact of large number of devices on the network*, reinforcing the need for dense network (DenseNets) deployments
2. *The integration of LTE and capillary networks as a scalable solution to support* the additional traffic generated by MTC devices
3. *Technology migration and deployment strategies*, considering the relevance of MTC as an integral part of the system

7.3.1 The impact of a larger number of devices

MTC communications are mostly associated with small and sporadic data transmission, coming from highly dense networks, with a number of MTC devices orders of magnitude higher than HTC devices [3,13]. The latest forecasts published by Cisco Systems indicate that approximately 4.9% of the global mobile traffic will be MTC-related by 2017. In the same report, it is estimated that MTC devices will represent 19.7% of the global number of mobile connections [13].

Besides the air interface optimizations discussed in Section 7.2.1, scalable connectivity and hierarchical network architectures are needed in order to support MTC traffic in the network [14]. Even though devices can be directly connected through cellular links, the implementation of aggregation points plays an important role to allow the connection of a larger number of devices and increase scalability. Such aggregation points are typically referred to as MTC gateways and are capable of collecting and processing data from MTC devices [14].

As explained in Ref. [14], an increase in the number of devices will force a more spread adoption of hierarchical solutions in cellular networks, combining macrocells with the deployment of small cells and the integration of other short-range technologies, such as Wi-Fi. The integration of multiple access radio technologies allows the usage of additional spectrum (licensed or unlicensed) and the off-loading of cellular communication links and can help increase data rates, improve capacity, reduce interference, and improve upon energy efficiency.

7.3.2 The integration of LTE and capillary networks as a scalable solution

The term capillary network was raised to give name to those short-range wireless communication technologies, e.g., ZigBee, Wi-Fi, and Bluetooth, which can be used to extend the coverage and functionality of main cellular communication links.

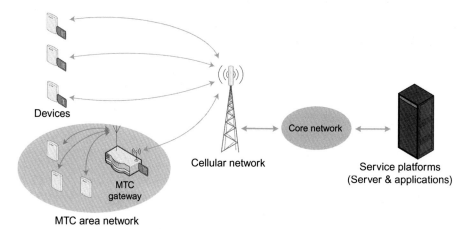

Devices

Core network

Cellular network

Service platforms
(Server & applications)

MTC
gateway

MTC area network

Figure 7.3 Capillary networks, represented as the area networks that are integrated to wide area networks by means of MTC gateways.

In many applications, these short-range technologies cannot constitute an integrated solution for an MTC application; however, they are ideal candidates to complement the functionality of cellular networks. Capillary networks are typically integrated with Wide Area Networks (WANs) through MTC gateways, thus creating the aforementioned hierarchical architectures; this is illustrated in Figure 7.3.

Next, an overview of the technical and architectural aspects of capillary solutions is presented, followed by a discussion considering the technical and economic implications for these technologies.

7.3.2.1   Overview of capillary networks

The IEEE and the IETF are two of the main Standards Developing Organizations (SDOs) that are working in this area. The IEEE focussed on the physical (PHY) and media access control (MAC) layer, while IETF focussed its work on the upper layers (excluding the application layer).

Arguably, the IEEE standards more relevant for MTC communications are the following:

1. IEEE 802.15.4 (which is used by the ZigBee Alliance) and all its subsequent amendments that are described below
2. IEEE 802.11 (which is used by the Wi-Fi Alliance) and all its subsequent amendments, among which it is worth highlighting the IEEE 802.11ah aimed at long-range low-power networks
3. IEEE 802.15.1 (which is used by Bluetooth)

The IEEE 802.15.4 has evolved in different amendments that aim at improving the standard for different applications [15]:

• *IEEE 802.15.4e*, which constitutes an extension to support industrial applications, e.g., factory automation and smart buildings

- *IEEE 802.15.4f*, targeted at active RFID systems for bidirectional communications
- *IEEE 802.15.4g*, for Smart Utility Networks
- *IEEE 802.15.4k*, aimed at critical infrastructure monitoring and ultralow-power operation

IETF's efforts in the area of capillary MTC are centered on

1. IETF 6LoWPAN (IPv6 over Low Power Wireless Personal Area Networks), which enables the use of IPv6 over constrained networks such as those based on the IEEE 802.15.4 Standard, by means of an adaption layer that performs, among others, header compression and packet fragmentation;
2. IETF ROLL (Routing Over Low power Lossy Networks), which addresses routing in low-power wireless networks;
3. IETF CoRE (Constrained RESTful Environments) [15], which aims at extending web services to devices in constrained environments, which cannot use a full IP protocol stack, by defining an application transfer protocol (CoAP) for M2M applications.

Therefore, the following layers could comprise a suitable protocol stack for an MTC deployment:

1. The IEEE 802.11.4-2006 PHY layer
2. The IEEE 802.11.4e MAC layer
3. IETF protocols: 6loWPAN, ROLL and CoAP integrated on top [16]

However, for applications that require higher data rates or seamless integration with existing products, a protocol stack made up with Wi-Fi would be more convenient. Indeed, recent studies have shown that the IEEE 802.11 can attain even better performance for MTC than any solution based on the IEEE 8021.5.4 Standards. For that reason, solutions based on Wi-Fi are deemed to gain market share in the coming years.

Regardless of the specific technology to be used, there are some cross technology challenges that still need to be addressed:

- *Interference management.* Most of communication technologies for the implementation of capillary networks operate on the license-free band at 2.4 GHz. This band is becoming more and more packed and cross system interference turns a severe challenge. For example, 802.15.4 networks running in the vicinity of 802.11 networks suffer from service denial due to the difference in the maximum transmission power. Therefore, the design of efficient interference management techniques is still an active area for further research.
- *Handling a large number of devices in the capillary domain.* Both the IEEE 802.11 and the 802.15.4 standards suffer from congestion when the number of devices is very high. This is due to the use of Carrier Sensing Multiple Access (CSMA) at the Medium Access Layer (MAC), which leads to very bad performance under heavy traffic loads. Therefore, the design of new ways of handling a great number of simultaneous connected devices becomes necessary.

The use of MTC Gateways or M2M Gateways [17] becomes fundamental when deploying these capillary solutions. They enable the interconnection of different networks operating under different radio technologies and can provide functions like protocol translations (between short-range technologies and LTE), resource management, device management, preliminary data processing and storage, and data compression and aggregation, among many others.

7.3.2.2   Techno-economic view on capillary networks

It is expected that a big share of MTC devices will lack of cellular connectivity and will only rely on capillary networks for their connectivity in local networks. It is then assumed that these devices will require to be connected to a gateway node, which could be connected to either the fixed network or cellular network infrastructure. In this section, we consider the feasibility of the capillary gateways to be connected to the MTC application servers via LTE as a backhaul technology.

There are several incentives to support this reasoning:

• Having cellular gateways reduces the deployment costs, since there is no need to provide fixed connectivity infrastructure.
• In remote areas, the access depends on cellular technologies since there is no other network infrastructure.
• It provides mobility capabilities to the devices. This case is applicable if the gateway also roams with the devices in the network.

The deployment scenario represents an important factor for the decision on the adoption of capillary solutions. If we consider an outdoor scenario, for the connection of smart reader in a suburban area, the deployment of capillary gateways would entail higher investments, not only for the equipment but also for the associated site leasing and maintenance costs.

A decisive factor in the decision-making is the number of devices to connect in the MTC application. When the deployment corresponds to a low number of devices, e.g., less than 20 devices, direct connectivity with cellular modems is the most cost-efficient solution; this is due to the fact that direct connectivity over cellular network excludes the costs associated with the deployment and maintenance of the capillary network and the required gateways. But as the number of devices in the network increases, the more cost-efficient the capillary solution will be, in terms of network deployment and connectivity fees. Therefore, it is desirable to opt for capillary solutions including gateways.

7.3.3   Technology migration and deployment strategies

Currently, there are two majorly debated views regarding the position of cellular networks in the MTC arena:

1. The first one corresponds to the establishment of dedicated legacy networks (i.e., GSM) for MTC. This could help separate MTC and HTC traffic to manage each of them independently. Nonetheless, the current amount of traffic generated by MTC devices might not provide enough revenues to justify the CAPEX and OPEX of maintaining an entire network up and running solely for MTC traffic. In addition, this implies the maintenance of licenses to use the dedicated spectrum for this exclusive purpose [2].
2. The second view proposes a migration from legacy networks to next-generation networks in order to allow spectrum refarming and the deployment of new networks capable of providing the required capacity and quality of service (QoS) for MTCs and HTCs seamlessly.

The most common strategy adopted by operators today is to deploy outdoor macro-cells and small cells to cope with the capacity requirements. Unfortunately, they fall short of providing the requested capacity and QoS for indoor-generated traffic.

In countries with high mobile broadband demand, MNOs are deploying denser heterogeneous 3G and 4G networks. Future deployment trends are having a strong focus on the feasibility of efficient network planning through denser network deployments in high demand areas in order to cope with the increasing capacity requirements. Research studies clearly state that the main enabler of data increase in the last decades is directly related to smaller cell deployments [18].

In order to reduce the cost-per-bit and allow a more effective reuse of bandwidth, smaller cell deployments have been studied. Results presented in Refs. [19–21] conclude that small cells and LTE femtocells bring benefits mainly on the reduction of deployment costs, depending on the pre-existing infrastructure of the operator, since higher savings are feasible in cases where the femtocells are not considered as part of the operator network; therefore, it does not constitute an additional cost for the operator.

Other research studies focus on the possibility of off-loading data traffic from the mobile network by means of Wi-Fi or femtocells [22]. The widespread adoption of Wi-Fi technology has motivated the integration of Wi-Fi with small cells, coining the term Integrated Femto-Wi-Fi (IFW) solution. The femtocell radio interface can be used for delay-constrained services, while throughput-demanding services can be served by the Wi-Fi radio interface.

MTCs are all about enabling services. For this reason, it is extremely important to understand the relationship between key players that will interact in the MTC ecosystem besides the MNOs. Some of the most relevant players are presented next [19,20]:

- *Mobile Network Operators (MNOs)*, which are the providers of wireless services, digital data pipes to transmit data. They are in charge of the control and operation of mobile networks and the management of customer relations, where customers refer to end users with a UE. Traditionally, MNOs have implemented complete vertical solutions and handle all different parts of the network. However, a shift is taking place, leading MNOs to reconsider their position and roles, moving from a data pipe provider to a complete service operator providing services, not just connectivity. This will be further discussed in Section 7.4.1.
- *Network Vendors (NVs)*, which are the manufacturers of the telecommunication equipment that deliver the technical assets for the MNOs. In recent years, NVs have expanded their traditional roles in the sector, and it is nowadays common to find NVs in charge of the operation and maintenance of the networks on behalf of MNOs. In this sense, MNOs outsource the operation and maintenance of the network in order to focus on the costumer relationship business. Related to the M2M ecosystem, important NVs in the market have developed vertical M2M solutions to handle device connectivity; their strong value relies on the fact the NVs have global and well-established relations with MNOs that give them a natural and beneficial position to promote the adoption of their solutions.
- *Managed Service Provider (MSP)*, which are third-party actors capable to reduce the operation cost of MNOs, offering services on top of the connectivity plane. The most representative case of MSP corresponds to the NVs that are in charge of the operation of the network; MNOs delegate these tasks to them mainly to reduce operation costs.

- *Data Aggregators*, which collect, store, analyze, and process the data generated by all those connected devices. Based on the sensitivity of the MTC data, MNOs or third-party actors are capable of handling this role but focussing on more detailed forms of analytical data processing needed. The challenges related to data aggregation in MTC correspond to privacy and data ownership, authenticity and security, and scarcity of data processing skills [23].
- *MTC Service Providers (SPs)*, which drive the MTC ecosystem and correspond to those companies that offer over-the-top (OTT) solutions based on MTC communications. These SPs normally prefer to adapt a service enablement kit in order to be able to offer their services over existing platforms unless the solution is either too complicated or implemented in a vertical mode that needs to be totally isolated regardless of the service enablement kits.

Today, it is still not clear how these players will interact in the future. Even when several solutions are already available in the market, there is no clear dominance of one solution, and the position and relationship among them vary depending on the application and the scenario in which they are implemented.

In the next section, the business implications of MTC in LTE are presented, highlighting the future challenges to overcome in order to enable the massive adoption of such solutions.

7.4 Business implications for MTC in LTE

The expected low traffic to be generated by most MTC devices will inevitably result in extremely low average revenues per device. For this reason, the main revenues are expected to come from the applications built on top of the transmitted data [24]. The major benefit of connected devices is directly related to seamless data availability. Moreover, in MTC applications, consumers are part of the aftermarket and deliver usage data and feedback to product manufacturers and MTC service providers (SPs), allowing cocreation of values [25,26]. Therefore, when the data represent valuable information and are processed appropriately, they can be exploited as a product.

It should be clear at this stage that there will not be one solution that fits all the technology requirements imposed by such a heterogeneous set of devices [1]. Following the same reasoning, both the technology and economic development of MTC communications will require a complex set of faculties that are unlikely to be covered by one single player in the industry. As discussed in Section 7.3.3, different players are pushing their view and solutions in order to position themselves in the market [19,20]. More importantly, new players are also entering the market and positioning themselves in the MTC ecosystem.

The shift in the relationship between players can be compared with the mobile broadband case, where the access to affordable and high-performance technology opened the market for OTT services and applications, resulting in a new set of actors that dominate the ICT market, reducing the traditional structure and control of MNOs [19,20].

MTCs are expected to be the next shifting point in the telecommunication sector, impacting on an extensive range of industries and services. The open debate is still on who will provide these services and what would be the relationship between the different actors [19,20].

7.4.1   Is there a need for a change in operators' mindset?

Previous studies, presented in Refs. [27,28], examined different cases related to MTC-related services in the market. Findings show how it is simpler to analyze the values and benefits of isolated solutions, but more importantly, it is highlighted how the implementation of successful services always requires radical changes from traditional business thinking, especially in terms of connectivity services, since telecommunication providers could not find a feasible structure to generate value from the M2M service. Therefore, alternative business relationships emerged in the market in order to successfully provide the services. In most cases, M2M solutions based on mobile communication are hampered due to the lack of a proper M2M business model, since the traditional vision for MNOs is based on a provider–consumer perspective like the one shown in Figure 7.4a.

As it has been explained in Section 7.3.3, M2M services are provided in a complex constellation of players, and the end consumer is rarely involved in a direct interaction with MNOs. The M2M ecosystem tends to be more like the one shown in Figure 7.4b.

7.4.2   The relationship between business challenges and engineers

Business studies on M2M have a strong focus on the analysis of the roles for incumbent and new actors needed in order to set up and manage M2M services and, more importantly, which are the key values that generate profits for the different stakeholders involved in each application. There are key challenges that still need to be addressed for M2M in the business domain, as pointed out by the OECD report [1]. For instance, system and network architectures proposed for M2M rarely consider the role and market power of existing players in the M2M ecosystem that could definitively affect the decision-making and future trends.

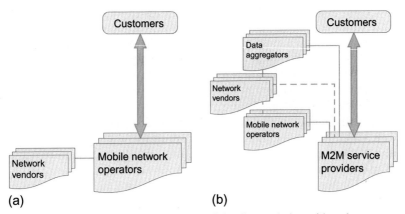

Figure 7.4 Change in the position of players and the direct relation with end customer. (a) Traditional provider–consumer perspective. (b) M2M ecosystem with complex constellation of players.

Also, current M2M services in the market remain as dedicated solutions developed and devoted to a single application, triggered by the high level of integration required with the targeted industry, i.e., solutions for energy efficiency are developed with direct focus on the energy sector, but connected vehicle services are developed with the car manufacture's perspective. Even within the same industrial sector, solutions are generally tailored for a specific company, leading to an increasing market fragmentation. Unavoidably, the consequence of these bespoke solutions is a higher development cost for M2M applications [29].

Another relevant aspect is the fact that M2M services are often provided in a complex and dynamic value chain, where the traditional provider–customer model adopted by MNOs does not apply [30], as presented in the previous section. Such complex interaction of players poses a challenge to the analysis of the real values and benefits for all the players involved. The real economic benefits are yet unclear for many applications. This has directly affected the adoption of large-scale solutions in M2M like smart cities in Europe, where there is no enough value for one player in order to take the lead and drive the ecosystem of players [28]. Specific M2M applications can be successfully deployed when their value is clear and the business thinking is adapted to the new market perspective. But the integration of solutions and large-scale application remain as open research challenges.

Two key challenges that directly limit any technical solution for M2M can be highlighted at this stage:

- It is difficult to change the position of traditional players that have significant market power.
- Players that are used to have the control position dominate some industrial sectors.

There are lessons to be learned by engineering companies that manage to take new positions in the market. The case refers to Ericsson, originally a network vendor (NV), which is now succeeding in their strategic move toward support of services based on connected devices. An excellent example is the "connected vehicle" case, presented in Refs. [19,20]. It describes the partnership between the car manufacturer Volvo and Ericsson, and it shows the new added value that involved different business thinking. The solution has the following value proposition:

- The car manufacturer will be able to offer value-added services either themselves or through third-party actors. Furthermore, they can gather valuable data on the operation of the car, which will improve maintenance and spare parts management.
- The drivers gain access to new range of applications and services.
- Third-party actors have a single channel to reach the end users.

Ericsson, who originally based their engineering business on selling network equipment to MNOs, has started to expand the set of activities and offers managed solutions to industrial actors interested in solutions based on M2M, providing an M2M service platform, on which the car manufacturer builds its services. In Figure 7.5, it is shown the change of position achieved by Ericsson, which was traditionally positioned only as a NV (following the description depicted in Figure 7.4); this shows the new type of business thinking applied in the M2M world, which is based on partnerships and rethinking of roles.

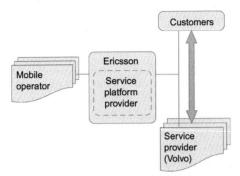

Figure 7.5 Redefined position for M2M service offering.

The challenge related to the control position of industrial player is evidenced in the study carried out in Ref. [28], which analyzes how the M2M projects related to smart cities and smart homes encountered similar barriers when dealing with different industrial sectors. Both projects pose the feasibility of having a common and shared network infrastructure in order to reduce the deployment and maintenance costs, lowering the entry barriers for new services based on M2M communications.

However, the interesting finding is that in both cases, dominant industry players like energy or media content companies showed unwillingness to share any type of infrastructure and are prepared to have their own system, and in some cases they even consider owning dedicated LTE networks for such purposes. There are several reasons for this: one is related to each company internal structure; they are used to operate in a certain manner and sharing assets will require deep restructuring, and since they already have a profitable business, there is no motivation to do so. Another reason is related to the direct customer and billing relationship, claiming that if the infrastructure is shared, it will be hard to differentiate the services and it will be more difficult to maintain their customer base. In any case, the barrier is more related to indisposition to change their current approach since the analysis presented in Ref. [27] concludes that even in cases where players are able to maintain the customer and billing control, and the communication infrastructure is only shared with companies that do not represent direct competition, the unwillingness to adopt a shared infrastructure remains invariable.

These challenges are open issues still to be addressed for services based on M2M communication, considering the position of important players, the increasing market power of new emerging companies, and differentiated market structure that varies in each country.

In the next section, business model options for M2M are discussed, focussing on the difficulties perceived by MNOs in order to be able to monetize from devices that do not necessarily generate the minimum required revenues.

7.4.3 Business models for M2M

It is agreed that M2M will expand the business possibilities in several industries, allowing new business model opportunities to take place; as it is stated on the OECD report [1], some examples of these new business models could be the following:

- *Pay as you drive insurance.* Allow charging drivers based on distances, location, and behavior. This could significantly reduce the cost associated with the insurance fee.
- *Products as services.* As explained in the OECD report (OECD, 2012), there are already companies providing light as a service (charging per lumen) or energy-saving, with charging according to the saving they generate. It is expected that M2M solutions use this type of business model to a large extent.

Nevertheless, defining effective business models for M2M in different industry sectors is a difficult problem that has not been solved in the market yet [31]. There are key challenges that need to be addressed. For example, many of the current solutions in the market have been developed as a complete vertical system, from connectivity to service provisioning [29]. Additionally, M2M services are frequently part of a dynamic value constellation where a provider–customer model does not apply [30].

All the discussion of new business models is centered on the M2M solutions, and little is mentioned about alternative business models for MNOs, especially in LTE networks, in relation to M2M communications. The reason for it is that when the traditional value network for the telecommunication sector is considered, the roles of MNOs are usually constrained to provisioning of technology to allow data transfer, shrinking the relevance and value of their service. The real limitation for MNOs is that revenues on M2M communications are simply too low to turn a profit just from the connectivity, even with the most efficient operation.

MNOs face an important challenge, since their traditional revenue stream is founded on charging based on a high amount of traffic per users. But M2M devices usually generate very low traffic and MNOs cannot base the charging on the traffic volumes; hence, alternative revenue models need to be considered [24].

In traditional HTC, users can be charged per device or small group of devices (SIM cards) such as smartphones and tablets and according to a subscription. The interaction between end users and MNOs is only related to billing and customer services in case of connectivity issues. In this sense, consumers have to adapt to the service offers given by the MNO or change to another operators.

The M2M case is entirely different, since the agreements are reached between MNOs and M2M SPs that have to manage large groups of devices with limited resources. International organizations like the 3GPP, ETSI, and OECD have identified several of the particular requirements from M2M users in order to effectively handle their devices [1,32]:

- Access to network status in order to differentiate network failures from device malfunctions.
- Ability to change the communication provider without replacing the SIM card inside each device and ability to have national roaming in case of network failure.
- Devices must be ready to operate out of the box but should be activated remotely once they are ready to be used.
- Roaming capabilities should be seamless in terms of cost.
- Guarantee network availability for long periods. In case of LTE, guarantee increasing coverage areas.

What this rationale tries to explain is the fact that MNOs should not only rely on their connectivity provision to monetize on M2M; on the contrary, solution-based business models, as recommended by Analysys Mason, could be more effective [33].

A solution-based business model implies that pricing on M2M should never be focussed on bytes per devices; instead, MNOs should focus on providing the necessary support they can provide in the operation of the services.

An example in the market of this approach is Telenor Connexion [33], which uses the solution-based business model. They base the price of the product according to the requirements of the M2M SP in terms of the level of service, type of information transferred over the network, managed service for devices, consulting and support in the integration process, and security requirements. The main objective should be to focus on service enablement, i.e., provide advanced M2M device management like remote software management, simplify communication between applications and devices, and simplify data collection [34].

7.5 Conclusions

LTE is today not suitable for supporting MTC in an efficient manner. Many technical challenges need to be solved, and optimizations must be carried out to make this technology the best option for supporting MTC. There are many reasons to believe that LTE and the subsequent family of standards (LTE-A) are the most suitable technology to facilitate the advent of the Internet of Things. All of them have been discussed in this chapter, where we have dealt with technical and business aspects related to the support of MTC over LTE networks. We have highlighted current works and described the different alternatives that are currently under consideration both in academia and in industry.

On the technical side, we have discussed open challenges and existing solutions to manage a high number of connected devices in highly dense networks, to reduce energy consumption, to reduce complexity and thus cost of devices, and to extend coverage for applications demanding extremely low data rates but aiming at very long life cycles.

On the techno-economic side, the emphasis has been put on the following aspects:

- *Small cells* and the use of short-range technologies (e.g., Wi-Fi) in combination with WANs are deemed to be the most cost-effective, scalable, and energy-efficient solution in ultra-dense scenarios. However, there is still a general unwillingness from many MNOs to deploy small cell or indoor solutions. This must change in the future so that true scalable and reliable M2M solutions can be deployed and operated.
- MNOs must face new competition in a highly dynamic market with many players searching for their clear position. The emergence of "independent" indoor network operators is just an example of the dynamics of the market and the arrival of new players to a traditionally static market. Also, NVs are becoming a strong force in the market since they are able to manage the connectivity of devices.
- Intermediary actors have a strong position; they control service platform and have relations with end users and SPs.

M2M is about *services* based on communication, not about *connectivity*. The heterogeneity of the solutions results in a complex ecosystem that involves many additional players that dynamically interact with MNOs.

The Internet of Things is around the corner, but we are still not there. Many challenges need to be faced at the technological and economic levels. This chapter aims at providing a holistic view of the problem, the current status, and future directions that must be taken to make the IoT a real revolution of the twenty-first century.

Acknowledgments

This work has been partially funded by M2MRISE, which is an EIT-ICT Labs project supported by the EIT (European Institute of Innovation and Technology); The European Commission through NEWCOM# (FP7-318306) and the Initial Training Network ADVANTAGE (FP7-607774).

References

[1] OECD. Machine-to-machine communications: connecting billions of devices, OECD Digital Economy Papers, No. 192, OECD Publishing. http://dx.doi.org/10.1787/5k9gsh2gp043-en, 2012.

[2] I. Streule, Growth in the M2M market will have long lasting mobile network effects, 2011. Available from: http://www.analysysmason.com/About-Us/News/Newsletter/Growth-in-the-M2M-market-will-have-long-lasting-mobile-network-effects/ [Accessed 4 May 2014].

[3] N. Lomas, Online Gizmos could top 50 billion in 2020, 2009. Available from: http://www.businessweek.com/globalbiz/content/jun2009/gb20090629_492027.htm [Accessed 4 May 2014].

[4] 3GPP. TS 22.011 v12.1.0 Technical Specification Group Services and System Aspects; Service accessibility, Release 12, 2014a. Available from: www.3gpp.org.

[5] 3GPP. TR 37.868 v11.0.0 Study on RAN Improvements for Machine-type Communications, Release 11, 2011. Available from: www.3gpp.org.

[6] 3GPP. TS 36.321 v12.1.0 Medium Access Control (MAC) protocol specification, Release 11, 2014b. Available from: www.3gpp.org.

[7] K. Zhou, N. Navid, S. Thrasyvoulos, LTE/LTE-A discontinuous reception modeling for machine type communications. IEEE Wireless Commun. Lett. 2 (1) (2013) 102–105, http://dx.doi.org/10.1109/WCL.2012.120312.120615.

[8] K. Wang, J. Alonso-Zarate, M. Dohler, Energy-efficiency of LTE for small data machine-to-machine communications. in: IEEE International Conference on Communications (ICC), 2013, 2013, http://dx.doi.org/10.1109/ICC.2013.6655207, vol., no., pp.4120,4124, 9-13.

[9] T. Nguyen, O. Berder, O. Sentieys, Energy-efficient cooperative techniques for infrastructure-to-vehicle communications. IEEE Trans. Intell. Transportation Syst. 12 (3) (2011) 659–668, http://dx.doi.org/10.1109/TITS.2011.2118754.

[10] 3GPP. TR 22.803 V12.2.0 Feasibility study for Proximity Services (ProSe), Release 12, 2013a. Available from: www.3gpp.org.

[11] 3GPP. TR 36.888 v12.0.0 Study on provision of low-cost Machine-Type Communications (MTC) User Equipments (UEs) based on LTE, Release 12, 2013b. Available from: www.3gpp.org.

[12] M. Beale, Future challenges in efficiently supporting M2M in the LTE standards. in: Wireless Communications and Networking Conference Workshops (WCNCW) 2012 IEEE, 2012, http://dx.doi.org/10.1109/WCNCW.2012.6215486, vol., no., pp.186,190, 1-1.

[13] Cisco. Cisco visual networking index: global mobile data traffic forecast update, 2013–2018, 2014. Available from: http://www.cisco.com/c/en/us/solutions/collateral/ser vice-provider/visual-networking-index-vni/white_paper_c11-520862.html [Accessed 4 May 2014].

[14] G. Wu, et al., M2M: from mobile to embedded internet, IEEE Commun. Mag. 49 (4) (2011) 36–43, http://dx.doi.org/10.1109/MCOM.2011.5741144.

[15] A. Bartoli, et al., Low-power low-rate goes long-range: the case for secure and cooperative machine-to-machine communications, in: NETWORKING 2011 Workshops, LNCS 6827, 2011, pp. 219–230.

[16] M. Palattella, et al., Standardized protocol stack for the Internet of (important) things, IEEE Commun. Surv. Tutorials 15 (3) (2013) 1389–1406, http://dx.doi.org/10.1109/ SURV.2012.111412.00158.

[17] W. Nitzold (Ed.), Expanding LTE for Devices, Final Report on LTE-M Algorithms and Procedures, 2012. Available from: http://www.ict-exalted.eu/fileadmin/documents/ EXALTED_WP3_D3.3_v1.0.pdf [Accessed 4 May 2014].

[18] M. Dohler, et al., Is the PHY layer dead? IEEE Commun. Mag. 49 (4) (2011) 159–165, http://dx.doi.org/10.1109/MCOM.2011.5741160.

[19] A. Laya, V. Bratu, J. Markendahl, Who is investing in machine-to-machine communications? in: 24th European Regional ITS Conference, Florence 2013, Florence, Italy, 2013.

[20] A. Laya, A. Widaa, J. Markendahl, Migration Strategies in Network Deployment to Support M2M Communications, Aalborg University, Copenhagen, 2013.

[21] Z. Frias, J. Perez, Techno-economic analysis of femtocell deployment in long-term evolution networks. EURASIP J. Wireless Commun. Networking 1 (2012) 288, http://dx.doi. org/10.1186/1687-1499-2012-288.

[22] A. Aijaz, H. Aghvami, M. Amani, A survey on mobile data offloading: technical and business perspectives, IEEE Wireless Commun. 20 (2) (2013) 104–112.

[23] Machina Research. Big Data in M2M: Tipping Points and Subnets of Things, White Paper, 2013.

[24] A. Daj, C. Samoila, D. Ursutiu, Digital marketing and regulatory challenges of machine-to-machine (M2M) communications, in: 9th International Conference on Remote Engineering and Virtual Instrumentation (REV), 2012, 2012.

[25] J. Heapy, Creating value beyond the product through services, Des. Manage. Rev. 22 (4) (2011) 32–39, http://dx.doi.org/10.1111/j.1948-7169.2011.00154.x.

[26] T. Mejtoft, Internet of things and co-creation of value. in: Internet of Things (iThings/ CPSCom), 2011 International Conference on and 4th International Conference on Cyber, Physical and Social Computing, 2011, pp. 672–677, http://dx.doi.org/10.1109/iThings/ CPSCom.2011.75, 19-22.

[27] J. Markendahl, A. Laya, Business challenges for Internet of things: findings from E-Home Care, smart access control, smart cities and homes, in: 29th annual IMP Conference, Atlanta, USA, 2013.

[28] A. Laya, J. Markendahl, The M2M promise, what could make it happen? A techno-economic analysis, in: IEEE 14th International Symposium and Workshops on a World of Wireless, Mobile and Multimedia Networks (WoWMoM), 2013, Madrid, Spain, 2013, http://dx.doi.org/10.1109/WoWMoM.2013.6583489, vol., no., pp.1,6, 4-7.

[29] IERC—Internet of Things European Research Cluster, IERC—Internet of things European research cluster, Cluster Book, third ed., Platinum, Halifax, UK, 2012.

[30] S. Leminen, M. Westerlund, M. Rajahonka, R. Siuruainen, Towards IOT ecosystems and business models, in: S. Andreev, S. Balandin, Y. Koucheryavy (Eds.), Internet of Things, Smart Spaces, and Next Generation Networking, in: Lecture Notes in Computer Science,

vol. 7469, Springer, Berlin, Heidelberg, 2012, pp. 15–26, http://dx.doi.org/10.1007/978-3-642-32686-8_2.

[31] S. Sharma, J.A. Gutiérrez, An evaluation framework for viable business models for m-commerce in the information technology sector. Electron. Mark. 20 (1) (2010) 33–52, http://dx.doi.org/10.1007/s12525-010-0028-9.

[32] 3GPP. TS 22.368 v12.3.0 Service requirements for Machine-Type Communications (MTC) Stage 1, Release 12, 2013c. Available from: www.3gpp.org.

[33] S. Hilton, M2M Insights for Mobile Network Operators, Analysys Mason Limited, London, 2013.

[34] V. Cackovic, Z. Popovic, Device connection platform for M2M communication, in: 20th International Conference on Software, Telecommunications and Computer Networks (SoftCOM), 2012, 2012, pp. 1,7, 11-13.

Part Two

Access, scheduling, mobility and security protocols

Traffic models for machine-to-machine (M2M) communications: types and applications

8

M. Laner[1], N. Nikaein[2], P. Svoboda[1], M. Popovic[3], D. Drajic[4], S. Krco[4]
[1]Vienna University of Technology, Vienna, Austria; [2]EURECOM, Biot, France;
[3]Telekom a.d., Belgrade, Serbia; [4]Ericsson d.o.o., Belgrade, Serbia

8.1 Introduction

Machine-type communication (MTC) or machine-to-machine (M2M) communication is regarded as a form of data communication that does not necessarily require human interaction [1]. This type of communication will play an important role in the information and communication technology (ICT) by enabling the future Internet of Things (IoT) and is expected to experience a significant growth within the next decade [2,3]. Moreover, supporting such a massive number of heterogeneous connected devices, many of them serving time-critical applications, is also an important part of 5G system requirement [4].

M2M communication services, in addition to conventional voice and Internet traffic, will be an integral part of the traffic transported by LTE/LTE-Advanced network. A very large number of devices can be attached to the operators' network, where the number of MTC devices could be of orders of magnitude greater than the number of cellular phones. This calls for new mechanisms to handle such a large number of devices with a low signaling and processing overhead.

While the literature focuses on enabling wide M2M deployment in LTE/LTE-A networks, the current mobile M2M traffic is in most cases conveyed through legacy networks. According to market predictions [5], connections over legacy networks will remain predominant worldwide in the following years, with significant growth of M2M connections on both 3G and 2G. Wide M2M application areas imply a variety of corresponding traffic patterns and QoS requirements, which in turn impose a substantial challenge for operators. While modern networks are mainly designed for human-type communication (HTC) and mostly downlink-dominant and bursty, M2M traffic is of generally different properties, mostly uplink-dominant, often periodic, and persistent. With the large number of connected devices, the knowledge of particular traffic patterns becomes substantial. Signaling congestion and network performance degradation affecting both voice and data services, as well as M2M services, proved to represent the worst-case scenarios [6]. The deployment of a new M2M service raises several questions for an operator: what are the traffic characteristics, what is

Machine-to-machine (M2M) Communications. http://dx.doi.org/10.1016/B978-1-78242-102-3.00008-3

the predicted number of connected devices, what is their spatiotemporal distribution, and what are their QoS requirements in terms of delay constraints? The operators need to assess whether new M2M services could jeopardize traditional HTC and whether the network can provide or guarantee the required latency [7]. Some system solutions proposed for LTE-A could be applicable to legacy networks but would have a deep implication in the network, making them generally unfeasible or expensive.

Understanding the properties of MTC traffic is therefore considered as the key for designing and optimizing future networks and the respective QoS schemes with the goal of provisioning adequate M2M communication services without compromising any conventional HTC services such as data, voice, and video. In particular, the success of 3GPP Rel-11 and Rel-12 Evolved Packet System (EPS) and its evolution toward 5G systems depends on the effectiveness of its class-based network-initiated QoS control scheme and the corresponding support of both MTC and HTC traffics.

Conventional HTC traffic and MTC traffic have two major differences: (i) HTC traffic is heterogeneous, whereas MTC traffic is highly homogeneous (all machines running the same application behave similarly), and further, (ii) HTC is uncoordinated on small timescales (up to minutes), while MTC may be coordinated, that is, many machines react on global events in a synchronized fashion. Some typical properties of MTC traffic may encompass

- massive number of devices (i.e., machines and users),
- few short packets to be transmitted per machine,
- low-duty-cycle traffic patterns (i.e., long periods between two transmission bursts),
- traffic patterns with small statistical variation produced by single devices,
- uplink-dominant traffic (i.e., uplink volume higher than downlink volume),
- real-time and delay-tolerant data bursts triggered by the same application,
- raw and aggregated packets (i.e., combining traffic of multiple sources into a single packet, relevant for specific nodes such as gateway),
- unsynchronized and synchronized packets (i.e., simultaneous access attempts from many devices reacting to the same/similar events),
- spatiotemporal-dependent traffic trigger.

Thus, well-known traffic models designed for HTC require adaptations for their application to MTC. A fundamental question is whether it is feasible to model the traffic of a large amount of autonomous machines individually. This approach is called source traffic modeling. It is in general more accurate than its counterpart, aggregated traffic modeling (i.e., treating the accumulated data from all MTC devices as a single stream). The aim of this chapter is to provide a thorough comparison of both approaches in the context of MTC.

This chapter is organized as follows. In Section 8.2, we present a generic traffic modeling methodology applicable to the majority of the application scenarios including MTC. In Section 8.3, a detailed description of different modeling strategies and a comparable analysis by applying the same traffic recorded from a fleet management in an operational cellular network are described. Finally, in Section 8.4, summary and concluding remarks are presented.

8.2 Generic methodology for traffic modeling

There are many approaches found in the literature to creating traffic models for network data traffic [8]. Figure 8.1 illustrates a generic traffic modeling methodology applicable to the majority of application scenarios including M2M communication. It defines the workflow to evaluate the performance of a system operating in the desired application scenario by extracting the traffic traces to build a precise traffic model and its respective key performance indicators as the performance evaluation metrics [9].

The first important step in the process of model selection is the definition of which parameters we want to model and how this should be done. Such parameters, for example, packet loss or data rate, will then be analyzed for statistical properties and an appropriate model that generates similar statistical patterns is selected. Please note that the parameter of interest may be composed of multiple subparameters that have to be taken into account. If detailed statistical properties are of interest, then an adequate sample size must be ensured in order to guarantee the statistical significance of the result. In the following, we will describe these steps in more detail.

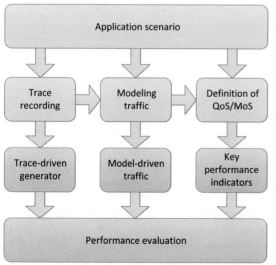

Figure 8.1 Generic traffic modeling methodology.

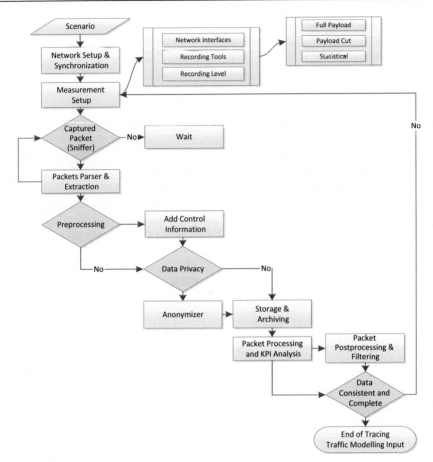

Figure 8.2 Packet tracing workflow.

8.2.1 *Trace recording*

The traffic analyzer tool generates a statistical dataset from a captured application scenario. It includes the following main building blocks: packet sniffer, parser, decoder, anonymizer, and analyzer, as well as data storage, visualization, and management. Figure 8.2 illustrates the methodology for traffic tracing and analysis. The first step is the scenario description of an experiment. It defines the application/flow configuration, network setup, measurement links, and the required statistical datasets. Such a scenario is used to allow the experiment to be reproduced. Based on the scenario, a network is setup and synchronized to allow for measurements at different network interfaces/ links with a common reference clock. Each measurement node needs to know the exact system time to provide time stamps on the measurements. If all links are symmetrical and identical, nodes can accurately estimate their clock offset by using round-trip time measurements. When the network is asymmetrical or time-varying, every node needs a time synchronization unit of its own. Typically, this is provided by Global Positioning System (GPS) providing pulse-per-second (PPS) reference signal indicating the start of a new second with respect to the Coordinated Universal Time (UTC). The next step is to determine the measurement links associated with network interfaces and define for each link the associated recording tools (e.g., Wireshark), packet formats (e.g., TCP, IP, and MAC), and recording levels. The recording level could vary from full payload (extract all user information), to payload cut (extract header information only), to statistical information (extract traffic characteristics only). It has to be mentioned that such measurement is normally done locally.

When a packet is captured (i.e., sniffed, traced, and recorded) on the measurement links under traffic analysis, the packet parser will immediately read the recorded packet and decode the protocol header information. This procedure will iterate for all recorded packets until the duration of the experiment is reached. Relevant information may be extracted or derived from the decoded packet header during the preprocessing step (e.g., location information). Some additional control information associated with the packet (e.g., time stamp and/or processing time) may also be added to facilitate data storage and management for fast search of specific traces with particular attributes.

The recorded traces are then stored and archived before being analyzed. As stated above, the stored data could be either full payload, payload cut, or statistical information depending on the measurement setup for each network interface. If privacy issues apply, the traces must be anonymized before the storage and archiving. Note that the anonymization should preserve user-packet association while hiding the user identity such that retrieving the user identify from anonymized trace becomes impossible. At this stage, the data are ready to be fully analyzed in order to produce output of statistical datasets in terms of protocol operations, packet information, and KPIs. Finally, the experiment may be repeated several times to adjust the measurement links and recording tools such that the produced datasets become consistent and complete.

8.2.2 Traffic modeling

Figure 8.3 depicts the workflow of traffic modeling in order to reproduce traffic flows or data sources in a communication network. In general, there are two main approaches to reproduce traffic: traffic emulation and statistical modeling. The emulation of application traffic is possible if the application internals and its finite state machine are known or could be fully or partially derived through functional analysis (e.g., a sensor generating regularly a message within a constant time). However, if part of the application is too complex or its behavior is unknown, a hybrid modeling approach combining traffic emulation and statistical modeling can be applied. For instance, in FTP modeling, the FTP protocol part could be emulated as its behavior is known, while the requested file size and the request rate per used could be modeled based on a statistical distribution.

Statistical modeling, on the other hand, analyzes application-specific recorded traffic traces by considering the underlying network under test, protocol interaction depending on at which OSI layer traffic is recorded, and other concurrent applications. The possible presence of interparameter correlation, for example, between packet size and rate or data rate and delay, is checked by applying the correlation metric between parameters. Depending on the presence of correlation and required accuracy, the parameters can be fitted to one-dimensional or multidimensional distributions.

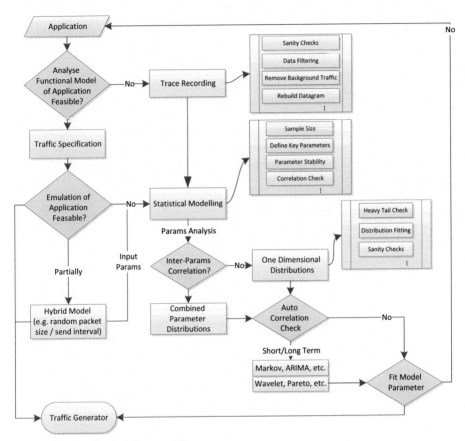

Figure 8.3 Traffic modeling workflow.

Another important property of the traffic is the autocorrelation properties inside the time series of the data packets, that is, how fast the values of the time series fluctuate and if two consecutive values in the time series are dependent on the history and therefore on each other. This property is called short-/long-range dependency of a time series. A time series can be considered as short range-dependent if the properties of the autocorrelation function (ACF) have a limited lag and can be reproduced in an ARMA (autoregressive–moving average), Markov, or similarly popular process. Long range-dependent time series on the other hand require specifically tailored models.

8.3 M2M traffic modeling

Traffic modeling means to design statistical processes such that they match the behavior of physical quantities of measured data traffic [8]. Traffic models are classified as source traffic models, for example, video, data, and voice, coming from one individual user, and aggregated traffic models for backbone networks or the Internet.

The typical MTC use case includes numerous simple machines assigned to one server or medium; therefore, the second modeling approach is more suitable.

For example, it is difficult to simulate thousands of discrete source traffic models in one scenario. An MTC scenario can be modeled as simple Poisson process; however, due to coordination (synchronizations) in MTC traffic, the respective arrival rate λ may be changing over time, $\lambda(t)$ (i.e., temporal modulation [10,11]). The more complex and individual the single MTC devices behave (e.g., video surveillance), the more questionable the approach of modeling them as aggregated traffic becomes assuming them to be all the same machine. The global data stream may exhibit high-order statistical properties, which are difficult to capture [12]. We further expect this effect to be enhanced by the synchronization of sources. In such a case, traffic modeling in terms of source traffic is preferable. Source traffic models that can capture the coordinated nature of MTC traffic are available [13,14]. However, they are designed for a low amount of sources and, thus, are too complex for MTC traffic (e.g., for N devices, an $N \times N$ matrix–vector multiplication is required in each time slot).

For multiple access and capacity evaluations, we do not need knowledge about the behavior of a single node, for example, simulate every link between a node and its base station. In this case, aggregated traffic models, such as homogeneous [15,16], with constant arrival rate, or inhomogeneous [11], with time-varying arrival rate, Poisson processes, are a satisfactory description of reality and therefore largely deployed. Respective setups are defined by 3GPP [2,3] and further discussed later. For the simulation of strongly scalable multiple access schemes in future networks (e.g., priority access, delay-tolerant devices, and QoS demands), mixed source traffic models have been adopted [17–19]. In those studies, the case of synchronized MTC devices has not been considered or only for a limited number of MTC devices.

We observe a divergence between traffic models deployed within different studies. On one hand, higher accuracy requires source traffic models, and on the other hand, reduced complexity claims for aggregated traffic models. The following subsections present different traffic modeling strategies of both worlds.

8.3.1 Use case: fleet management

We consider one specific use case as an example for studying several modeling strategies introduced below. This use case is a "fleet management" scenario of 1000 trucks run by a transportation company in Central Europe. It relies on measured M2M traffic from an operational 2G/3G network. This traffic has been captured at the Gn interface, which is the interface with the highest data aggregation in the mobile core network. The resolution of this dataset is on packet granularity. The dataset does not feature payload but rather a set of parameters per packet. Those are, among others, the time stamp, the packet size, the direction, IP addresses, port numbers, pseudo identifiers of the sending devices, and the Access Point Name (APN). The capturing period extended over one week, containing approximately 27 million packets originating from 1000 devices for the considered application, have been observed. Due to the knowledge of the identifiers, single devices can be traced reliably over the whole observation period. The accumulated data rate was on average 1.89 kB/s, yielding a rate of 2 B/s/device. On average, 4% of the MTC devices of the traced class are active. Further, they exhibit coordinated behavior. We have observed 100 time instants, out of which more than 20% of all devices were trying to transmit data simultaneously. Consequently, this dataset incorporates coordinated and uncoordinated phases and allows for (separate or joint) modeling of both cases.

8.3.2 Source traffic modeling

For every M2M application, there are four common basic stages of communication as follows:

1. Collection of data
2. Transmission of selected data thought a communication network
3. Assessment of the data
4. Response to the available information

As discussed subsequently, that yields different traffic patterns and associated states, which can then be modeled by specific processes.

8.3.2.1 M2M traffic states

Analyzing the functionality of such M2M applications has revealed that MTC has three elementary traffic patterns [20]:

1. Periodic update (PU): This type of traffic occurs if devices transmit status reports of updates to a central unit on a regular basis. It can be seen as an event triggered by the device at a regular interval. Typically, PU is non-real-time and has a regular time pattern and a constant data size. The transmitting interval might be reconfigured by the server. A typical example of the PU message is smart meter reading (e.g., gas, electricity, and water).
2. Event-driven (ED): In case an event is triggered by an MTC device and the corresponding data have to be transmitted, its traffic pattern conforms to this second class. An event may either be caused by a measurement parameter passing a certain threshold or be generated by the node acting as server to send commands to the device and control it remotely. ED is mainly real-time traffic with a variable time pattern and data size in both uplink direction and downlink direction. An example of the real-time ED messages in the uplink is an alarm/health emergency notification and in the downlink could be the distribution of a local warning message, for example, in case of Tsunami alert. In some cases, ED traffic is non-real-time, for example, when a device sends a location update to the server or receives a configuration and firmware update from the server.
3. Payload exchange (PE): This last type of data traffic is issued after an event, namely, following one of the previous traffic types (PU or ED). It comprises all cases where larger amount of data is exchanged between the sensing devices and a server. This traffic is more likely to be uplink-dominant and can be either of constant size as in the telemetry or of variable size like a transmission of an image or even of data streaming triggered by an alarm.

Real-world applications are often a combination of the abovementioned traffic types. Hence, using the three elementary types above for traffic modeling enables building models with high degree of complexity and accuracy. For example, a device may enter the power saving mode and trigger a PU pattern at regular intervals. Another example is that a PE is only triggered after the ED to provide further details about the events. It has to be mentioned that the PU and the ED can be regarded as the short control information type of traffic (very low data rate), while PE may entail bursts of data traffic.

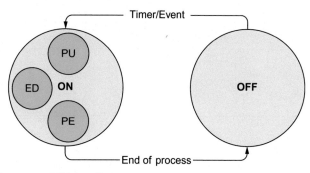

Figure 8.4 Elementary M2M traffic state structure.

For a convenient modeling of MTC traffic (by deploying the above-described traffic types), we propose an ON–OFF structure, as depicted in Figure 8.4. Together with the three distinct traffic patterns mentioned above, this can be integrated in a Markov structure with four different states: OFF, PU, ED, and PE. The classification of the states into several ON and one OFF states facilitates the handling of the almost vanishing data rates, for example, long periods of no data between phases of activity. The OFF state is thereby equivalent to an artificial traffic type, where no packets are transmitted neither from nor to the respective machine. This corresponds to situations such as the terminal being in idle/sleep mode. This enables the assignment of meaningful side information to each state, such as respective QoS parameters. For example, the attribute "latency <100 ms" may be added to the state ED, in order to ensure fast forwarding of alarms.

8.3.2.2 Source modeling via semi-Markov models

For modeling the data streams within single states i, we deploy renewal processes (Ref. [21], p. 254). They consist of a random packet interdeparture times (IDT) D_i and a random packet sizes (PS) Y_i. Both random processes D_i and Y_i are independent and identically distributed (i.i.d.), with arbitrary marginal probability density functions (PDFs) $f_{D,i}(t)$ and $f_{Y,i}(y)$. Two special cases are periodic patterns, for example, fixed interdeparture time, and Poisson processes, for example, exponentially distributed IDT.

In order to model state transitions, we define a semi-Markov model (SMM) (Ref. [21], p. 352). A SMM defines transition probabilities p_{ij} between states, with $p_{ii} = 0$ transition probability to the current state. The transition probabilities are arranged in the transition probability matrix P. Further, a random sojourn time or holding time T_i is introduced per state, with arbitrary independent distribution $f_{T,i}(t)$. SMM models are advantageous for MTC modeling for several reasons: they (i) allow capturing a broad spectrum of traffic characteristics [22], especially the almost vanishing data rate, (ii) enable augmented modeling if side information is available (e.g., the exact number of states is known) [8], and (iii) enable advanced fitting mechanisms to be established [23], which allow for good fitting quality, even if nothing but raw traffic measurements are given.

The input parameters for the model are summarized in Table 8.1, where "·" represents parameters to be fitted to a desired MTC traffic pattern and the completed items

Table 8.1 Input parameters of the SMM approach

States	$f_{D,s}(t)$	$f_{Y,s}(y)$	$f_{T,s}(t)$	P			
OFF	Deg(∞)	Deg(0)	·	0	·	·	·
PU	Deg(∞)	·	Deg(ΔT)	·	0	·	·
ED	Deg(∞)	·	Deg(ΔT)	·	·	0	·
PE	·	·	·	·	·	·	0

are state-specific constants. Deg($\cdot$) represents the degenerate distribution, corresponding to a constant value, and ΔT represents the minimum temporal resolution of the model. Note that the state-specific constants conform two special cases, namely, (i) no traffic is generated within a state, for example, OFF state and (ii) the sojourn time is very short and only one chunk of data is transmitted, for example, PU and ED states.

An example use case is the modeling of a fleet management scenario, which is used as reference for the presented modeling techniques. The resulting model parameters for the outlined use case are the following (the range of distributions has been restricted to parametric distributions with at most two parameters for simplicity):

- $f_{T,\mathrm{OFF}}(t) = \mathrm{Deg}(397\mathrm{s})$
- $f_{Y,\mathrm{PU}}(t) = \mathrm{Deg}(197\mathrm{B})$
- $f_{Y,\mathrm{ED}}(t) = \mathrm{Deg}(120\mathrm{B})$
- $f_{D,\mathrm{PE}}(t) = \mathrm{Exp}(6.65\mathrm{s})$
- $f_{Y,\mathrm{PE}}(y) = \mathrm{Exp}(43\mathrm{B})$
- $f_{T,\mathrm{PE}}(t) = \mathrm{Exp}(6907\mathrm{s})$
- $P = \begin{pmatrix} 0 & 1 & 1 & 1 \\ 0.915 & 0 & 0 & 0 \\ 0.058 & 0 & 0 & 0 \\ 0.027 & 0 & 0 & 0 \end{pmatrix}$

8.3.3 Aggregated traffic modeling

Because of its popularity, we first provide an overview of the 3GPP model developed in 2012. The 3GPP model consists of two scenarios called model 1 and model 2. The first one treats uncoordinated events triggering data traffic and the second one coordinated events triggering data traffic. Both scenarios are defined by a distribution of packet arrivals (or, equivalently, access trials) over a given time period T, see Table 8.2. This is shown in Figure 8.5, where the probability density functions (PDFs)

Table 8.2 Parameters of the 3GPP model

Characteristics	Model 1	Model 2
Number of devices N	1000, 3000, 5000, 10,000, 30,000	
Distribution $f(t)$ over [0,1]	Uniform	Beta(3,4)
Period T	60 s	10 s

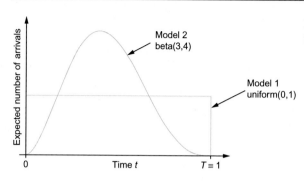

Figure 8.5 Access intensity of the 3GPP model.

of both distributions are depicted. The distributions $f(t)$ are both defined on the interval [0, 1], which has to be rescaled to the time interval [0, T] to yield $f_T(t)$. In order to simulate arrivals, it is sufficient to draw N samples from the given distribution and sort them in time, where N is the expected number of MTC devices, see Table 8.2. This number may reach up to 30,000, which is the maximum amount of smart meter devices expected to be served by one cell in a densely populated urban area [2,3].

In general, it is undesirable to generate a full traffic pattern over T beforehand as this requires a large amount of memory on the simulation machine. In the present case, this may not be an issue, however, as basic problems, such as undefined run length T or large amounts of generated data, may require a sequential drawing of samples. This issue is discussed in [11], where it is pointed out that the 3GPP model is equivalent to a modulated Poisson process. Thereby, the modulation is achieved by averaging the PDF of the arrival distribution $f_T(t)$ for time bins ΔT. This is depicted in Figure 8.6, where the mean arrival rate $\lambda(t)$ of a Poisson process is modulated in each time bin Δ_T by a beta distribution (cf. Table 8.2 as reference). For infinitesimal ΔT, both curves coincide. Consequently, sequential sampling is performed by the generation of a Poisson distributed number of arrivals in each time bin ΔT with mean arrival rate $\lambda(t)$. In order to obtain an expected outcome of N samples within the period T (i.e., one sample per machine), the arrival rate has to be normalized according to

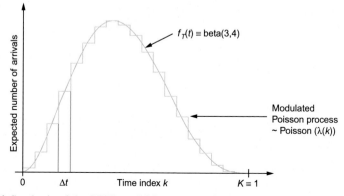

Figure 8.6 Synthesis of the 3GPP model from a modulated Poisson process.

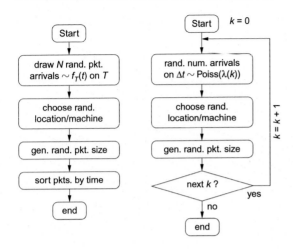

Figure 8.7 Generation of samples from the 3GPP model. Left, nonsequential; right, sequential.

$\lambda(t) = f_T(t) \cdot \Delta_T \cdot T \cdot N$. The two different sampling strategies are summarized in Figure 8.7.

The 3GPP model however is not well suited for further M2M-specific requirements, such as:

(i) the amount of machines becomes small, so that a data source has to be associated with a fixed location;

(ii) multiple packets come from the same machine;

(iii) the synchronous traffic (model 2) influences uncoordinated traffic (model 1); and

(iv) the network has an influence on the traffic patterns.

Due to the small set of parameters, fitting of the 3GPP model to traced data is very simple. There are only two unknown parameters: (i) the number of MTC devices in the uncoordinated case and (ii) the number of MTC devices in the coordinated case. Applied on the use case of fleet management, presented above, both parameters are estimated to $N = 1000$.

8.3.4 Source modeling for coordinated traffic via Markov-modulated Poisson processes

To circumvent the limitations of the 3GPP model, a source modeling approach is adopted here. Each machine is thereby represented by a separate model. This approach is only feasible if a trade-off between mutual couplings among data sources (synchronization) and a tolerable complexity for large amounts of devices is found. Generic traffic models couple devices (i.e., multiple random processes) by bidirectional links between them. This is however too complex for the present purpose. Instead, one background process is proposed, modulating all MTC device entities.

Markov-modulated Poisson processes (MMPPs) are presented in the following, which model a single MTC device. Due to their simplicity, however, the operation of large amounts of device models in parallel is computationally feasible. Further, the coupling to the background process requires only low complexity.

8.3.4.1 MMPPs: the basics

Markov models and MMPPs are commonly deployed in traffic modeling and queuing theory. They allow for analytically tractable results for many use cases [10,21]. MMPP models consist of a Poisson process modulated by the rate $\lambda_i[k]$, which is determined by the state of a Markov chain $s_n[k]$. Thereby, k denotes the time index, obtained by migrating from continuous time t to discrete time by $k = \dfrac{t}{\Delta_T}$, where Δ_T denotes an arbitrary but constant time interval, constituting the "heartbeat" of the system. Further, $i = 1 \ldots I$ denotes the index of Markov state and $n = 1 \ldots N$ the index of the machine-type device. The principle of a modulated Poisson process is depicted in Figure 8.8, where $p_{i,j}$ are the transition probabilities between the states of the chain. In the present source modeling approach, each MTC device n out of N is represented by a separate Markov chain and a corresponding Poisson process. The state transition probabilities form the state transition matrix P and the state probabilities π_i form the state probability vector π.

In the stationary case, both parameters are related by the balance equation $\pi = \pi\, P$. Hence, π is an eigenvector of P to the eigenvalue of 1. Further, the overall perceived rate of the MMPP is $\lambda_g = \displaystyle\sum_{i=1}^{I} \lambda_i \pi_i$, where I is the total number of states. A basic example for an MTC device modeled by a MMPP is a two-state MMPP with the first state representing regular operation and the second representing the sending of an alarm. This is in analogy to the 3GPP model (see above), where two models (model 1 and model 2) are capturing coordinated behavior and uncoordinated behavior.

8.3.4.2 Coupling multiple MMPP processes

The state transition matrix P has to be determined such that each device model resides a prescribed amount of time in each state. From the perspective of a single device, this is straightforward; from a global perspective, the devices transit from the regular to the alarm state in a strongly correlated manner in both time and space. To imitate this behavior for multiple MMPP models, some coupling is required.

In the context of pattern recognition, coupled Markov chains are well known [24,25]. They are realized as multiple chains that mutually influence their transition probability matrices $P_n(t)$. In terms of discrete time, they are given by $P_n[k]$, corresponding to the notation deployed in the following. The matrices are influenced by the respective multiplication of weighting factors, which depend on the past states $s_m[k-1]$ of neighboring chains m.

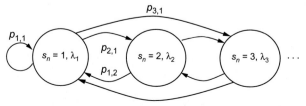

Figure 8.8 Markov chain driving the MMPP model.

We only consider a unidirectional influence from a background process (master) $\Theta(t)$ to the MTC device MMPP models in the presented framework; an example of the said background process is a fire in a factory triggering several fire sensors to response or a traffic jam on the road triggering telemetry sensors in trucks to transmit in a correlated fashion. This approach is named coupled Markov-modulated Poisson processes (CMMPP). Again, we perform the transition from continuous to discrete time by sampling at time instant k: $\theta[k] = \Theta(k \, \Delta_T)$. The separate tuning of each of the weighting parameters for each machine is avoided by defining the following framework [26]:

Let there be two transition matrices P_C and P_U globally valid for all N MMPP models, representing the strictly coordinated and strictly uncoordinated behavior, and a background process $\Theta(t)$ producing samples $\theta[k]$ within the interval [0, 1]. Further, a parameter $\delta_n \in [0, 1]$, constant over time, is associated to each MTC device n yielding $\theta_n[k] = \delta_n \cdot \theta[k]$. Then, the state transition matrix $P_n[k]$ can be calculated for machine n at time t according to the following expression:

$$P_n[k] = \theta_n[K] \cdot P_C + (1 - \theta_n[k]) \cdot P_U \tag{8.1}$$

This convex combination of both transition matrices yields again a valid transition matrix. The matrices P_C and P_U are transition matrices for the case of perfectly coordinated device behavior and uncoordinated device behavior, respectively. The parameter δ_n corresponds to a measure of closeness (distance) to the epicenter (point in space, on which the expected coordination is maximum). The closer $\theta_n[k]$ to zero, the more uncoordinated the respective machines behaves; the closer $\theta_n[k]$ to one, the stronger the coordination. The background process $\Theta(t)$ is allowed to have an infinite number of states, yielding $\theta[k]$ to be a continuous process.

The synthetic generation of data traffic according to the CMMPP model is described in the flow diagram in Figure 8.9. Two nested loops are required, for both devices n and time index k, respectively. The transition matrix $P_n[k]$ is calculated anew for any iteration according to Equation (8.1). From a complexity perspective, this is feasible since the number of states is usually low and the required convex combination can be computed efficiently. The random state update from $s_n[k-1]$ to $s_n[k]$ is performed afterward. Finally, a number of arrivals and packet sizes are generated according to the current state $s_n[k]$.

The abovementioned fleet management application, for example, yields the following model parameters:

- $\Delta_T = 1\,\text{s}$
- $\lambda_0 = 0.15\,\text{B/s}$
- $\lambda_1 = 6.5\,\text{B/s}$
- $\lambda_2 = 24.7\,\text{B/s}$
- $\theta[k] = 1 \quad @\, k = 0, \quad 0 \quad \text{otherwise}$
- $\delta_n = 1$
- $P_U = \begin{pmatrix} 1 - 6.75 \times 10^{-5} & 1.47 \times 10^{-4} & 0.39 \\ 6.75 \times 10^{-5} & 1 - 1.47 \times 10^{-4} & 0 \\ 0 & 0 & 0.61 \end{pmatrix}$
- $P_C = \begin{pmatrix} 0.66 & 0 & 0 \\ 0 & 0.66 & 0 \\ 0.33 & 0.33 & 1 \end{pmatrix}$

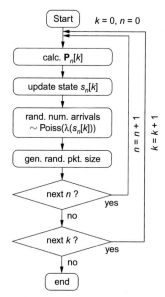

Figure 8.9 Generation of samples from the CMMPP model; the indices n and k stand for the device and time instant, respectively.

8.4 Model fitting from recorded traffic

The question remains on how to obtain the parameters for above models. Typically, the fitting procedures of the three modeling approaches outlined above rely on the time series of the data rate produced by actual MTC devices. The properties of the traffic streams at higher granularity (e.g., packet level) are not of particular interest.

The detailed fitting procedures for the modeling approaches are outlined in the flow diagram in Figure 8.10. All procedures are covered by a subset of the three main building blocks, B1, B2, and B3. The SMM model requires blocks B1 and B3, the aggregated traffic model requires B2, and the CMMPP model requires B1 and B2. The functionality of the building blocks is the following:

- B1: a Markov model for the behavior of an individual device is derived.
- B2: an aggregated traffic model is built for the global data rate.
- B3: each state obtained in B1 is described as general renewal processes.

The three building blocks are presented in detail in the following subsections.

8.4.1 B1: modeling individual devices as Markov chains

This block is required to fit transition probability matrices to traced traffic. In the case of SMM modeling, it is the matrix P of the embedded Markov chain; in the case of CMMPP modeling, this is the matrix P_U. Further, data rates λ_i associated to the states i are obtained, either for direct use in the model (i.e., CMMPP) or for further modeling steps (i.e., SMM and B3). For the aggregated traffic model (i.e., 3GPP model), this block is not required, since the model has no notion of individual machines.

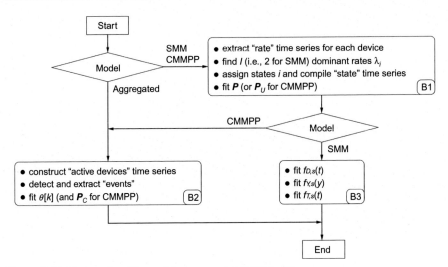

Figure 8.10 Fitting of traffic models to recorded MTC traffic.

We first extract the time series of the data rate of individual MTC devices. A temporal resolution of one second is proposed, in order to capture all relevant effects. This resolution may however lead to strong variations in the data rate due to sparse arrivals of packets. In order to smooth these variations, a sliding average ought to be used. In the present case, we deployed a sliding average of 30 min and a cosine-shaped window. Figure 8.11 shows a corresponding time series of an MTC device over the duration of 4 days. The device corresponds to the fleet management application described in Section 8.3.1. It is clearly visible that there are roughly two main rates used for communication: (i) 0.5 B/s and (ii) 7 B/s.

From the obtained samples of the data rate, we derive the histogram (empirical distribution). This distribution should ideally consist of few regions of high density (common data rates), which constitute the Markov states of the final model. It has to be decided upon how many states shall be deployed for modeling. This is equivalent to fixing the number I for CMMPP models. For SMM models, this number is always $I=2$, since only two different rates can be achieved by the respective approach. This is due to the restrictions imposed by definition on the four individual states; see

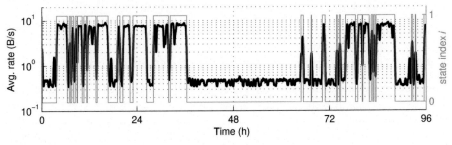

Figure 8.11 Rate time series of a single MTC device.

Table 8.1. Namely, the states OFF, PU, and ED do not exhibit the same flexibility as the PE state.

When I is fixed, one has to place boundaries between the rate regions associated to specific states. For instance, in the case of SMM models, a single threshold rate is defined, yielding all instances with data rate above that threshold to be associated to state PE and all others to a combination of the states OFF, PU, and ED. This task requires either human efforts or a set of elaborate rules on how to place these boundaries in an opportunistic or deep-learning manner. We achieved a satisfactory performance using the k-means algorithm on logarithmically compressed rate samples. The rates λ_i, the output of this modeling block, can be computed by averaging over all rates within one region (state i).

A time series of state sojourns can now be constructed, from which the state transition probabilities (i.e., P for SMM models and P_U for CMMPP models) are directly obtained. For SMM models, the state transitions to other states are of primary interest, and for CMMPP models, transitions to the state itself have to be further considered. For example, Figure 8.11 shows a respective time series where the number of states was chosen to $I = 2$ due to the two dominant data rates.

To evaluate one individual device, it is enough to find values for P and λ_i from a single device time series. However, to obtain statistically significant modeling parameters, several devices should be considered. Based on the obtained modeling parameters, one then has to decide whether (i) an individual model is fitted to each device and afterward combined to an average model or (ii) the data are combined to an average device, to which a single model is fitted.

8.4.2 B2: modeling aggregated traffic

This building block is required to fit numerical values to the aggregated traffic stream, for both regular operation and event-based operation scenarios. For the aggregated traffic model (i.e., 3GPP model), the unknown parameter is only the number of devices N (for both kinds of operational scenarios); however, the suggested distributions (e.g., beta distribution) could also be tuned to better fit the traced traffic (e.g., adjusting the parameters of the beta distribution). For the CMMPP model, the unknown parameters are the background process $\theta[k]$, as well as the transition probability matrix P_C for the coordinated case. For the SMM model, this block is not required; nevertheless, results obtained here can be used to refine the probability of occurrence of an event or, equivalently, the transition probabilities to the PE state.

For modeling aggregated traffic, it is required to construct the time series of the global number of active devices (i.e., N). Thereby, a resolution of one second is chosen, which is not critical for the present purpose, but can be adapted to specific use cases. Further, smoothing is required to reliably detect events (defined below). For this purpose, we deploy the same moving average filter as outlined in the section above (i.e., cosine window with 30-min duration).

Comparing the original and the smoothed time series of the number of active devices allows for the detection of events. Thereby, the term event has to be defined by a set of rules. For the use case outlined in Section 8.3.1, for example, we define the occurrence of an event when

• the number of active users in one second is ten times higher compared to the sliding average (normal operation),
• no such change in the number of active users has occurred within the last 30 s.

With this policy, events are reliably detected and the frequency of occurrence of events can be calculated (useful for the refinement of transition probabilities of SMM models).

In the context of the 3GPP and CMMPP models, the simulation of coordinated traffic as a single isolated event is targeted. For this purpose, it is convenient to define a "typical" event by averaging over all detected events. As per Figure 8.12, the typical event for the use case is depicted through a black solid line, where 88 events have been detected. From this curve, the average number of active users (e.g., 21.4) has been subtracted in order to avoid any bias (which is modeled separately and in parallel). During a typical event, 315 devices become active within the same second (peak second). The most extensive event during the tracing period triggered roughly 500 devices simultaneously, the lightest only 180 MTC devices. Comparing these numbers to the total device population (i.e., 961) shows that an event triggers 33% of the devices, whereas only 2.2% of the devices are active per second during uncoordinated operation.

Fitting the 3GPP model to the typical event requires to sum over all communication activities during the duration of an event. The typical event contains 976 activities, yielding a number of $N = 1000$ MTC devices the best fit of the 3GPP model in the coordinated case. Note, however, that the 3GPP model assumes all activities to originate from different devices. This is not the case for the fleet management scenario,

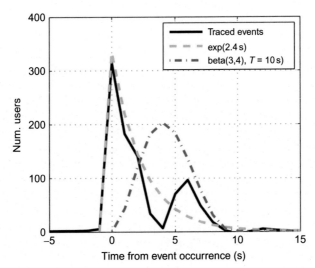

Figure 8.12 "Typical" event for the use case (see Section 8.3.1); modeled by the original 3GPP model and an exponential distribution with $\mu = 2.4$ s.

where each active device has on average 3 activities per event. In uncoordinated operation, there are on average 21.4 activities per second, amounting to 1284 activities per minute. The closest value adhering to the 3GPP model is here again $N = 1000$ devices.

Figure 8.12 shows moderate accordance of the 3GPP model and the modeled use case (especially regarding the duration of the event); note that only one parameter (i.e., N) has been fitted. This confirms that the beta(3,4) distribution and the duration $T = 10$ s are decent choices for coordinated M2M behavior for the given use case. A point of criticism is that the model assumes the event to start smoothly and reach its peak after some seconds. For the traced event, this is not the case. We observe that the event starts within one second and ends smoothly. This could be modeled more accurately by an exponential distribution, as shown in Figure 8.12. Thereby, the average duration is determined to be $\mu = 2.4$ s.

The determination of the parameters $\theta[k]$ and P_C of the CMMPP model allows for a more accurate representation of reality. An additional state j is introduced to the Markov chains of single devices, which represents coordinated operation, whereby the associated mean rate λ_j must be determined. In the case of the presented fleet management application, the rate calculates to $\lambda_j = 24.7$ B/s. It is reasonable to assume the background process $\theta[k]$ to resemble a unit impulse function (i.e., one at index zero, zero otherwise), since all devices involved into an event exhibit activities within the first second. Thereby, $\theta[k] = 0$ corresponds to uncoordinated operation and $\theta[k] = 1$ to the coordinated case. Accordingly, P_C is constructed such that 33% of the devices fall into state j (event), the rest remains in the prior state (uncoordinated operation). Finally, the transition probabilities in P_U, which cause devices to return from state j, must be adjusted such that the exponential decay shown in Figure 8.12 is matched. For example, assuming a time slot duration of 1 s, this can be achieved by setting the probability of remaining in state j to $p_{U,jj} = 0.61$.

8.4.3 B3: modeling single Markov states

This building block is required only for the SMM approach. The desired output parameters are the distributions for packet size, packet interarrival times, and sojourn times associated with each of the four states. Thereby, several of these distributions are fixed a priori, such that only few have to be derived from traced data (see Table 8.1).

For this purpose, the time series of state sojourns for individual devices (obtained from the building block B1) can directly be reused. Those series exhibit two states: (i) state 0 with low data rate λ_0 and (ii) state 1 with high data rate λ_1. As already mentioned, state 0 is matched by the combination of states OFF, PU, and ED; state 1 is resembled by PE. Accordingly,

- $f_{T,\text{OFF}}(t)$ is set to the constant value of the average interpacket time in state 0,
- $f_{Y,\text{PU}}(y)$ is set to the distribution of the packet size in state 0,
- $f_{Y,\text{ED}}(y)$ is set to the distribution of the packet size during events,
- $f_{D,\text{PE}}(t)$ is set to the distribution of the packet interarrival time in state 1,
- $f_{Y,\text{PE}}(y)$ is set to the distribution of the packet size in state 1,
- $f_{T,s}(t)$ is set to the distribution of the sojourn time in state 1.

The extraction and fitting of these parameters is straight forward. We encourage the usage of simple distributions (e.g., unimodal distributions) and simple fitting procedures (e.g., method of moments). By fitting the fleet management scenario, for example, only degenerate (i.e., constant) and exponential distributions are adopted. Note that all introduced distributions are restricted to strictly positive support.

8.5 Conclusions

Recent studies have confirmed the impact of M2M traffic on network performance and accessibility through a measurement analysis in operational networks [6]. Thus, supporting the coexistence of M2M uplink-dominant traffic with conventional downlink-dominant user traffic without any service degradation coupled with the massive number of connected devices to the cellular infrastructure is a primary requirement for the next-generation wireless network design.

The study of MTC application cases showed that it is possible to dissect the traffic states of an M2M node into three generic states, namely, event-driven, periodic update, and payload exchange. These three states can be implemented as a source traffic model using an SMM. The source traffic modeling considers MTC application-specific traffic as a single stream. The parameters for such a model were extracted for different use cases leading to source traffic models for each node. The prediction for M2M nodes per cell assume numbers of up to 10,000 devices at a time. Traffic patterns of such a high number of users can only be modeled by working with aggregated traffic streams in a feasible way. The aggregated approach has a lower complexity and is capable of capturing the coordinated traffic. A hybrid solution can be achieved using a CMMPP approach where the state of the nodes is coupled via a second Markov process. This allows a low-complexity modeling of the traffic without neglecting the feature of correlated event-driven traffic.

Table 8.3 presents a comparative summary of the 3GPP aggregated modeling approach with the SMM and CMMPP source modeling approaches.

Table 8.3 Input parameters of the SMM approach

Metric	Aggregated	SMM	CMMPP
Modeling device granularity		✓	✓
Modeling coordinated devices	✓		✓
Spatiotemporal coordination			✓
Modeling packet		✓	
Modeling data rate	✓	✓	✓
Random run time feasible		✓	✓
Device location		✓	✓
Modeling QoS constraint		✓	✓
Coupling of traffic states			✓
Complexity (N MTC devices)	$O(1)$	$O(N)$	$O(N)$

To provide comparable results, the same-recorded network traffic produced by a fleet management use case has been applied to all the modeling strategies. The measurement results show that in fact, the devices trigger in a correlated fashion in the time domain. While basic models like the one proposed by 3GPP fail to produce the same activity peak in the number of active users, the CMMPP model proved to be accurate under these conditions reproducing the peak of active users with a linear growing modeling complexity.

References

[1] ETSI, Machine-to-Machine Communications (M2M); M2M Service Requirements, s.l.: ETSI TS 102 689, 2010.
[2] 3GPP, Service Requirements for Machine-Type Communication, TR 22.368, s.l.: s.n., 2012.
[3] 3GPP, Study on RAN Improvements for Machine-Type Communications, Technical report, TR 37.868, s.l.: s.n., 2012.
[4] Ericsson. 5G Radio Access: Research and Vision, s.l.: Ericsson, 2013.
[5] GSMA, The Mobile Economy 2013, s.l.: GSMA, 2013.
[6] M. Popović, D. Drajić, S. Krćo, Evaluation of the UTRAN (HSPA) performance in different configurations in the presence of M2M and online gaming traffic, Trans. Emerg. Telecommun. Technol. (2013), http://dx.doi.org/10.1002/ett.2738.
[7] M.Z. Shafiq, A First Look at Cellular Machine-to-Machine Trafffic – Large Scale Measurements and Characterization s.l, in: 12th ACM SIGMETRICS/PERFORMANCE Joint International Conference on Measurement and Modeling of Computer Systems, 2012.
[8] A. Adas, Traffic models in broadband networks, IEEE Commun. Mag. 35 (7) (1997) 82–89.
[9] LOLA. D3.5 Traffic Models for M2M and Online Gaming Network Traffic, s.l.: s.n., 2012.
[10] H. Heffes, D. Lucantoni, A Markov modulated characterization of packetized voice and data traffic and related statistical multiplexer performance, IEEE J. SAC 4 (6) (1986) 856–868.
[11] R.C.D. Paiva, R.D. Vieira, M. Säily, Random access capacity evaluation with synchronized MTC users over wireless networks, in: Proceedings of VTC Spring, 2011, s.l., s.n.
[12] G. Casale, E.Z. Zhang, E. Smirni, Trace data characterization and fitting for Markov modeling, Elsevier Performance Eval. 67 (2) (2010) 61–79.
[13] M. Laner, et al., Users in Cells: A Data Traffic Analysis. s.l., IEEE WCNC, 2012.
[14] M. Laner, P. Svoboda, M. Rupp, Modeling randomness in network traffic, in: Sigmetrics, ACM, London, UK, 2012, s.l.
[15] K. Zhou, et al., Contention based access for machine-type communications over LTE, in: Vehicular Technology Conference, 2012, s.l.
[16] R. Ratasuk, J. Tan, A. Ghosh, Coverage and capacity analysis for machine type communications in LTE, in: Vehicular Technology Conference, 2012, s.l.
[17] S.-Y. Lien, K.-C. Chen, Toward ubiquitous massive accesses in 3GPP machine-to-machine communications, IEEE Commun. Mag. 47 (4) (2011) 66–74.
[18] Y. Jou, et al., M2M over CDMA2000 1x Case Studies. s.l., IEEE WCNC, 2011.
[19] Y. Zhang, et al., Home M2M networks: architectures, standards, and QoS improvements, IEEE Commun. Mag. 49 (4) (2011) 44–52.
[20] N. Nikaein, et al., Simple traffic modeling framework for machine type communication, in: IEEE ISWCS, 2013, s.l.

[21] R. Nelson, Probability, Stochastic Processes, and Queueing Theory, Springer-Verlag, New York, 1995, s.l.
[22] S. Yu, Z. Liu, M.S. Squillante, C.H. Xia, L. Zhang, A hidden semi-Markov model for web workload self-similarity, in: IEEE Proceedings of the Performance, Computing, and Communications Conference, 2012, s.l.
[23] S.-Z. Yu, Hidden semi-Markov models, Elsevier J. Artif. Intell. 174 (2) (2010) 215–243.
[24] M. Brand, Coupled hidden Markov models for modeling interacting processes, s.l.: MIT Technical Report, 1997.
[25] M. Brand, N. Oliver, A. Pentland, Coupled hidden Markov models for complex action recognition, in: Proc. of the IEEE Conf. on Computer Vision and Pattern Recognition, 1997, pp. 994–999, s.l.
[26] M. Laner, P. Svoboda, N. Nikaein, M. Rupp, Traffic models for machine type communications, in: IEEE ISWCS, 2013, s.l.

Random access procedures and radio access network (RAN) overload control in standard and advanced long-term evolution (LTE and LTE-A) networks

9

N.K. Pratas, H. Thomsen, P. Popovski
Aalborg University, Aalborg, Denmark

9.1 Introduction

With the advent of machine-to-machine (M2M) communications, denoted in 3GPP as machine-type communications (MTC), and the densification of cellular networks (i.e., cell sizes comparable to local area networks), there will be a massive increase in the number of terminal devices connected to the network. One of the main outcomes of this growth will be the overload of the signaling plane of the cellular networks, if no proper load mechanisms and optimized signaling procedures are put in place. With that goal in mind, in 3GPP, there have been in recent standard releases several study items [1] where many of the MTC-related challenges to the existing network architecture have been studied. These challenges include the support of lower complexity LTE devices, efficient triggering support, efficient transmission of user packets with small amounts of data, simplification and reduction of signaling flows [2], and overload control mechanisms in both the radio access network and the core network.

Figure 9.1 depicts the Evolved Packet System (EPS) architecture model for MTC, showing the entities present in the control and user plane. The control plane is where the signaling flows take place, e.g., the activation and registration of user in a network, the establishment of voice call, and the sending of an SMS. While, the user plane is where the actual user data flows take place, e.g., data payload of a voice call and data files transfer. The cellular system composed of the Long-Term Evolution (LTE) and the System Architecture Evolution (SAE), jointly denoted as Evolved Packet System (EPS), supports only packet-switched services, providing IP connectivity between the user equipment (UE) and the packet data network (PDN). The radio network part is denoted as Evolved Universal Terrestrial Access Network (E-UTRAN), and the core network (CN), which encompasses all nonradio aspects, is denoted as Evolved Packet Core (EPC). Finally, the External Services includes all entities not under the standardization umbrella of 3GPP, which include application servers, either locally or through other external PDNs, such as the Internet. More recently, it has been proposed a cellular-centric M2M architecture [5].

Machine-to-machine (M2M) Communications. http://dx.doi.org/10.1016/B978-1-78242-102-3.00009-5

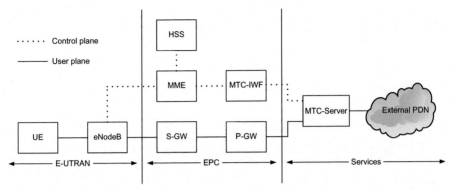

Figure 9.1 EPS MTC architecture model [3,4].

The uplink overload in LTE/LTE-A, and in fact any cellular network, is caused by a very large number of UEs attempting access to the network. To understand the impact of a massive amount of users transmitting small packets to the network, consider Figure 9.2, which depicts the messages exchanged between the UE and the EPC network (eNB, MME, and S-GW/P-GW). The first four message exchanges denote the access reservation protocol (ARP) and occur within the E-UTRAN domain. The remaining ones occur within the EPC domain, although relayed through the E-UTRAN.

The overload can occur at both the E-UTRAN and EPC levels, although due to different causes. At the E-UTRAN level, the overload occurs whenever the number of UEs attempting access to the network is much larger than the amount of radio resources available. At the EPC level, the overload occurs whenever the amount of UEs in a region (traffic incoming from multiple eNodeBs) attempting access is much larger than the processing capacity of the MME entity as well as the link(s) that connects it to the network. In the EPC, the overload can also occur in any of the other entities, although in [3], the MME was identified to be the more susceptible one.

Load control can be accomplished through proper downstream triggering of devices. Especially, if the MTC application is mostly based in collecting information from the MTC devices through triggering, then the application should distribute the triggering in such a way that the amount of users accessing the network does not exceed the amount of resources available. This can only occur if the application is aware of the network load at each of its points. Therefore, while referring to the deployment scenarios of an MTC application in 3GPP network, the indirect and hybrid model would allow such information to be available to the application. An alternative approach would be to let the network decide when each of these devices should be triggered. In this chapter, we consider solely the challenges and mechanisms associated with overload control in the 3GPP context in the upstream direction, i.e., from the user equipment (UE) to the network.

We note that although in this chapter we emphasize the load control mechanism in MTC context, the same mechanisms are applicable to other devices and traffic profiles.

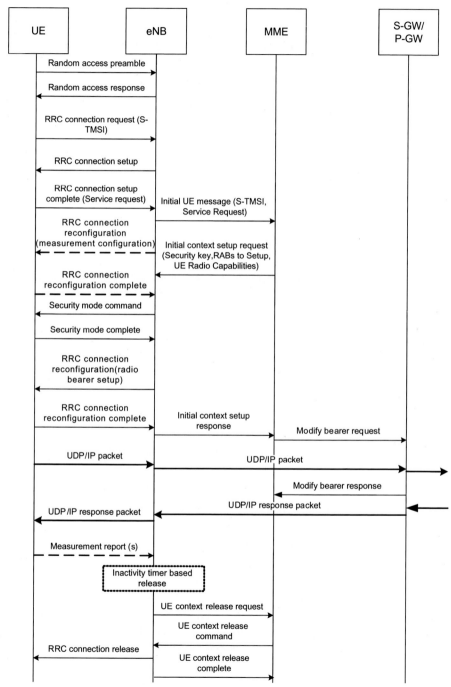

Figure 9.2 3GPP Release 11 message sequence for the transfer of a single IP packet [6].

In the remaining part of this chapter, we describe in detail the LTE/LTE-A access reservation protocol and the extended access barring scheme, followed by an overview of alternative load control principles not selected to be implemented in the E-UTRAN. Then, we give an overview of the challenges associated with overload in the EPC, the mechanism put in place to control it, and how these trigger the extended access barring in the E-UTRAN. Then, we give a short synopsis of the past work done in 3GPP in the context of load control, as well as the future directions. Further, we introduce the concept of system re-engineering, which allows cellular communication systems such as LTE/LTE-A to be more resilient to overload conditions. Finally, we finalize the book chapter with a recap of the chapter contents and future outlook of the developments of MTC load control in cellular networks.

9.2   E-UTRAN access reservation protocol

In this section, we describe the baseline access reservation protocol (ARP) employed in the E-UTRAN. We describe the random access procedure itself and the ancillary subjects that are relevant to the protocol, such as power ramping and preamble configurations.

The access reservation protocol employed in the LTE/LTE-A standard can take on two forms, *contention-based* and *contention-free* [7]. The contention-based method is used in the connection establishment, while the non-contention-based method is used when users have dedicated resources. This latter case is used in, e.g., handover purposes.

The contention-based access reservation is a four-step procedure, which as shown in the message flow is depicted in Figure 9.3, where four different messages are exchanged between the UE and eNodeB [7].

In [8], the current proposals by the scientific community to allow the ARP to support efficiently machine-type traffic are discussed.

9.2.1   Random access preamble

The process begins with the UE selecting uniformly at random one preamble sequence from the set of preamble sequences used for contention-based access and transmitting it at the next physical random access channel (PRACH) opportunity. There are 64 preamble sequences in total, for both contention-based random access and contention-free random access [9]. Furthermore, the preambles used for contention-based random access are divided into two groups. In that way, by choosing a preamble from a particular group, the UE transmits a single bit of information in the RACH. The two groups are denoted as groups A and B, respectively. From which group the preambles will be transmitted depends on the UE's message size.

The preambles used in the random access procedure are derived from Zadoff–Chu (ZC) sequences [10]. This is done by cyclically shifting a base sequence (also called a root sequence), to obtain the preambles. The autocorrelation function of a ZC

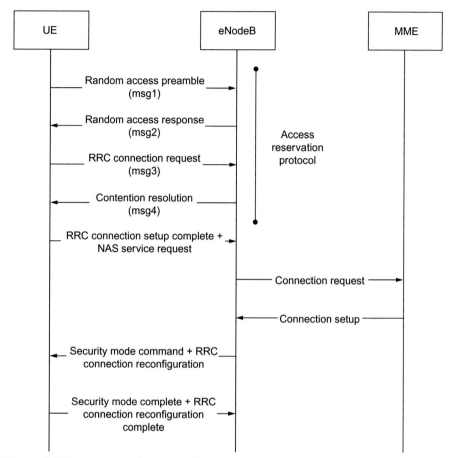

Figure 9.3 Access reservation protocol in E-UTRAN.

sequence is periodic, which makes such sequences useful for preamble detection and timing. They also have very good cross correlation properties.

The cyclic shifts of the preambles need to be sufficiently wide to take uplink timing uncertainty into account. In LTE, there are four possible preamble format configurations. These are shown in Figure 9.4. The formats can have either short or long cyclic prefix (CP). The long CP, used in formats 1 and 3, enables an increased tolerance for timing uncertainty. By aggregating two preamble sequences, as done in formats 2 and 3, there can be better compensation for path loss, which also means that these two formats are better for cells of large radius.

The possible PRACH configurations are shown in Figure 9.5, for the case of preamble configuration 1 and frame structure type 1 [9]. The PRACH configuration is selected by the eNodeB and depends on the load, i.e., on the amount of users attempting to connect to the network [11].

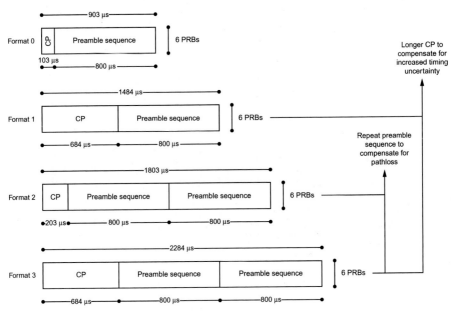

Figure 9.4 PRACH preamble formats.

9.2.2 *Random access response*

After the preamble transmission, the UE listens on the physical downlink control channel (PDCCH) for a Random Access Response message [9]. The UE starts to listen to the PDCCH, three subframes after the subframe containing the end of the preamble, and listens for ra-ResponseWindowSize subframes, where ra-ResponseWindowSize is a parameter set by configuration [12]. If the eNodeB is able to detect that a preamble was transmitted within this period, it can reply with the time and frequency where the preamble was detected. This is called the Random Access Response (RAR) message. If several UEs transmitted the same preamble, they will all receive the RAR. In the case where the eNodeB is able to discern that multiple UEs selected the same preamble, then it can reply with a back-off message. This is only possible whenever the cell size is more than twice the distance corresponding to the maximum delay spread; the BS may, in some circumstances, be able to differentiate the transmission of the same preamble by two or more users, provided that the users are separable in terms of the power delay profile [11]. If the UE is not able to receive the RAR within the specified time period, it will increase the counter for the number of preamble transmissions by one. The UE may use power ramping, when the eNodeB does not receive the preamble transmission.

In step (1) of the random access procedure, contending UEs can use power ramping. This principle is shown in Figure 9.6.

The UE starts by transmitting a preamble (1), as explained in the previous section. In step (2) of the figure, the preamble is not detected at the eNodeB, due to, e.g., collision, fading, or some other means. The UE then increases its preamble transmission

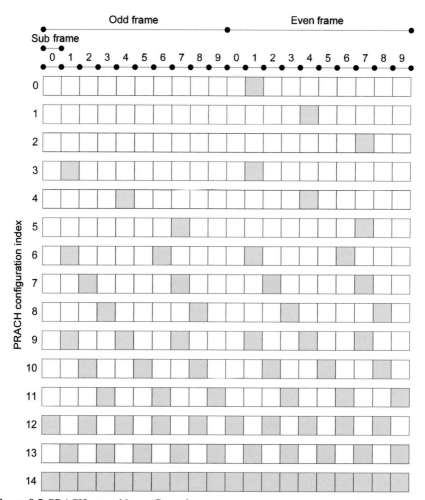

Figure 9.5 PRACH preamble configurations.

counter by one. At the next PRACH resource, the UE transmits another preamble (3), with the transmission power increased, compared to the first transmission. The eNodeB then receives the preamble (4) and responds with a RAR. In step (5), the UE transmits its data (layer 2 or 3 message) on the physical uplink shared channel (PUSCH).

9.2.3   RRC connection request

The UEs that got the RAR in step (3) each send a contention resolution message. This message includes radio resource control (RRC) request and scheduling request. Hybrid automatic repeat request (HARQ) is used in the transmission. If several UEs sent the same preamble in step (1), they will transmit their RRC connection request in the same PUSCH scheduled resources and therefore collide again.

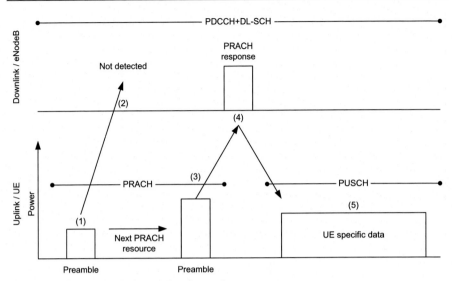

Figure 9.6 Power ramping operation for random access.

9.2.4 Contention resolution

Also denoted as request acknowledgment, here, the eNodeB acknowledges any message received in the third step. This message, as for the one in the previous step, also uses HARQ.

More detailed information about each of the steps of the access reservation procedure and the format and contents of each of the exchanged messages can be found in [7,12].

9.3 Extended access barring protocol

In this section, we provide a description of the extended access barring protocol, standardized in 3GPP to handle overload situations in the E-UTRAN.

Contrary to the GERAN overload control mechanism "implicit reject" [13] that stops the UEs connection attempt after the UE has already made a request to the base station and therefore has spent radio resources, the extended access barring (EAB) mechanism can restrict completely the access of the UEs. Therefore, the extended access class barring (EAB) mechanism works by stopping the radio access of UEs configured as EAB.

The extended access class barring applies only to the UE (user equipment) that is configured with low access priority and extended access barring as defined in [14]. This configuration is targeted primarily for usage by UEs that can tolerate being deferred when competing with other UEs for accessing network resources, as is the case during congestion situations.

The network distinguishes between UEs configured with low access priority and therefore eligible to extended access barring by the indication of low priority when establishing a connection with the UTRAN. We note that this low-priority flag is used by the load control mechanisms present at the CN entities, since the eNodeB when broadcasting the EAB flag does not require to know how many UEs are affected. The UEs themselves, when configured with low access priority, abstain from accessing the channel whenever the EAB flag is active.

An UE, configured for low access priority and extended access barring, may be allowed to override the restrictions imposed as long as its configuration allows it. This exceptional behavior is primarily for usage by applications that most of the time can tolerate being deferred due to low access priority when competing with other UEs for accessing network resources but which occasionally require access to the network when the low-access-priority configuration would prevent getting access. For activating this behavior, the UE requests the activation of packet data network (PDN) connection without indicating that it is configured as low access priority.

The permission for overriding low access priority and extended access barring restrictions by the application still needs to be handled with care since as long as such a PDN connection without low access priority is active, the UE is not affected by any access restriction conditions that the network may set for access with low access priority. As the 3GPP system cannot determine whether any overriding of access restrictions by such UE is justified, the operator has to establish an overlaying mechanism that prevents abuse of such privileges, e.g., through specific tariffing to avoid excessive usage of overriding the low access priority.

The EAB mechanism is shown in Figure 9.7, and it works as follows. The eNodeB starts by paging the devices. The indication of EAB is done through the EAB system information block (SIB) [15]. Only devices configured to EAB are allowed to read this block. The duration of the paging cycle is typically 2.56 seconds.

After the broadcasting of the paging information, the eNodeB sends a barring bitmap, consisting of 10 bits numbered 0–9, representing different access classes. The eNodeB can also bar devices that are roaming, i.e., devices not in their home network [16].

The devices do the barring check by comparing the detected bitmap to the access class (AC) value of the device. If the AC value and the broadcasted bitmap match, the device will not initiate any communication until the EAB SIB is changed. The network rotates access classes, by broadcasting a different EAB bitmap each time. The duration of barring varies. If the bitmap matches, the device proceeds with the access attempt.

9.4   Alternative E-UTRAN load control principles

This section reviews the various alternative load control principles considered by 3GPP prior to the adoption of the EAB protocol [17,18].

These principles are the following:

- *Access class barring*—This method consists of separating users into groups, also called access classes. This method is used by the eNodeB to control the load. It does so by blocking

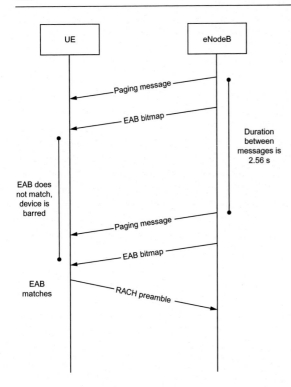

Figure 9.7 Extended access barring mechanism used in 3GPP LTE/LTE-A networks.

one or several user classes. The number of classes is optional and depends on the required granularity. The EAB protocol follows this principle.

- *Separate RACH resources for MTC*—The PRACH resources can be separated into two groups, one for H2H traffic and the other for M2M traffic.
- *Dynamic allocation of RACH resources*—In this solution, the network can adjust the amount of PRACH resources for M2M traffic based on the load, in a dynamical manner.
- *Back-off schemes*—In overload conditions, the network can request the M2M devices to back off. These devices will then attempt transmission at a later stage. This scheme is less suitable for massive batch arrivals than other schemes [19].
- *Slotted access*—Here, resources are assigned by the network for exclusive use by either individual M2M devices or groups of such devices. An M2M can be assigned an access slot based on its ID [17].
- *Pull-based schemes*—This scheme is a paging scheme, where the eNodeB triggers the M2M devices to transmit. Such triggering request is enabled by the core network. This method is only used under favorable traffic conditions.

9.5   Overview of core network challenges and solutions for load control

In this section, we describe solely the core network (CN) mechanisms in place that trigger the load control in the UTRAN, and although the GERAN has also an overload control mechanism denoted as "implicit reject" [13], this is not described here. The overload control general principle in 3GPP networks is that whenever the network starts to become overloaded, the network starts to discard/reject the access requests from low-priority devices and only if still overloaded then starts to discard the higher priority devices.

The introduction of devices classified as low priority was triggered by recognizing that many of the M2M application use cases considered by 3GPP are assumed to be delay-tolerant [20]. This led to the introduction of low-access-priority indication (LAPI) in the control plane signaling and at the radio resource control (RRC) level.

The classification of low-priority device is in general applicable to all the applications running in that same device [16], although there were introduced exceptions after 3GPP Release 11, where the device is able to override the low-access-priority configuration in case it requires emergency access. This facility is a consequence of the regulatory requirements that dictate such behavior in normal cellular systems. This overriding facility will be available for a certain number of restricted applications in these low-access-priority devices.

We now describe the different overload protection mechanisms introduced to protect the 3GPP networks, although with a focus on the EPS. In Figure 9.8, a simplified view of the Evolved Packet Core (EPC) and the entities that are able to perform overload protection mechanisms is depicted.

To enact the overload protection mechanism, it is first necessary to detect if there is an overload occurring at any node or nodes of the network. There are two overload detection and minimization strategies that can be put in place in a 3GPP network [16].

The first is an external one, where the operator can deploy traffic analysis tools to measure the load conditions at each point and node of the network, which makes the strategy rather comprehensive. Upon detection or in the case of prior knowledge of outside world events (such as large population gatherings), the operator may act directly by sending commands to the RAN or CN entities so to off-load some of the traffic being generated. These operator originating commands will then externally trigger the overload control functionalities standardized by 3GPP at each of the network nodes. This strategy is traditionally denoted as operations and management (O&M).

The second strategy is the local detection and triggering at the 3GPP network nodes, which is in line with the self-organizing network (SON) paradigm [21]. Upon

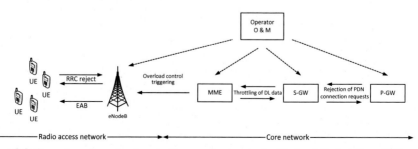

Figure 9.8 Overview of overload protection mechanism in the Evolved Packet Core [16].

overload detection, the PDN gateway (P-GW) may start rejecting new connections and indicate a back-off time for a certain access point name (APN) to the mobility management entity (MME). The MME then rejects all incoming PDN connection requests, for the specified Access Point Name (APN) during the "PDN-GW back-off time." Upon receiving such rejection including a back-off time, the device starts a back-off timer and will not try to access that APN until the back-off timer has expired.

When the MME detects an overload, it can act on it by rejecting non-access stratum (NAS) requests and include back-off time values to off-load the CN. An MME may also reject downlink data notification requests for low-priority traffic for devices in idle mode. Also, to further off-load the MME, the MME can request the S-GWs to selectively reduce the number of downlink data notification requests sent for downlink low-priority traffic for devices in idle mode. If the MME experiences excessive load, it can issue an overload message with a load level indication toward the RAN and whether low-access-priority devices are to be rejected. The RAN will then start blocking some of the traffic incoming toward the CN node.

When off-loading due to an overloaded CN, the RAN can reject the UEs' RRC access attempts while informing them of a waiting timer, during which the UE will not access the network until the timer expires, unless the access attempt is for an emergency or high-priority access request. Unfortunately, even when the UE moves out from the cell where its access request was rejected, it will still wait for the timer to expire before reattempting the access request, preventing therefore the UE to reconnect to the network. Even though this waiting time is available in devices configured for low access priority, the timer value cannot exceed 30 min [22].

Note that the RRC rejection approach is not efficient from a radio resource utilization perspective, since at that point, the UE has already wasted the radio resources associated with the access reservation request (or in the case the packet switch has already established the resources associated with the request). This is why the extended access barring (EAB) mechanism is so important, since it stops the access reservation from occurring, i.e., the UE will not even send the random access preamble message as depicted in Figure 9.2.

Full description of the overload control mechanism at each of the 3GPP network nodes can be found within [22,23].

9.6   Ongoing 3GPP work on load control

The work in 3GPP in the context of network improvements for the handling machine-type communications started with the study reported in [1]. From that report, a possible set of requirements to be agreed on were highlighted, although no specification phase was started. In Release 10 [24], a specification phase was started where the goal was to provide a high-level functional description of the network improvements, the introduction of device with low-priority indication, and finally of overload and congestion control at different levels of the CN and RAN. The results of the findings were reported on [3] and network improvements at the CN captured in the specifications [22,23,25]. Still, at the Release 10, no RAN mechanism was specified.

In Release 11 [26], the findings of the study item on improvements on the RAN for machine-type communications with emphasis on load control were documented on [17]. From the several identified load control mechanism candidates, extended access barring (EAB) was selected for specification and the necessary RAN improvements captured on [12,20,27].

In Release 12 [28], the study on overload control mechanism was resumed since the overload control solutions found in the previous releases act only on UEs that are on RRC_IDLE state (i.e., in the case of overload it prevents new users to become activated in the network). The observed trend is to have a high number of smartphones "always on" in the network, which results in having a majority of these devices in RRC_CONNECTED mode, making the access control mechanism in plane ineffective, since they only apply to devices trying to switch from RRC_IDLE to RRC_CONNECTED mode. This leaves an overloaded network susceptible to become even more overloaded, since, especially in situations of, e.g., disaster, people will try to send requests to the network again and again, and since their UEs are in RRC_CONNECTED mode, the network is forced to process those same requests. The end result is that the network starts discarding even emergency calls and high-priority calls, which has to be avoided. The current considered solutions are going in the direction of defining an application-specific access barring mechanism, based on specific service requirements for the system that shall be able to allow or prohibit the communication initiation of particular applications, as defined by the operator and subject to regional regulations that can change according to context, i.e., in case of disaster. Further, in [29,30], overload control solutions at the CN level, focus on tightening the control of generated signaling streams originating from multiple radio access technologies (RATs) are studied.

9.7   Resilience to overload through protocol re-engineering

As mentioned previously, the majority of MTC applications are seen to be delay-tolerant, require very low data rates, and activate the radio connection very sporadically. Therefore, it is of no surprise to verify that the sum of the total data rate required by all MTC connections in a cell should be, typically, well supported by the maximal data rate available in the cell.

Consider the following illustrative example: Assuming that if there is a single UE in a cell, the base station can support a connection of at most 10 Mbps in the downstream and upstream directions. While if there are 10^4 UEs requiring 1000 bps each in the uplink and downlink, the cellular system will not be able to handle the request traffic, mainly due to the lack of resource granularity due to the signaling overhead associated with the protocol of establishing and maintaining the connection of each of the devices to the network.

Although the physical layer in place would theoretically be enough to serve the incoming traffic, the overhead of the upper-layer protocol put on top of it introduces a limitation to the system capacity. This leads us to the following question: Is it possible to re-engineer the protocol of an existing communication system while keeping its physical-layer specifications essentially intact, so that the system can support a massive number of devices with low-rate connections? The rationale is that by reusing the physical layer, we are reusing a major part of the infrastructure and device structure.

Similar efforts have been observed in the standardization work toward adapting the LTE system to support low-cost MTC devices [16] and other EU research projects (e.g., EXALTED); however, the changes introduced to the LTE system are limited and may only satisfy the MTC requirements in the short term.

A concrete example of the LTE protocol re-engineering has been put forward in [31,32], denoted as coded reservation. Herein, the LTE's access reservation protocol was re-engineered so as to increase the capacity of the phase associated with the transmission of the random access preamble. The core difference between this scheme and the standard one is on how the UEs start the access reservation procedure, while in the standard LTE, the UEs do so through the transmission of a random access preamble in a PRACH slot; in the coded reservation, the UEs now do so by transmitting over several PRACH slots up to one random access preamble.

To exemplify the coded reservation scheme, consider the illustration in Figure 9.9, which depicts the random access preamble phase in the LTE standard access reservation protocol and in the coded reservation. In Figure 9.9a, corresponding to the LTE standard scheme, the UE1 and UE2 select the same preamble, here denoted as "A," in the first PRACH slot, while UE3 selects to not transmit any preamble (not initiating the ARP procedure); this is identified by the idle symbol "I." The eNodeB would then detect that the preamble "A" was activated and then proceed with the remaining

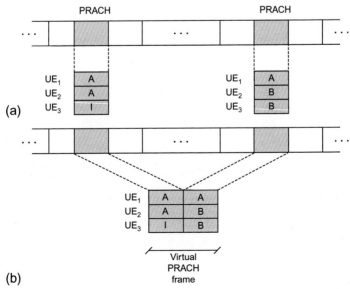

Figure 9.9 Example of re-engineering of the "random access preamble" phase using the coded reservation principles. (a) Standard LTE ARP; (b) coded reservation LTE ARP.

phases of the ARP, resulting on UE1 and UE2 colliding in the third phase of the ARP procedure. In the subsequent PRACH slot, UE1 selects the preamble "A," while UE2 and UE3 select the preamble "B," leading UE2 and UE3 to collide in the ARP's third phase. In Figure 9.9b, the same contention example is shown but now re-engineered using the coded reservation principle. The two PRACH slots are now combined in one virtual PRACH frame and the UEs are allowed to transmit up to one preamble in each slot of the frame. The users no longer contend in a single slot but at the frame level, with a contention code word. The eNodeB observes the whole contention frame (i.e., the preambles transmitted at each RACH slot) and from there generates the list of possible contention code words: AI, AA, AB, IA, and IB, where the idle symbol "I" accounts for the case where the UEs remained silent in the respective slot. In this example, all three UEs no longer collide in the third phase of the ARP procedure, as the contention code words AA, AB, and IB correspond, respectively, to UE1, UE2, and UE3.

Through the coded reservation scheme, the first phase of the LTE ARP procedure is re-engineered resulting in the expansion of the available contention resources following the law $(M+1)^L - 1$, where M is the number of preambles and L the length of the virtual frame. The term "$M+1$" accounts for the idle symbol and the term "-1" accounts for exclusion of the contention code word composed solely by idle symbols. The available contention resource scales then exponentially with L, which is a substantial improvement over the linear increase in the standard LTE with L and M achieved by selecting different PRACH configurations, as depicted in Figure 9.5.

In Figure 9.10, three throughput curves with respect to the LTE baseline and coded reservation are shown, when $L=4$ contention slots are used. It can be seen that even when using a rather small number of preambles in the latter case, it is possible to achieve a larger contention space and, consequentially, increase the throughput.

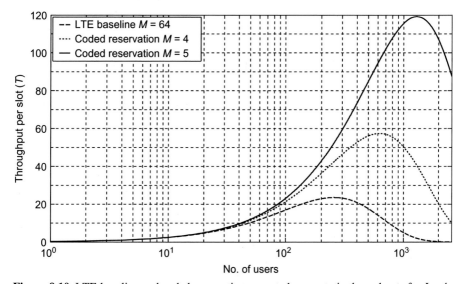

Figure 9.10 LTE baseline and coded reservation expected asymptotic throughputs for $L=4$.

Therefore, this approach lends itself readily to the creation of several QoS classes by proper partitioning of the contention space in the code domain, as discussed in [31,32].

The re-engineering approach has also been applied in a GSM ARP context [33], although in the third phase of the ARP.

Existing cellular system can therefore be made resilient to overload conditions, by proper re-engineering of the existing system. This is of practical interest since it allows reusing the existing cellular technologies and their deployed infrastructure.

9.8 Conclusion

In this chapter, we provided a detailed description of the LTE/LTE-A access reservation procedure and of the load control mechanism in place in the E-UTRAN and in the EPC. Further, we have introduced how a cellular system such as the LTE/LTE-A can be re-engineered into a system that is resilient to overload situations.

It is interesting to note that the future efforts, in the context of load control, by 3GPP are not on enhancing the existing load control mechanism for controlling the incoming traffic of nonactive UEs (in RRC_IDLE), but instead on providing load control for users already active in the network (in RRC_CONNECTED). This comes in sequence with the LTE/LTE-A being completely packet-switched, while the previous cellular generations were predominantly circuit-switched. Therefore, with the possibility of a single UE generating several application streams, due to multiple apps running on it, it is now necessary to extend the load control to be application-aware, which was not a priority in previous 3GPP releases. This therefore enables the definition of MTC to be extended from sensor-type devices attempting to connect to the network, to also include normal UE devices that are running in the background several apps that require short communication bursts with the PDN, emulating the behavior of a sensor network.

Acknowledgments

The work presented in this chapter was supported by the Danish Council for Independent Research (Det Frie Forskningsråd) within the Sapere Aude Research Leader program, Grant No. 11-105159 "Dependable Wireless Bits for Machine-to-Machine (M2M) Communications" and by the framework of the FP7 project ICT-317669 METIS. The authors would like to acknowledge the contributions of their colleagues in METIS, although the views expressed are those of the authors and do not necessarily represent the project.

References

[1] 3GPP Technical Report 22.868, Study on facilitating machine to machine communication in 3GPP systems. Available from: www.3gpp.org.
[2] 3GPP TS 24.368, Non-Access Stratum (NAS) configuration Management Object (MO). Available from: www.3gpp.org.

[3] 3GPP Technical Report 23.888, System improvements for Machine-Type Communications. Available from: www.3gpp.org.

[4] The LTE Network Architecture, Strategic White Paper.

[5] A. Lo, Y. Law, M. Jacobsson, A cellular-centric service architecture for machine-to-machine (M2M) communications. Wireless Commun. IEEE 20 (5) (2013) 143–151, http://dx.doi.org/10.1109/MWC.2013.6664485.

[6] 3GPP Technical Report 23.887, Architectural Enhancements for Machine Type and other mobile data applications Communications. Available from: www.3gpp.org.

[7] 3GPP TS 36.321, Medium Access Control (MAC) protocol specification. Available from: www.3gpp.org.

[8] A. Laya, L. Alonso, J. Alonso-Zarate, Is the random access channel of LTE and LTE-A suitable for M2M communications? A survey of alternatives, Commun. Surveys Tutorials IEEE 99 (2014) 1–13, http://dx.doi.org/10.1109/SURV.2013.111313.00244.

[9] 3GPP TS 36.211, Physical channels and modulation. Available from: www.3gpp.org.

[10] 3GPP TS 36.201, LTE physical layer; general description. Available from: www.3gpp.org.

[11] S. Sesia, I. Toufik, M. Baker, LTE—The UMTS Long Term Evolution: From Theory to Practice, Wiley, New York, 2011.

[12] 3GPP TS 36.331, Evolved Universal Terrestrial Radio Access (E-UTRA); Radio Resource Control (RRC); Protocol specification. Available from: www.3gpp.org.

[13] 3GPP TS 44.018, Mobile radio interface layer 3 specification; Radio Resource Control (RRC) protocol. Available from: www.3gpp.org.

[14] 3GPP TS 22.011, Service accessibility. Available from: www.3gpp.org.

[15] 3GPP TS 25.304, User Equipment (UE) procedures in idle mode and procedures for cell reselection in connected mode. Available from: www.3gpp.org.

[16] P. Jain, P. Hedman, H. Zisimopoulos, Machine type communications in 3GPP systems. Commun. Mag. IEEE 50 (11) (2012) 28–35, http://dx.doi.org/10.1109/MCOM.2012.6353679.

[17] 3GPP Technical Report 37.868, Study on RAN Improvements for Machine-type Communications. Available from: www.3gpp.org.

[18] A. Lo, Y.W. Law, M. Jacobsson, M. Kucharzak, Enhanced LTE-advanced random-access mechanism for massive machine-to-machine (M2M) communications, in: In the 27th World Wireless Research Forum (WWRF) Meeting, Düsseldorf, Germany, October 18–20, 2011.

[19] M.Z. Shafiq, L. Ji, A.X. Liu, J. Pang, J. Wang, A first look at cellular machine-to-machine traffic: large scale measurement and characterization, ACM SIGMETRICS Performance Evaluation Review 40 (1) (2012) 65–76, http://dx.doi.org/10.1145/2318857.2254767.

[20] 3GPP TS 22.368, Service requirements for Machine-Type Communications (MTC). Available from: www.3gpp.org.

[21] 3GPP TS 32.500, Telecommunication management; Self-Organizing Networks (SON); Concepts and requirements. Available from: www.3gpp.org.

[22] 3GPP TS 23.401, General Packet Radio Service (GPRS) enhancements for Evolved Universal Terrestrial Radio Access Network (E-UTRAN) Access. Available from: www.3gpp.org.

[23] 3GPP TS 23.060, General Packet Radio Service (GPRS); Service description; Stage 2. Available from: www.3gpp.org.

[24] 3GPP Overview of 3GPP Release 10. Available from: www.3gpp.org.

[25] 3GPP TS 23.236, Intra-domain connection of Radio Access Network (RAN) nodes to multiple Core Network (CN) nodes. Available from: www.3gpp.org.

[26] 3GPP Overview of 3GPP Release 11. Available from: www.3gpp.org.

[27] 3GPP TS 25.331, Radio Resource Control (RRC); Protocol specification. Available from: www.3gpp.org.

[28] 3GPP Overview of 3GPP Release 12. Available from: www.3gpp.org.
[29] 3GPP Technical Report 23.843, Study on Core Network (CN) overload solutions. Available from: www.3gpp.org.
[30] 3GPP Technical Report 29.809, Study on Diameter overload control mechanisms. Available from: www.3gpp.org.
[31] N.K. Pratas, H. Thomsen, C. Stefanovic, P. Popovski, Code-expanded random access for machine-type communications, in: Globecom Workshops (GC Workkshops), 2012 IEEE, 2012, pp. 1681–1686, http://dx.doi.org/10.1109/GLOCOMW.2012.6477838.
[32] H. Thomsen, N.K. Pratas, C. Stefanovic, P. Popovski, Code-expanded radio access protocol for machine-to-machine communications, Transactions on Emerging Telecommunications Technologies Special Issue: Machine-to-Machine: An Emerging Communication Paradigm 24 (4) (2013) 355–365, http://dx.doi.org/10.1002/ett.2656.
[33] G.C. Madueno, C. Stefanovic, P. Popovski, How many smart meters can be deployed in a GSM cell?. in: IEEE ICC 2013—Second IEEE Workshop on Telecommunication Standards: From Research to Standards, Budapest, Hungary, June, 2013, http://dx.doi.org/10.1109/ICCW.2013.6649431.

Packet scheduling strategies for machine-to-machine (M2M) communications over long-term evolution (LTE) cellular networks

A. Alexiou, A. Gotsis
University of Piraeus, Piraeus, Greece

10.1 State of the art in M2M multiple access in legacy cellular systems

A steady growth is forecast worldwide for the machine-to-machine (M2M) cellular market. By providing an established global network infrastructure and cost-efficient communication modules, cellular systems play a critical role in the broad adoption and the successful deployment of M2M. Recent market reports predict more than 500 million embedded M2M connections by 2015 [1]. They also indicate the very high number of M2M devices per cell as one of the most demanding challenges. For example, the number of smart-metering devices per cell in a typical urban environment is expected to be in the order of tens of thousands [2]. Diverse smart-metering, health monitoring/alerting, and intelligent transportation scenarios are foreseen to consist the M2M applications portfolio [3]. To this end, both the 3rd Generation Partnership Project (3GPP) and the IEEE standardization organizations have initiated M2M study items and working groups, targeting the support of such application scenarios (see, e.g., [4,5]) through the evolving systems standardization releases.

Today, most M2M applications use the general packet radio service (GPRS) network infrastructure and specific services such as the short message service (SMS). This is a manageable, cost-efficient way to address M2M deployment, as long as the number of devices remains relatively small. GPRS is based on the packet radio principle used for carrying end users' packet data to/from GPRS terminals and/or external networks. Nevertheless, GPRS is designed for bursty traffic, usually encountered in applications such as Internet browsing or e-mail. Resources for GPRS traffic can be reserved in a static or dynamic fashion [6].

In this way, GPRS offers a promising cellular infrastructure for the support of the also bursty M2M applications traffic. Nowadays, the vendors and providers tend to be in favor of selecting GPRS for the support of M2M on the basis of the following arguments:

- Immediate M2M business entry.
- Low cost and convenient deployment.

Machine-to-machine (M2M) Communications. http://dx.doi.org/10.1016/B978-1-78242-102-3.00010-1

- Ubiquitous and international operability.
- Roaming between mobile operators.
- GPRS is a proven real-world tested technology, which is also open and standardized.

Despite the advantages of being a low-cost and well-established technology, the suitability of GPRS for future M2M applications raises several concerns: although the capacity of a GPRS cell depends on several parameters, such as the frequency reuse pattern, the spectral efficiency of a GPRS cell usually cannot exceed the 100–150 kbps/cell/MHz. When voice users are also assumed to be active, the number of supported data users becomes limited (<30) [7]. Considering these limitations, GPRS will soon prove inadequate for the support of the envisioned M2M applications and services with thousands of devices per cell. Another issue is that a connection is initiated by the device in the GPRS network [8], which may be a serious obstacle in M2M scenarios.

10.2 Signaling and scheduling limitations for M2M over LTE

The limited capacity of second-generation cellular systems indicates the need to consider higher capacity systems, such as the 3GPP long-term evolution (LTE) especially targeting the exploitation of flexible radio resource management (RRM) capabilities. LTE has been, however, designed for broadband applications, while most M2M applications transmit and receive very small volumes of data, leading to an unreasonable split between payload and required control information. Moreover, the requirements for low-energy consumption, low latency, and high reliability, which are critical in M2M communications, are not by default addressed in broadband cellular systems.

In order to overcome the deficiencies of LTE with respect to M2M communications, 3GPP has introduced the machine-type communications study item [9]. Besides the necessity of supporting a very large number of devices, the efforts also focus on addressing the vast dynamic range of M2M service characteristics, the need for low energy consumption and the importance of driving cost low, in order to be able to introduce a viable business model/opportunity.

Packet scheduling constitutes the key RRM mechanism for optimizing the overall system resource utilization efficiency while guaranteeing individual quality-of-service (QoS) requirements. Particularly in LTE, the flexible resource allocation of time–frequency resource elements or physical resource blocks (PRBs) renders efficient scheduling an even more decisive system performance factor. In order to optimally allocate the PRBs to user equipment (UEs) and/or machine-type communication devices (MTCDs), the scheduler should exploit channel state information and traffic dynamics knowledge on a fast timescale, ideally per transmission time interval (TTI). The uplink (UL) and downlink (DL) signaling channels for carrying channel quality, traffic, and allocation information are necessary for facilitating scheduling. The unique characteristics of M2M traffic, namely, the large number of devices and the bursty low rate load nature, perplex scheduling as both complexity and signaling are heavily increased. It is worth mentioning that, besides the flexible resource allocation capabilities of LTE, features such as relaying in LTE-Advanced (LTE-A) may offer additional benefits for M2M communications, such as coverage extension for low-energy devices.

This chapter aims at presenting the challenges of M2M scheduling over existing and evolving cellular networks as well as proposing novel scheduling techniques. In what follows, signaling and scheduling limitations of LTE will be identified in the context of M2M communications. We will then review existing approaches and present two new scheduling schemes and their performance merits, and, finally, we will elaborate on long-term perspectives.

10.2.1 Signaling in LTE scheduling

When it comes to the most critical M2M application scenarios, such as smart metering, e-health, and intelligent transport systems, involving the aggregation of a considerable volume/number of data streams with varying QoS requirements, the UL is expected to constitute the most critical link. For this reason, this chapter focuses mainly on the UL scheduling.

LTE provides a random-access (transport) channel (RACH) in the UL, mainly utilized for initial radio-link establishment, handover, and synchronization. RACH may also carry UL scheduling requests if no dedicated physical uplink control channel (PUCCH) has been assigned. This is done through a contention-based mechanism (as opposed to the contention-free scheduling). Each UE competing for access may use one of the 64 available orthogonal sequences per cell (a number significantly lower than the expected volume of MTCDs). If more messages should be transmitted, collisions are likely to occur.

In LTE, UL scheduled access is performed at the base station (eNB) in a centralized manner, and resource allocation decisions are communicated to the UEs through appropriate control channels. More specifically, each UE sends scheduling requests to the eNB via L1/L2 control signaling, that is, the PUCCH, requiring access to the uplink shared channel.

Each UE is assigned one PUCCH, which, in the presence of a large number of MTCDs, as it happens in many M2M scenarios, results in shortage of PUCCH resources. Moreover, in order to cater for channel-aware scheduling, the per-device channel quality information is needed to be sent to the eNB; this information is also carried by a UL control channel. As the number of devices grows, so does the associated signaling load.

Based on the scheduling requests, the eNB decides the PRB-to-UE (or MTCD) allocation on each TTI and sends the scheduling grants to the MTCDs through the corresponding physical downlink control channel (PDCCH). The PDCCH is loaded into the first 1–3 orthogonal frequency division multiplexing symbols of the DL time–frequency grid, consuming system resources. According to 3GPP specifications, up to 10 UEs (or MTCDs) may be loaded in a single subframe. Hence, PDCCH is unable to support hundreds of MTCDs simultaneously requesting access to the shared channel, as envisioned in many future M2M scenarios.

Considering either random or scheduled access, supporting UL becomes prohibitive in many M2M scenarios, as LTE mechanisms often prove inadequate to efficiently support several MTCDs in LTE cells. Modifications of existing approaches, as well as the design of new solutions for reducing the required signaling overhead are of utmost importance for the realization of M2M-enabled LTE and beyond systems.

10.2.2   The LTE scheduling framework

Although scheduling algorithms, and RRM procedures in general, are not part of standardization, but rather an implementation-specific issue, the associated signaling itself needs to be standardized. Thus, any scheduling algorithm proposal should be compatible with feasible control signaling requirements. To this end, a generic packet-scheduling framework has been proposed in 3GPP [10] according to which (Figure 10.1): (1) a set of UEs is picked by a time-domain scheduling entity for packet transmission on a certain frame and (2) a frequency-domain scheduling entity allocates the PRBs to the selected UEs. Both entities make allocation decisions based on metrics reflecting channel quality, experienced packet delay, and QoS tolerance. Making use of this flexible framework, several schemes have been devised for dynamically allocating resources to UEs with heterogeneous QoS requirements, covering a large dynamic range of requirements from non-real-time to real-time services

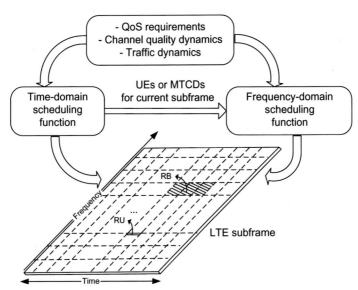

Figure 10.1 The frequency–time domain LTE packet scheduling framework for supporting machine devices.

(e.g., File Transfer Protocol down-/up-loading, web surfing, and video streaming) [11]. A set of 9 QoS classes have been prescribed in 3GPP specifications (QoS class identifiers, QCIs) classifying services (or *radio bearers*) based on resource type (Guaranteed bit rate (GBR)/non-GBR), priority order, packet delay budget, and packet loss rate characteristics (Figure 10.2).

Nevertheless, in most M2M scenarios, scheduling entities have to deal with extremely diverse QoS criteria. Delay tolerance, for instance, may take values ranging from tens of milliseconds (vehicle collision) to several minutes (environmental monitoring). Error rate tolerance may scale accordingly and span a dynamic range of several orders of magnitude. Thus, forming specific QoS classes tailored to a variety of M2M scenarios is not a trivial task. As far as scheduling that exploits channel and traffic dynamics is concerned, the presence of a huge number of MTCDs may induce further processing delays and impose storage/buffering constraints. Developing practical scheduling schemes that support large numbers of MTCDs without deteriorating at the same time the performance of standard LTE services is therefore a challenging objective.

QCI	Resource type	Priority	Packet delay budget	Packet error loss rate	Example services
1	GBR	2	100 ms	10^{-2}	Conversational voice
2		4	150 ms	10^{-3}	Conversational video (live streaming)
3		3	50 ms	10^{-3}	Real time gaming
4		5	300 ms	10^{-6}	Non-conversational video (buffered streaming)
5	Non-GBR	1	100 ms	10^{-6}	IMS signalling
6		6	300 ms	10^{-6}	Video (buffered streaming) TCP-based (e.g., www, e-mail, chat, ftp, p2p file sharing, progressive video, etc.)
7		7	100 ms	10^{-3}	Voice, Video (live streaming) Interactive gaming
8		8	300 ms	10^{-6}	Video (buffered streaming) TCP-based (e.g., www, e-mail, chat, ftp, p2p file sharing, progressive video, etc.)
9		9			

Figure 10.2 QoS class identifiers (QCIs) characteristics standardized by 3GPP.

10.3   Existing approaches for M2M scheduling over LTE

M2M scheduling over cellular (LTE and beyond) systems is a research area that has only recently attracted considerable interest by academic researchers, operators, and infrastructure vendors. In order to efficiently design and successfully deploy M2M over cellular systems, one has to take into consideration three major constraints:

    i. Capacity scaling requirements, that is, sustaining per M2M device QoS as the number of M2M devices increases dramatically.
   ii. Signaling, standardization, and backward compatibility issues associated with the adoption of novel M2M scheduling approaches.
  iii. A graceful evolution path that would guarantee the feasibility of M2M adoption by operators and service providers while addressing (i) and (ii) at the same time.

In this section, some promising initial approaches for M2M scheduling are presented.

10.3.1   Group-based scheduling

A reduced complexity approach for managing radio resources and scheduling is proposed in ref. [12]. It is based on the formation of MTCD groups or "clusters," where each cluster is associated with a prescribed QoS profile. Then, MTCDs are transparently connected to and managed by the LTE network, since the eNBs control the cluster entities. Scheduling prioritization is imposed on a cluster basis, a policy that significantly reduces complexity and overhead. Cluster formation is dictated by the packet arrival rate and maximum tolerable jitter. Nevertheless, quantization of varying QoS requirements by means of associating a single QoS profile to each cluster results in performance degradation that heavily depends on the actual QoS requirements dynamic range and the degree of adaptability of clustering decisions.

10.3.2   Time granularity of scheduling

In LTE, full dynamic channel- and QoS-aware scheduling per TTI is optimal in terms of system performance. In that respect, LTE should be able to support full dynamic scheduling for M2M communications as well. However, the larger number of MTCDs by several orders of magnitude (compared to legacy UEs) imposes severe constraints on the applicability of such highly dynamic approaches. A similar problem has been identified in the LTE standard, regarding the provision of Voice over Internet Protocol (VoIP) services: LTE should in principle be able to support hundreds of VoIP UEs, each generating small amounts of periodic data traffic. In the literature, semipersistent scheduling schemes have been proposed for efficiently dealing with such special traffic characteristics (see, e.g., [13]). Semipersistent scheduling makes an allocation decision for a longer time period; thus, it is not needed to inform the UEs on a TTI basis. Due to the similarity of the traffic nature of VoIP with M2M traffic, semipersistent scheduling is a possible candidate for M2M scenarios. However, during semipersistent scheduling, various parameters, such as resource allocation and link adaptation, remain fixed. Should radio-link conditions change, a new allocation would need to be initiated. Semipersistent scheduling is therefore more suitable for periodic communications and relatively static radio-link conditions.

10.4   Novel approaches for M2M scheduling over LTE

In what follows, new performance results based on two recently proposed novel M2M over LTE scheduling solutions are presented, and some longer-term M2M evolution ideas and directions are discussed.

10.4.1   QoS classes and LTE scheduling

Group-based scheduling associates each MTCD with one QoS profile out of a fixed number of QoS classes. Although the group-based proposals lower the signaling overhead burden, the QoS classes' definition for MTCDs, as well as the introduction of new MTCD-QCIs on top of the ones adopted by LTE, raises several issues, mainly stemming from the fact that the number of different classes would have to be practically infinite, in order to avoid performance degradation due to quantization effects as the number of MTCDs increases and their QoS requirements vary. This means that both the number of QoS classes and the range of each class must be carefully chosen in order to successfully capture the diverse characteristics of M2M applications. In contrast to current policies, a dynamic formation of QoS classes according to a particular application scenario may be more appropriate for M2M communications, given that the MTCD topology and individual characteristics are not a priori known.

Before one devises a novel analytic (or combined analytic/empirical) model for MTCD-QCI formation, we first need to explain/assess the effect of QoS classes granularity on the system performance. To this end, two simple scheduling algorithms were proposed in ref. [14], which assign resources to MTCDs based on individual delay budget. Both the channel conditions and the maximum allowed delay of each device that requests transmission are taken into account. By exploiting the channel quality of each device, the cell throughput increases through the multiuser diversity gain. Moreover, because of the large number of MTC devices per cell, the delay tolerance of each one, which can range from a few milliseconds to several hours, is a key factor for efficiently allocating the limited resource blocks. In contrast to other scheduling algorithms, no fixed predefined classes of devices are formed in this approach. The exact delay constraint of each device is considered instead, addressing the M2M communication requirements in a realistic fashion. This results in increased number of effectively served requests, along with further improvements taking advantage of the exact delay constraints, for example, optimizing sleep mode duration.

The first algorithm focuses more on the channel quality of each user while the second one on the maximum delay tolerance. The first algorithm is basically an extension of the conventional channel-aware scheduler, but now, the maximum delay tolerance of each device is taken into account. The second algorithm gives priority to devices with low delay tolerance and then tries to find the best resource block in terms of channel quality in order to assign the resources to that device. The channel quality metric utilized is based on the signal-to-noise ratio (SNR) over a certain resource block as seen by a device. The signal-to-interference-and-noise ratio could be used instead in order to provide a metric that more accurately reflects the interference conditions in the network.

This work assessed and quantified the importance of having an adequate number of different M2M classes and spanning the whole range of the M2M QoS requirements. Figure 10.3 depicts the performance of a priority-based scheduler (MTC devices with lower delay tolerance are served first), when different numbers of classes are used.

The performance results are extracted via numerical simulations, following a simplified LTE system model. A 5 MHz UL single-cell environment is considered, where the number of MTC devices ranges from 0 to 1000, with each one having an average SNR that uniformly ranges from 0 to 10 dBs. Perfect decoding is also assumed. Each device requests a predefined number of LTE physical resource blocks, uniformly distributed in such a way that the total load is controlled (results are depicted for high load conditions). The maximum delay tolerance of each device uniformly ranges from 10 ms to 10 min. Regarding the LTE-specs-based scheduler, classes are formed following the levels of {50, 100, 150, 300} ms, whereas for the other schemes, the whole delay span is split uniformly into corresponding delay ranges. It can be seen that using

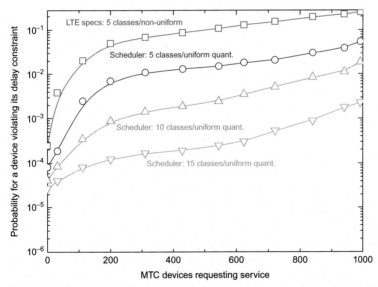

Figure 10.3 LTE scheduler's performance as function of the number/type of available M2M classes. The impact of QoS classes formation on delay performance: increasing the number of classes and modifying the quantization type leads to enhanced performance.

For a more detailed description of the algorithms and simulation procedures, refer to ICT-EXALTED Project Deliverable D3.3, Final report on LTE-M algorithms and procedures. Available from: www.ict-exalted.eu/.

the standard QCI classes (as specified for the existing LTE services), the percentage of MTC devices failing to satisfy their constraints is significantly higher than in the case where the various classes span the whole dynamic range of requirements. More specifically, simply by using classes resulting from the quantization of the whole range of delay tolerances (e.g., uniform quantization between the minimum and the maximum required delay), the performance of the scheduler significantly improves. Ensuring the availability of an adequate number and dynamic range of required QCIs for M2M communications through standardization, is expected to play a decisive role in the adoption of such an approach.

10.4.2   Low-complexity scheduling and M2M bandwidth estimation

With the objective of devising a low-complexity solution for scheduling MTCDs (single and grouped), the approach proposed in ref. [12] has been extended in ref. [15] and the time-controlled M2M feature was adopted. According to this feature, M2M device groups are scheduled at specific periodic intervals, similarly to the semipersistent VoIP paradigm, while Poisson-like-generated traffic is assumed. Scheduling is modeled as an M/D/1 queuing problem, and, following the ideas of such a queuing framework, an analytic model is derived, which relates the scheduling period, the average offered traffic load, and the QoS requirements, in terms of packet delay budget and dropped packet rate as in LTE.

Considering M2M devices sharing homogeneous traffic characteristics and QoS requirements, an exact analytic model is first derived that combines statistical QoS performance indicators (in terms of a delay threshold violation probability) with the allocated grant period. This model could be utilized either for assessing the performance of M2M over LTE without the need of running time-consuming simulations or for estimating the minimum grant period to meet prescribed statistical QoS requirements. The latter is in fact an accurate bandwidth estimation method for M2M-type traffic. Building on the previous model, the multiple-M2M-class scenario is investigated, where lower priority M2M devices may postpone their transmissions in the presence of a higher priority M2M device in the same grant period. For this scenario, an approximate analytic model is derived for assessing the QoS performance loss due to this "grant collision" phenomenon.

Finally, a simple modification to the considered periodic-like fixed grant scheduling algorithm is proposed, in order to improve the performance of low-priority (LP) M2M devices, along with an approximating model that predicts the improvement levels. According to this modification, the scheduler monitors the running average of the grant period, and when it detects that it is lower than the target one, it allocates the next available (if any) resource block to this M2M class. Such an action distorts the original pattern, but the extra allocated resource blocks improve the QoS performance. Although not optimal, this modification introduces a very simple and effective way to improve the performance of LP classes as illustrated through both simulation experiments and queueing analysis [15].

The developed analytic models, which are based on concepts from queueing theory and validated through extensive system-level simulation experiments, may be utilized for:

1. tuning the scheduling decisions, that is, decide which is the minimum scheduling period for an M2M cluster in order to meet the probabilistic QoS targets and
2. estimating the minimum LTE bandwidth reserved for M2M in order to support different M2M loads with prescribed QoS requirements.

In Figure 10.4, indicative results for the packet delay distribution corresponding to variable M2M traffic loads and scheduling periods are illustrated. One may easily observe how, for a specific scheduling and load scenario, the 90th percentile of the packet delay budget may be predicted. Besides the analytic tractability of this approach, such a periodic-like scheduling scheme is very efficient in terms of signaling, since each MTCD needs to be informed only about the period to be active.

This approach constitutes an initial attempt to shed some light on the viability of such scheduling schemes for M2M-enabled LTE systems. Nevertheless, hybrid full dynamic/semipersistent schemes for traffic mixes including legacy LTE UEs and MTCDs could be more appropriate. A thorough study of these issues is necessary for identifying the inherent performance–complexity trade-offs of various scheduling schemes. Moreover, studies targeting a different mix of loads are necessary for assessing the real performance bounds of LTE or other future cellular networks aimed to support M2M services.

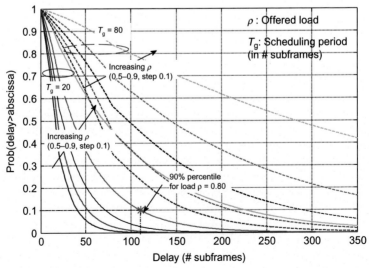

Figure 10.4 Predicted packet delay budget distribution for various M2M traffic intensity and scheduling period setups [15].

10.5 Technology innovations and challenges for M2M scheduling over wireless networks beyond 2020

The issues presented in this chapter mainly call for modifications on the existing protocols. However, in order to efficiently support M2M over future cellular systems, more radical changes may be needed in the foreseen future, regarding the radio access network architecture and procedures. An important question that the future wireless system designer needs to answer is whether an evolutionary or a revolutionary approach would be more appropriate in the advancement process of cellular systems like LTE, in order to provide M2M communications in a cost-efficient, scalable, and reliable manner.

As a first example, the initial access procedure of LTE (Figure 10.5) involves a lot of steps that increase overhead and latency and may be avoided for M2M data transmission. Simpler access procedures (evolutionary approach) or completely separated standard LTE- and M2M-oriented control and data channels (revolutionary approach) may be needed. Second, we have assumed so far that M2M data are loaded on the current LTE frame. However, the LTE frame is designed for data-hungry applications;

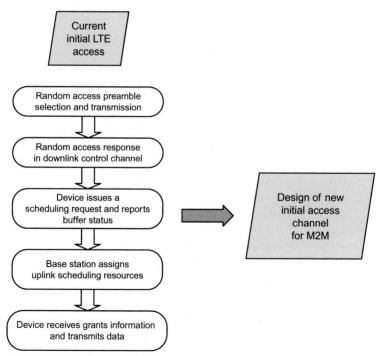

Figure 10.5 From current LTE initial access to a new random access channel for M2M data transmission.
See also ref. [16].

thus, the minimum resource element is 180 kHz, significantly larger than the requirements for typical M2M applications. Alternative physical layer-oriented solutions will thus be needed, such as aggregating data from multiple devices into a single resource block or employing advanced hierarchical modulation techniques. Finally, in order to minimize layer 3 signaling and latency, a *hybrid contention- and scheduling-based scheme* may be applied incorporating the merits of low-complexity random access together with those of high-performance centralized scheduled access. Such a hybrid contention/reservation protocol has been proposed in ref. [17].

10.5.1 Hybrid contention/reservation multiple access protocol for future M2M over cellular systems

Having a large number of M2M devices trying to forward their data to a gateway node leads to increased traffic intensity, congestion, and increased packet loss probabilities. The key point in coping with this problem is to redesign the medium access control (MAC) protocols in order to be able to handle various levels of traffic intensity and at the same time satisfy specific performance optimization criteria. It has been theoretically proven and evidenced by simulation/experimental results that the low-complexity (contention-based) random access is bound to become inefficient in a region beyond a certain threshold of traffic load/density, where scheduled access would rather be used instead. Several hybrid schemes have been proposed—mainly in the framework of wireless sensor networks—that combine the merits of random and scheduled access in terms of complexity and performance, attempting to achieve the best of both worlds. Several random/scheduled access hybrid schemes have been proposed, such Zebra-MAC, Funneling-MAC, and CSMA/TDMA schemes, suggesting the importance of employing hybrid schemes designed to adapt to traffic intensity; to introduce scheduling strategies for improving the network capacity, fairness, and packet loss performance; and to dynamically assign part of the contention access period to TDMA (to address traffic congestion scenarios). A common observation in all these approaches is that contention-based access is used for relatively low traffic conditions, while contention-free slot assignments become dominant as traffic density increases. However, it is not a trivial task to identify the *switching point between the contention-based access and contention-free access*, which optimizes the system performance in terms of throughput and energy efficiency.

In an effort to address the hybrid contention/reservation MAC protocol design challenge, a novel hybrid protocol was presented in ref. [17], which decides on the access mode to be used, based on the trade-off between the expected throughput and protocol complexity. The expected throughput can be predicted by exploiting an analytic framework based on the queueing theory principles, which evaluate the performance of both contention-based and contention-free access schemes.

More specifically, exact and approximate expressions are derived for evaluating the performance of both access schemes under consideration, in terms of packet loss probability and throughput. These expressions can be found quite useful for predicting the performance of contention-free or contention-based access schemes and thus deciding on the more suitable one (or understanding the impact of various involved parameters). Furthermore, based on these theoretical expressions, a switching criterion between the contention-based access and reservation-based access is proposed, enabling the introduction of a hybrid approach for accessing the medium. The switching criterion that was found to optimize the hybrid scheme performance utilizes the ratio between the throughput values for the contention-based and reservation MAC protocols. The throughput values can be theoretically evaluated (predicted) using the derived analytic expressions, and the corresponding value of λ (packet arrival rate of the assumed Poisson process) can be communicated to the node taking scheduling decisions, whenever there is a considerable variation in the generated traffic, using a few reserved bits in the data packets. Using such an approach, a near optimum throughput performance is guaranteed, while depending upon the requirements, for example, increasing demands for the throughput, the switching criterion threshold can be modified accordingly.

10.6 Conclusions

Supporting M2M communications over advanced next-generation cellular infrastructures, such as LTE, LTE-A, and systems beyond, opens up a wide range of new applications beneficial for both individuals and the whole society. An important enabler for M2M is the design of efficient and dynamic packet-scheduling schemes. In this chapter, we first explained why second-generation systems are not expected to be able to carry the increasing M2M load and then recognized the limitations of the existing scheduling approaches with respect to the distinctive M2M traffic characteristics. Then we reviewed proposals appearing in the literature for supporting this type of traffic in an advanced cellular network, such as device clustering and semipersistent/fixed grant scheduling. We presented new results on possible scheduling solutions and explained why there are still significant challenges to deal with. We finally argued that contrary to the presented evolutionary changes, revolutionary paths may be also needed for fully enabling M2M capabilities over LTE, LTE-A, and systems beyond 2020.

References

[1] Machine-to-machine (M2M), The rise of the machines, Juniper Networks White Paper (2011).
[2] Vodafone, R2-102296, RACH intensity of time controlled devices, 3GPP Technical Report, April (2010). Available from: www.3gpp.org.
[3] EXALTED Deliverable 2-1, Description of baseline reference systems, scenarios, technical requirements & evaluation methodology (2011). Available from: http://www.ict-exalted.eu.
[4] 3GPP TS 22.368, Service requirements for machine-type communications (MTC); Stage 1 (Release 11), v. 11.3.0, (2011–09). Available from: www.3gpp.org.

[5] IEEE 802.16's Machine-to-Machine (M2M) Task Group, "Enhancements to Support Machine-to-Machine Applications for WirelessMAN-Advanced," DRAFT Amendment to IEEE Standard, 2011.

[6] ETSI GSM Specification 02.60, Digital cellular telecommunications system (Phase2+), General Packet Radio Service; Service description, Stage 1 (1999).

[7] T. Halonen, J. Romero, J. Melero, GSM, GPRS and EDGE Performance—Evolution Towards 3G/UMTS, John Wiley & Sons, Chichester, 2002, ISBN 0470 84457 4.

[8] M. Martsola, T. Kiravuo, J. Lindqvist, Machine to machine communication in cellular networks, in: Mobile Technology, Applications and Systems, 2nd International Conference on 15–17 November, 2005, Guangzhou.

[9] 3GPP Technical Report 23.888, System improvements for machine-type communications (Release 11), v. 1.6.0 (2011–11). Available from: www.3gpp.org.

[10] 3GPP R1-072277, Downlink interference coordination, Technical Document, Nokia & Nokia Siemens Networks, Kobe-Japan, May (2007). Available from: http://www.3gpp.org.

[11] G. Monghal, K.I. Pedersen, I.Z. Kovacs, P.E. Mogensen, QoS oriented time and frequency domain packet schedulers for the UTRAN long term evolution, in: Vehicular Technology Conference, VTC, Spring, 11–14 May, IEEE, 2008, Marina Bay, Singapore, pp. 2532–2536.

[12] S.-Y. Lien, K.-C. Chen, Y. Lin, Toward ubiquitous massive accesses in 3GPP machine-to-machine communications, IEEE Commun. Mag. 49 (4) (2011) 66–74.

[13] D. Jiang, H. Wang, E. Malkamaki, E. Tuomaala, Principle and performance of semi-persistent scheduling for VoIP in LTE system, in: International Conference on Wireless Communications, Networking and Mobile Computing, 2007 WiCom, 21–25 September, 2007, pp. 2861–2864.

[14] A.S. Lioumpas, A. Alexiou, Uplink scheduling for machine-to-machine communications in LTE-based cellular systems, in: Proc. IEEE Globecom 2011 Conference, Houston, Texas, USA, 2011.

[15] A.G. Gotsis, A.S. Lioumpas, A. Alexiou, Analytical modeling and performance evaluation of realistic time-controlled M2M scheduling over LTE cellular networks, Trans. Emerging Tel. Tech. 24 (4) (2013) 378–388, Special issue on Machine-to-machine: an emerging communication paradigm.

[16] A.G. Gotsis, A.S. Lioumpas, A. Alexiou, M2M scheduling over LTE: challenges and new perspectives, IEEE Vehicular Technol. Mag. 7 (3) (2012) 34–39.

[17] P.S. Bithas, A.S. Lioumpas, A. Alexiou, A hybrid contention/reservation medium access protocol for wireless sensor networks, in: IEEE Globecom 2012—International Workshop on Machine-to-Machine Communications, Anaheim, CA, December, 2012.

Mobility management for machine-to-machine (M2M) communications

A. Elmangoush[1], A. Corici[2], A. Al-Hezmi[2], T. Magedanz[1]
[1]Technische Universität Berlin, Berlin, Germany; [2]Fraunhofer FOKUS Research Institute, Berlin, Germany

11.1 Introduction

The connected world is extending exponentially including physical objects besides computers and smartphones in a global Internet of Things. In the past 5 years, the number of connected devices has increased by 300%. Currently, more than nine billion devices around the world are connected to the Internet, and the estimations show that by the end of 2020, there will be one trillion connected devices worldwide [1]. The motivation for this new trend in ubiquitous networking is twofold: technical and economic. On the one hand, the advancement of semiconductor industry shrinking lithography continues to reduce chipset cost and power consumption. On the other hand, the technology evolution in the Internet and advanced wireless networks make it possible to provide broadband data services at a significantly lower cost per bit transferred than in the past.

Linking the communication and the information technology infrastructures forms the new machine-to-machine (M2M) communication trend, which is one of the key enabler technologies for smart cities. M2M communications will enable every physical and virtual object to integrate seamlessly into a large-scale infrastructure without human intervention. This applies and reinforces a convergence of heterogeneous technology families in various layers (application, transportation, and physical) to accommodate the technical communication and service requirements. Unlike traditional human-to-human (H2H) or human-to-machine services, which mainly involve multimedia sessions, messaging, web browsing, and remote control, M2M services have different features and communication requirements [2,3]. M2M provides the opportunity of deploying new services and engaging various kinds of connected objects with various domain-specific applications. This leads to new types of traffic patterns emerging in future networks. As seen in Figure 11.1, the traffic generated from connected devices will have a huge variety in size and rate. The dissimilarity between M2M and H2H communication requirements and traffic is expected to bring more complexity to the networking architecture.

Current communication platforms are optimized for H2H communication, concentrating on the optimization of the communication between devices, which are under direct human control. On the one hand, M2M presumes the independent communication

Machine-to-machine (M2M) Communications. http://dx.doi.org/10.1016/B978-1-78242-102-3.00011-3

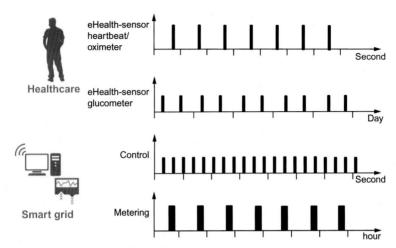

Figure 11.1 Various M2M traffic profiles.

of large amounts of data of heterogeneous types and sizes over the network independently of the human interaction between a large number of devices—sensors and actuators—and service platforms. On the other hand, the principle of limiting human interaction in M2M systems and integrating billions of devices with heterogeneous computational capabilities into them demands a fully network and application agnostic system.

M2M networks require varying levels of mobility support due to the diverse nature of the connected physical objects: humans, cars, robots, etc. This mobility should be accommodated at specific communication layers like Media Access Control (MAC), network, and session layers in order to ensure session continuity for mobile M2M nodes. In this regard, the user should be able to retrieve the desired sensing information or send commands to a specific actuator irrespective of the mobility of these objects. While most M2M traffic patterns can be estimated, device mobility is complex to predict.

There are different types of device mobility in M2M communication. In the field domain, the mobility can be achieved on the MAC or network level like Wi-Fi or wireless sensor network (WSN), respectively. However, there are several M2M applications that require global mobility, which can be supported with either cellular or satellite access technologies. Figure 11.2 shows a high-level M2M architecture with related mobility concepts. Terminology related to network mobility support is discussed in Ref. [4]; the document differentiates between two types of mobility handover. A horizontal handover occurs when a mobile node (MN) moves between access points (APs) that use the same technology, while a vertical handover takes place when an MN changes the point-of-attachment technology used. Also, the concept of host and network mobility support and various types of network mobility are discussed. As the M2M applications have various requirements on the connectivity, in particular on the global connectivity, this chapter focuses mainly on cellular mobility.

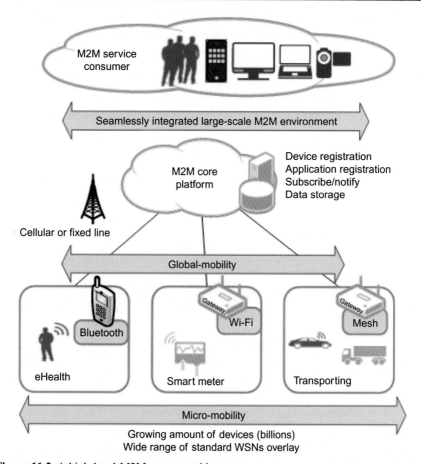

Figure 11.2 A high-level M2M system architecture.

In this chapter, we will discuss the motivation and challenges of managing mobility for M2M nodes. Section 11.2 gives an overview of some M2M use cases and discusses the need to support mobility as an essential feature. Section 11.3 summarizes the challenges of supporting mobility in M2M communication; Section 11.4 discusses the infrastructure requirements and considerations in M2M, with a brief review of existing standards in this area. In Section 11.5, we present the main solutions and research work in supporting M2M mobility while referring to the main challenges addressed. Finally, Section 11.6 summarizes the chapter and highlights opportunities of M2M communication evaluation.

11.2 Use cases for M2M mobility

M2M communication aims to enable connecting everyday existing objects and allows "nonhuman" content providers to feed the Internet with data in various formats frequently. The networking world is stepping toward a fully integrated Future Internet, which significantly increases the need for sophisticated smart applications. Smart city services should have the capability to analyze data and provide instant real-time solutions for many applications in order to promote positive outcomes. There are a lot of domains and environments in which M2M communication is likely to improve the quality of our lives, such as smart environments (home, office, and plant), transportation, and health care. Recognizing the need for reliable network infrastructures, various standards developing organizations have recently promoted standardization activities in the M2M domain and started their work by analyzing potential use cases and service requirements [5–8]. The concept behind M2M has been in existence for many years; many communication systems in the industrial area (e.g., fleet management, toll collect systems, and goods tracking) are operational using today's networking infrastructure such as SMS over GSM systems. The main driver for the evolution of M2M frameworks is the expected increase in data volume and the number of connections, combined with the desire to enable interoperability on a global scale. Table 11.1 clarifies the diverse requirements of different M2M use cases. Typical use cases for M2M communications that require mobility support are detailed in the following subsections.

11.2.1 eHealth

In the health sector, the utilization of M2M technology will lead to a more cost-effective health care system, allowing people with health problems to better manage their health, without disturbing their daily work and life. eHealth applications

Table 11.1 Communication requirements of some M2M use cases

Use Case	eHealth	Smart Home	Transportation	Smart Metering
Mobility	Low mobility	Low/no mobility	High-speed mobility	No mobility
Latency tolerance	Milliseconds (emergency case) to hours (remote monitoring)	Minutes	Milliseconds to minutes	Seconds to hours
Traffic pattern	Random	Random	Random	Regular
Message size	Few bytes	Bytes to Kilo bytes	Bytes	Small
Device density	Medium	Low	High	Medium
Required throughput	10 kbps (blood pressure monitor) to 91 kbps (H.263 video-encoding streams) [9]	kbps to Mbps (depending on installed devices)	1–3 kbps [10]	9.6–56 kbps [11]

aggregate a set of biomedical measurements from sensors connected to patients' smartphones, to continually monitor the status of the user. Through applying high-level analysis on these measurements, it will allow the early discovery of any critical health condition and actuate different actions according to the detected situation. Examples of such actions include triggering the medical sensors to increase the sampling rates so as to cope with the possible dangerous situation in a timely manner, displaying warning messages to the patient and his/her relatives with some urgent advice, and sending an ambulance request to the nearest health care provider with information about his/her location and situation. Generally, medical sensors use wireless body area networks such as Bluetooth or ZigBee, to forward measurements to a gateway device that has a WAN connectivity. Using mobile capable gateways, such as smart mobile devices, PDAs, and tablet PCs, gives the user more flexibility to carry on with their daily activities while monitoring their status constantly in real time. ETSI provides some eHealth use cases in Ref. [8] describing the requirements and information flow for each. The existing research work related to cloud-based eHealth care worldwide addresses advances in the field of medicine such as providing reliable and cost-effective personal health care [12–14].

11.2.2   Transportation

Recently, carmakers have been integrating more sensors in vehicles for driver comfort and safety applications on roads. It is now possible to create cars, trucks, and aircraft that are completely or partly autonomous [15]. It is foreseen that M2M communication technology will play a major role in developing intelligent transportation systems (ITS) of people and goods. ITS aim to offer great opportunities in maximizing utilization of roadways while minimizing fuel consumption, congestion, and the environmental impact of vehicle traffic. Current available technologies enable various value-added services in the transportation domain such as the integration of real-time traffic into GPS navigational aids, parking spot reservation on demand, and vehicle diagnostics and maintenance. The integration of M2M platforms into the transport domain will result in developing several sophisticated services. oneM2M have listed some in Ref. [5]. Recently, cellular communication technologies (such as 3G, Long-Term Evolution (LTE), and LTE-advanced (LTE-A)) are considered good candidates for car-to-x communication, which has strong requirements on latency and mobility [16]. The 802.11p standard defines enhancements to 802.11 with the objective of fulfilling the requirements of the ITS applications. This includes data exchange between high-speed vehicles and between the vehicles and the roadside infrastructure [17].

11.2.3   Smart buildings

Automating building management is one of the key M2M applications aiming to implement more comfortable, convenient, and secure environments. With integrating proper sensors and actuators, different objects inside the building could be controlled to handle time-consuming operations or manage energy efficiency. For example, inside fires can be detected by monitoring smoke sensors. Fire from the neighboring buildings can be detected by applying advanced video recognition patterns to the outside video cameras. Motion sensors can enable earthquake detection. All these data could be aggregated by an M2M gateway that can trigger corresponding actions such as:

- playing a dedicated audio alarm to alert the inhabitants,
- sending notification to the emergency authorities including the event type and location,
- acting upon preset actions like shutting down the gas and electric appliances and turning on electric lights with accumulators, and
- following recommendations from the owner or the emergency authorities.

Additionally, the M2M gateway could communicate with central smart city platforms to receive information from different stakeholders in the smart city. This information would be translated by the M2M gateway into recommended actions, and the owner of the household can be notified about possible operations to decide upon. An example is to receive weather status warnings from related agencies and respond accordingly, such as in case of an incoming dust storm by actuating all the window shutters to close automatically. The M2M gateway may also have mobility policies so that it is able to perform handovers between the local Wi-Fi AP and the public operator network in case of AP failure during an emergency, and the operator network is considered more reliable.

11.3 Challenges of M2M mobility

Mobility support refers to the ability of the communication architecture to enable connected entities (e.g., M2M nodes) to move across different access networks (i.e., changing the point of attachment) while performing transactions with other entities. Currently, several connectivity options are available to connect M2M devices and sensing objects to each other and to control servers. Defining the requirements for the future mobile networks, 3rd Generation Partnership Project (3GPP) presumes that a large number of the wireless connected devices are low-mobility devices, that is, devices that do not move, move infrequently, or move only within a certain region. Thus, M2M devices are widely considered as low-mobility devices that will not require ubiquitous mobility support. However, this is not the case in some M2M use cases as described in the previous section. On the one hand, smart devices are connected through the networks to service platforms in a self-controlled system. On the other hand, the current communication networks are designed to support human-centered communication and optimized for devices under direct human control. Generally, the common research challenges faced by the previously mentioned initiatives toward an M2M infrastructure can be summarized as follows:

1. *Scalability*: Mainly, scalability refers to the ability of the solution to support large numbers of entities or transactions among system entities. A mobility scheme is said to be scalable if its performance does not drop as the number of nodes increases. Due to the rapid increase in the number of M2M connected objects coupled with heterogeneity in sensor networks, scalability is a major technical challenge for enabling ubiquitous access in large-scale deployment of smart environments. Existing networks are implemented to support connecting large numbers of mobile devices; nevertheless, only a small portion of them are active at

a given time and the traffic flow is mainly in the download direction. Existing technologies still need enhancements to enable higher uplink traffic flow from M2M nodes.

2. *Location management*: If an MN offers services or data to other nodes, it must be able to be located by these nodes as it moves, as well as keeping the privacy of its topological location. M2M devices can be connected to M2M platforms directly using WAN connections or through a gateway using low-cost radio protocols such as IEEE 802.11 or IEEE 802.15. Connecting through a gateway is preferred to collect and process data from simpler M2M devices and manage their operation. However, some M2M applications might require peer-to-peer connectivity between end devices; such connectivity can be supported at various levels of hierarchy depending on latency requirements.

3. *Addressing*: In order to support the different communication requirements of M2M services (e.g., mobility, performance, and security), hierarchical deployments of multiple communication topologies will be needed [18]. In this regard, mobility management must work well with Internet Protocol (IP) routing such as acquiring a new topologically correct IP address upon moving.

4. *Security*: Generally, security is a crucial issue in M2M communication due to the massive information exchange related to various daily activities. M2M mobility mechanisms should be implemented while considering the resource-constrained nodes.

5. *Reliability*: There are several applications having restricted requirements on the end-to-end delay of transmitting a packet, which should be maintained during the process of the handover from one cell to another and from one access technology to another access technology, the so-called horizontal and vertical mobility, respectively. A long delay might result in violation of quality-of-service (QoS) objectives of an M2M service. It is important to quantify the handover delay in terms of various system parameters in order to design the associated wireless communication systems efficiently. This can be a challenge for small-coverage wireless networks (e.g., Wi-Fi network), since mobile objects might spend only a short period of time covered by each AP.

11.4 Infrastructure considerations for mobility in M2M

The main goal of M2M platforms is to connect efficiently billions of objects and enable the association of gathered data to a set of smart services. The heterogeneity of integrated communication technologies, targeted service domain, and data representation highlights the need to study the requirements of platforms required to enable the instrumented, interconnected, interoperated, and intelligent smart city. In our previous work [19,20], we address the requirements and design aspects of a reference M2M communication platform as an enabler for smart cities. In this section, we discuss the different infrastructure considerations to support M2M mobility and related standard work in this area.

11.4.1 Overview of M2M network reference architecture by standard organizations

Standardization is essential to remove the technical barriers and ensure interoperable M2M services and networks. In this section, we preview the M2M reference architectures proposed by standard organizations working in the M2M domain.

11.4.1.1 IEEE 802 LAN/MAN standards

Several IEEE task groups (TGs) are addressing the impact of M2M communication on the IEEE 802 radio access networks. The IEEE 802.16p TG aims to enhance the mobile WiMAX base standards IEEE 802.16e and IEEE 802.16m for M2M, identifying a number of requirements for mainly MAC-related functions such as network entry and group and device addressing [21]. Other working groups addressing M2M communications include IEEE 802.11 and IEEE 802.15.4. The ZigBee alliance (http://www.zigbee.org/) has recently developed further network and application layer protocols using small low-power radio devices based on IEEE 802.15.4. The target applications include smart energy, health care, remote control, and consumer electronics equipment. IEEE 802.11p defines enhancements to IEEE 802.11 required to support ITS applications.

11.4.1.2 3GPP MTC reference architecture

The 3GPP refers to M2M communication as machine-type communication (MTC) starting from the evolved packet core (EPC) architecture. 3GPP started standardization activities on MTC in September 2008 as part of 3GPP Rel-10 specifications. The service requirement working group (3GPP SA WG1) had specified a number of use cases and scenarios and derived a set of service requirements accordingly [22]. In 3GPP Rel-11, some of the proposed MTC features were finalized, such as addressing and device triggering. Most important is the enhancement of the 3GPP architecture to support MTC applications [23], by introducing a machine-type communication interworking function (MTC-IWF) to interact an external MTC server with the mobility management entity (MME), Home Subscriber Server (HSS), and Mobile Switching Center (MSC). The access network discovery and selection function (ANDSF) allows application servers (e.g., M2M platform) to trigger the M2M end devices to select certain wireless access technologies according to defined policies.

In Rel-12, new enhancements for small data transmission and minimizing overheads are considered in addition to enhancements of device triggering methods by using reference points between MTC-IWF and serving nodes (i.e., Serving GPRS Support Node (SGSN), MME, and MSC). The standard will also intend to optimize the user equipment (UE) power consumption to prevent battery drain and enable group-based features that allow multicast communication to an MTC group of devices sharing one or more MTC features.

11.4.1.3 ETSI M2M reference architecture

Aiming at an efficient end-to-end delivery of the M2M services, ETSI has defined a set of requirements [24]. These requirements mainly address features related to security and communication management as well as the functional requirements for a horizontal middleware oriented toward M2M communication in which the communication with various sensors and actuators is executed in a convergent and consistent manner for multiple applications. The ETSI goal is to define a middleware service capability layer (SCL) that interacts with M2M nodes over open interfaces: mIa, dIa, and mId.

These interfaces offer generic and extendable mechanisms for interactions with the SCLs at both device and gateway domain and network domain. The ETSI M2M reference architecture consists of three parts:

1. *M2M area network*: This consists of heterogeneous endpoint devices, such as sensors and actuators, connected through a sensor network based on various technologies, for example, ZigBee, M-Bus, and Bluetooth. This part of the network ends with an M2M gateway, which hides the complexity of the area network from the rest of the communicating entities. The gateway provides a set of service capabilities to the M2M applications in this domain, including the generic communication (GC) capacity that handles transport and session management functionality.

2. *M2M middleware core*: The M2M core implements functionality to facilitate the communication between the devices (in the M2M area network) and the network application domain. The M2M core provides features such as device management, reachability, and GC mechanisms over the communication network. Additionally, the M2M core handles the data exchange between devices and applications. On the one hand, it aggregates the information received from the device and transmitted to applications, which shows interest of that information by means of subscribing to its resource. On the other hand, it orchestrates the information received from applications, such as actuation commands or parameter updates that are transferred to the devices, depending on the urgency of the communication and on the momentary network conditions, as well as on the parameters of the device.

3. *Application domain*: The M2M middleware allows the connection of multiple applications addressing very heterogeneous use cases in different industries, such as energy, automotive, health, and transportation. The main function of any M2M application is to aggregate data presenting measurements from the surrounding environment, perform some calculations on them prior to decision making, and finally send commands to act according to that decision.

ETSI does not provide specification of M2M area networks, nor does it specify the access and core network; the ETSI framework provides the generic SCL for devolving M2M services independently from the underlying network. However, the Network Telco Operator Exposure (NTOE) capability is defined by ETSI to allow ETSI M2M applications to trigger certain policies on the network side or to retrieve any network information. Policies can be used to define the operation of data processing and transmission such as store-and-forward handling due to mobility or traffic offloading.

11.4.1.4   oneM2M

In 2012, oneM2M [25] was established with the aim of consolidating the standardization work in M2M communication. oneM2M is a consortium of seven standards development bodies working in M2M communication standardization. More than 260 participating partners and members have joined oneM2M to participate in the standardization of M2M communication systems, including ETSI and Open Mobile Alliance (OMA). The participating organizations intend to transfer all standardization activities in the scope of M2M service layers to oneM2M. oneM2M specifies a high-level architecture at both the field and infrastructure domains, to support end-to-end M2M services. The oneM2M functional architecture comprises the following entities:

1. *Application entity* (*AE*): responsible for providing an end-to-end M2M logic solution, that is, eHealth, logistic, and smart energy.
2. *Common services entity* (*CSE*): comprises a set of common service functions (CSFs) that are common to M2M environments and exposed to other entities through four reference points Mca, Mcn, Mcc, and Mcc'. oneM2M specified 12 different CSFs; some of them can be optionally implemented at a given CSE depending on the implementation domain and device, supported networks, etc. A CSE could be implemented on different kinds of nodes such as middle node (i.e., M2M gateways) at the field domain and infrastructure node (i.e., M2M server infrastructure) at the infrastructure domain.
3. *Underlying network services entity* (*NSE*): to provide services to the CSEs such as device management, location services, and device triggering.

oneM2M is specifying four reference points:

1. *Mca reference point*: for interaction communication between an AE and CSEs that enables the AE to use the exposed services from the CSE.
2. *Mcn reference point*: to allow the CSE to use services provided by the underlying NSEs.
3. *Mcc reference point*: to enable the interworking between CSEs. Any CSE could use some functionality provided by another CSE in order to provide service to other entities.
4. *Mcc' reference point*: the Mcc' shall be implemented on CSEs at infrastructure nodes to enable inter-domain communication between CSEs at different service provider domains.

Comparing with ETSI NTOE capability, oneM2M specifies the Network Service Exposure, Service Execution and Triggering (NSSE) CSF for managing communications with the Underlying Networks for accessing network service functions over the Mcn reference point. The network service functions provided by the underlying network include service functions such as device triggering, mobility management, data transmission, and data path policy rule setting. In this regard, this component can be considered to have the main capability to optimize mobility management for M2M traffic.

11.4.2 Managing mobility with 3GPP EPC

Currently, the most advance mobile core network architecture is represented by the 3GPP EPC that is able to manage the connectivity through a large variety of 3GPP and non-3GPP access networks including LTE, Universal Mobile Telecommunications System (UMTS), General Packet Radio Service (GPRS), Code Division Multiple Access (CDMA) 2000, Wi-Fi, and WiMAX. EPC includes convergent mechanisms for authentication and authorization, mobility support, and policy-based resource reservations and charging, enabling ubiquitous seamless IP connectivity for a large number of heterogeneous mobile devices. For providing the connectivity service, EPC includes a set of components, each maintaining some state information related to the connected mobile devices. In Figure 11.3, EPC architecture is depicted in a simplified form, including only the LTE access network. For the other access networks, similar functionality is available.

   In the case of the LTE access, EPC includes an MME for controlling the radio connectivity and the radio resource allocation and handovers as well as a policy and charging rules function (PCRF) and a charging system for allocating resources for

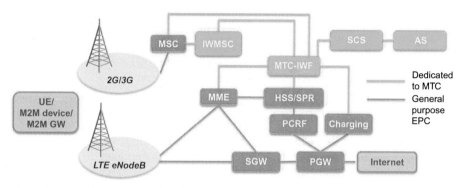

Figure 11.3 EPC architecture for LTE access network.

specific data streams of each subscriber and for their metering and charging. The EPC also features a HSS database including the static subscription profile as well as the dynamic information on the current mobile device network location, the allocated IP addresses, etc.

Handling the radio resource management and connectivity, the LTE eNodeB is connected to a serving gateway (SGW) that is connected to an additional packet data network gateway anchoring the data traffic and ensuring transparent mobility and convergent resource control and charging. The gateways are handling both the control and user plane for routing the IP traffic of the UE or M2M device. The actual data traffic is encapsulated in GPRS Tunneling Protocol (GTP) messages enabling the transparent location changes of the mobile devices without changing the IP address. For this, the gateways are signaling the network location changes using GTPv2 messages. Additionally, they communicate with the PCRF via Diameter protocol for receiving the specific QoS, for gating and charging rules, and with the charging system for transmitting charging data records. The ANDSF performs data management and controls functionality to assist the UE/M2M device on the selection of the optimal access network by exchanging discovery information and intersystem mobility policy [26].

In idle mode, the device is connected only using a default bearer and when in active mode a device can have dedicated bearers apart from the default bearer. Bearers represent routing and QoS resources to be used for communication. Additionally, a UE, in our case an M2M device or gateway, may perform a horizontal handover between two cells from the same access network (2G to 2G, 3G to 3G, and LTE to LTE) when in active or idle mode. Concerning horizontal handovers performed in LTE, the access network technology having the most advanced support for M2M communication, the handover could be done by changing between two SGW or two MMEs. The SGW is a hybrid component including not only data routing functionality but also controlling/signaling functionality. The MME is a pure controlling entity. After a handover is performed, the entire connection context is transferred on the new SGW or MME and the routing rules from the core network are changed according to the control plane signaling. The M2M device or gateway may connect to the network or could be triggered to connect by the M2M server; the MTC-IWF can determine the optimal trigger delivery method. The M2M server should be aware of the connectivity status of the devices in order to know if device triggering is needed or normal transmission can be used directly. For mobile networks, the M2M server can be aware of the connectivity status (or the signaling path) of the M2M devices or gateways using the Diameter Rx interface toward the PCRF [27] without maintaining a Transmission Control Protocol (TCP) or keep-alive connection.

11.4.3   Software-defined networks

The complexity of the EPC comes from the ever-increasing number of subscribers using the core network, the amount of protocols and procedures in use, and the fine-grained level of processing at flow level applied on the data path. Policy decisions, routing, and matching based on the final destination of UE/M2M device traffic, and the communication using separate diameter sessions for each data session of the subscriber both for policy control and for charging are not scalable; thus, a new data path handling designed for the mobile core network is required.

One of the most promising options is coming from software-defined network (SDN), separating completely the control from the data plane through having a centralized network controller. The controller schedules applications and controls routing and traffic engineering and a set of distributed user plane forwarding entities (FEs), which only route the data traffic according to the rules received from the controller, thus having smaller functionality and reduced delay. The basic SDN concept is depicted in Figure 11.4.

The SDN global view on the network enables the decision takers to enforce optimal routing paths from the perspective of the specific application, while the distribution of

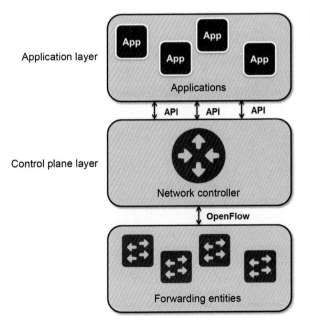

Figure 11.4 SDN general concept.

the user plane allows the creation of optimal data paths with reduced delay. The communication between the network controller and the FEs could be realized using the OpenFlow protocol [28]. As the SDN world was rather limited to traffic engineering and routing, in order to get it accepted by the industry, further applications have to be considered such as mobile core network support. Coming to meet this advanced SDN feature, OpenFlow 1.4.0 includes novel features such as the flow monitoring mechanisms enabling the consistency between multiple controllers and bundle messages from controllers to multiple switches. These new mechanisms make the OpenFlow-based communication more appropriate for deployment in production operator environments.

However, only a limited number of initial architecture developments are available in the literature such as the ones presented in Refs. [29] and [30] that present basic implementation developments based on OpenFlow in the EPC. The majority of OpenFlow-based frameworks (Onix [31] and HyperFlow [32]) or modifications to the basic OpenFlow architecture (DIFANE [33], DevoFlow [34], or Kandoo [35]) are addressing the basic limitations of routing and traffic engineering rather than the development of an appropriate OpenFlow-based mobile core network architecture that requires additionally support for frequent node network relocation due to handovers, IDLE mode management, authentication and authorization, and policy and charging control.

11.4.4   Management and control of M2M devices

M2M networks connect sensors and actuators as well; thus, device management protocols are essential here. Although device management components are present in both ETSI and oneM2M technical architecture, the device management protocol that is supposed to be used by the components is left out of scope. Different options are usable for controlling devices, such as SMS-based protocols from the legacy systems. The OMA is providing a platform-independent device management (OMA DM) protocol for general devices [36]. The DM protocol defines an interface between the DM server and the DM client to manage and configure devices on top of the HTTP transport protocol. Recently, OMA introduced the Lightweight M2M (LWM2M) [37], a device management protocol matching the constraint requirement for M2M domain by using Constrained Application Protocol (CoAP) [38] as a transport protocol. The payload can be encoded as plain text, JavaScript Object Notation (JSON) object, or Tag-Length-Value (TLV)-encapsulated, making the encoding very efficient. Security can be achieved by using Datagram Transport Layer Security.

The standard is suitable for any M2M solution as it is independent of the network, being applicable to both cellular networks and WSN. Apart from the management protocol, LWM2M supports service enablement with the same architecture and underlying transport protocol. The architecture comprises a DM client residing on the M2M device and a DM server located on the M2M service provider or the network provider, as shown in Figure 11.5. The DM server enables exchanging DM object information according to the resource models for handling access control, connectivity monitoring, connectivity statistics, device, firmware update, and location management.

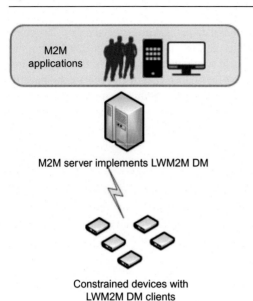

Figure 11.5 LWM2M DM system concept.

Unfortunately, the related object models for mobility aspects, connectivity monitoring and connectivity statistics, provide information about the current connectivity (mobile operator country code and network code, APN, cell ID, IP address, link quality, and transmission rates). They do not currently enable the LWM2M server to inform the LWM2M client about the recommended access network to use or exchange more complex policies related to urgency of the transmission and the mobility parameters in cases where the link quality is too low.

11.5   State-of-the-art solutions

Mobility solutions can be provided by either improving the networking architecture or revising the architecture to reflect the changing environment and requirements. In this section, we review some opportune solutions to support M2M mobility based on state-of-the-art communication networking systems.

11.5.1   Mobility support for IPv6

The IPv6 supports a large address space and a better autoconfiguration mechanism than IPv4, which allow assigning a unique address to each connected device in the Internet space, in addition to the better support of security and QoS. The implantation of IPv6 in M2M communications is widely considered as a necessity to support the huge amount of connected objects and enabling an efficient mobility support [39]. The IPv6 standardization at the Internet Engineering Task Force (IETF) has reached its final stage and its implementation is already in a stable state.

To enable efficient transmission of IPv6 datagrams over constrained links such as 802.15.4 links, the IETF has opened the IPv6 over low-power WPAN (6LoWPAN) charter. 6LoWPAN introduces an adaptation layer between the IP stack's link and network layers [40] and allows the Internet extension to small and smart devices by providing three basic services: IPv6 header compression, fragmentation, and mesh under routing support. As 6LoWPAN supports mesh and star topology and ad hoc networks, a certain degree of node mobility is possible. Different types of 6LoWPAN mobility protocols and available solutions are discussed in Ref. [41].

Traditional IP mobility support mechanisms, i.e., Mobile IPv4 and Mobile IPv6, are host-based mechanisms, where the MN is assumed to be aware of its mobility and responsible for performing some management operations in order to be able to maintain the ongoing communication sessions. In the case of resource-constrained nodes in M2M networks, such mechanisms are not optimal for power-constrained devices. This raises the need to propose network-based mobility mechanisms. The IETF Network-Based Mobility Extensions (netext) working group introduced several solutions to support network-based flow mobility, including the Proxy Mobile IPv6 (PMIPv6) [42], which is the base protocol for network-based localized mobility management. With the PMIPv6, there is no need to implement any mobility stack on the MN, as it will not be involved in mobility-related signaling nor change its IPv6 address. The network devices are responsible for performing the signaling with the home agent and performing the mobility management on behalf of the MN attached to the network. In a further work, fast handover for PMIPv6 was standardized [43] to reduce handover latency and data loss in PMIPv6. Additionally, the IETF has standardized the Host Identity Protocol [44], which decouples node identity from its address by introducing a secure cryptographically namespace as a node identifier. This will support securing mobility and a multihoming M2M overlay network. Chandavarkar and Reddy [45] reviewed mobility management protocols and mechanisms of heterogeneous IP networks, focusing on the challenging issues for the researchers.

11.5.2   Cognitive M2M communication

Cognitive radio technology is going through a phase of active research activity, addressing challenges related to spectrum sensing and allocation. Cognitive radio focuses on the tuning of parameters at the physical and link layers to provide efficient spectrum sharing. Some recent research work has taken place to empower the M2M domain by utilizing this new paradigm of networking [46,47]. Such technology can support vertical mobility among heterogeneous cognitive radio networks for efficient information delivery by allowing the opportunistic utilization of more spectrum frequencies. Tragos and Angelakis [47] proposed a new model of M2M communications based on the cognitive radio technology that considers the energy consumption for all spectrum-related decisions. Another proposal of applying the cognitive technology in M2M is presented in Ref. [48]; the proposed solution uses the TV white space for smart grid applications.

The idea of reusing licensed spectrum to support ubiquitous communications is prompted by various technical issues [48]:

1. Due to the massive number of connected objects, its efficacy to use larger parts of the spectrums to support large-scale data transmission, with low cost and less energy consumption.
2. The intensive interference between heterogeneous wireless signal inference and external electromagnetic interference from electronic equipment may degrade the performance of M2M communications. Leveraging the software configurability of cognitive radio allows M2M nodes to rapidly switch between different wireless modes and avoid the interference with the external radio environment.
3. M2M networks comprise constrained devices that are designed to work for several years without battery replacement. Cognitive machines in a secondary network are capable of adaptively adjusting their transmission power levels based on the operating environments, without interfering with the primary network and at the same time not causing spectrum pollution.

11.5.3 Delay-tolerant networking

Delay-tolerant networking (DTN) is an approach to address technical issues of interconnecting networks that lack some fundamental connectivity assumptions such as the following:

• A reliable end-to-end path between the source and destination during communication session.
• Retransmission based on timely and stable feedback from data receivers is an effective means for repairing errors (for reliable communication).
• All routers and end stations support the TCP/IP protocol suite.
• Endpoint-based security mechanisms are sufficient for meeting most security concerns.
• Packet switching is the most appropriate abstraction for interoperability and performance.
• Selecting a single route between sender and receiver is sufficient for achieving acceptable communication performance.

Due to the lack of a complete routing path in such networks, the mobile ad hoc network protocols cannot be used. Instead, the "store-carry-and-forward" manner is applied here to support data delivery and device mobility in DTN. Two main approaches are widely used; the first approach defines an overlay layer on top of transport layers that uses persistent storage to solve network interruption-related problems and provides functionalities similar to the Internet layer. The overlay network approach is represented by the bundle protocol (BP) [49] and the Licklider Transmission Protocol (RFC5326). Some open-source implementations for the DTN BP are available [50]. The second approach is based on collecting data from nodes using data mobile ubiquitous LAN extensions, that is, mobile relay nodes that collect data from static sensor nodes to an infrastructured AP [51]. Although the latter approach comprises higher transmitting latency, it is more efficient to use with energy-constrained sensors. For example, in environment monitoring scenarios, sensors are used to collect noncritical data for long periods. Recent research work has investigated routing models in DTN [52,53].

11.6   Summary and conclusions

Embedding more powerful and programmable sensors into smartphones (e.g., GPS, camera, and accelerometer) and extending different types of sensors and actuators with cellular capabilities offer a plethora of opportunities for the development of novel business cases and significantly increasing the current ecosystem. This will result in many changes in the conventional communication infrastructure to support M2M communication and huge amounts of data being produced in different domains. In this chapter, we have discussed new challenges that need to be addressed to enable seamless reliable M2M services and M2M mobility while referring to requirements of popular M2M applications. Furthermore, various standardization bodies are in the process of defining specifications for M2M communications and improving the existing systems accordingly. We presented the M2M standard-based infrastructure and the impact of M2M service's requirements on the existing network frameworks.

Discovering and managing connections between ubiquitous devices and applications are essential capabilities for M2M systems. The mobility management scheme in M2M infrastructure must be scalable and efficient to support different types of traffic patterns, which emerge due to the heterogeneity of connected objects/devices. Advantageous solutions are discussed on how to support reliable data exchange between servers and M2M MNs.

Acknowledgments

The authors would like to thank Ronald Steinke and Alejandra Escobar Rubalcava for their related discussions and review comments that helped to improve the quality of this chapter.

References

[1] J. Manyika, M. Chui, J. Bughin, R. Dobbs, P. Bisson, A. Marrs, Disruptive Technologies: Advances That Will Transform Life, Business, and the Global Economy, vol. 180, McKinsey Global Institute, San Francisco, May 2013.
[2] 3GPP-TS22.368, Service requirements for Machine-Type Communications (MTC) Rel-12. 3GPP, 2013. Available from: www.3gpp.org.
[3] ETSI TS 102 689 V1.1.2, Machine-to-Machine communications (M2M); M2M service requirements, 2011. Available from: http://www.etsi.org/standards.
[4] J. Manner, M. Kojo, RFC 3753—mobility related terminology, 2004. Available: https://datatracker.ietf.org/doc/rfc3753/.
[5] OneM2M TR 0001, oneM2M Use cases collection, 2013. Available from: www.onem2m.org.
[6] ETSI-TR102.898, Machine to Machine communications (M2M); use cases of automotive applications in M2M capable networks, 2013. Available from: http://www.etsi.org/standards.
[7] ETSI-TR102.857, Machine to Machine communications (M2M); use cases of M2M applications for connected consumer, 2010. Available from: http://www.etsi.org/standards.
[8] ETSI-TR102.732, Machine-to-Machine communications (M2M); use cases of M2M applications for eHealth, 2013. Available from: http://www.etsi.org/standards.
[9] L. Skorin-Kapov, M. Matijasevic, Analysis of QoS requirements for e-health services and mapping to evolved packet system QoS classes, Int. J. Telemed. Appl. 2010 (January) (2010) 1–18.

[10] P. Papadimitratos, A.D. La Fortelle, M. Paristech, K. Evenssen, R. Brignolo, S. Cosenza, Vehicular communication systems: enabling technologies, applications, and future outlook on intelligent transportation, IEEE Commun. Mag. 47 (November) (2009) 84–95.

[11] V.C. Gungor, D. Sahin, T. Kocak, S. Ergut, C. Buccella, C. Cecati, G.P. Hancke, A survey on smart grid potential applications and communication requirements, IEEE Trans. Industr. Inform. 9 (1) (2013) 28–42.

[12] Fi-Star project. Available from: https://www.fi-star.eu/home.html.

[13] L. Fan, W. Buchanan, C. Thummler, D. Bell, DACAR platform for eHealth services cloud, in: 2011 IEEE International Conference on Cloud Computing (CLOUD), 2011, pp. 219–226.

[14] A. Kailas, C.-C. Chong, F. Watanabe, From mobile phones to personal wellness dashboards, IEEE Pulse 1 (1) (2010) 57–63.

[15] E. Guizzo, How Google's self-driving car works, IEEE Spectr. (2011). Available from: http://spectrum.ieee.org/automaton/robotics/artificial-intelligence/how-google-self-driving-car-works (accessed 14 April 2014).

[16] P. Papadimitratos, A.D. La Fortelle, M. Paristech, K. Evenssen, R. Brignolo, S. Cosenza, Vehicular communication systems?: enabling technologies, applications, and future outlook on intelligent transportation. IEEE Commun. Mag. 47 (11) (2009) 84–95, http://dx.doi.org/10.1109/MCOM.2009.5307471.

[17] I. Rozas-Ramallal, T.M. Fernández-Caramés, A. Dapena, J.A. García-Naya, Evaluation of H.264/AVC over IEEE 802.11p vehicular networks, EURASIP J. Adv. Signal Process. 1 (2013) 77.

[18] P. Bhagwat, C. Perkins, S. Tripathi, Network layer mobility: an architecture and survey, IEEE Pers. Commun. 3 (1996) 54–64.

[19] M. Corici, H. Coskun, A. Elmangoush, A. Kurniawan, M. Tong, T. Magedanz, S. Wahle, OpenMTC: prototyping machine type communication in carrier grade operator networks, Globecom Workshops (GC Wkshps), 2012 IEEE, 3–7 December 2012, Anaheim, CA, pp. 1735–1740, http://dx.doi.org/10.1109/GLOCOMW.2012.6477847.

[20] A. Elmangoush, H. Coskun, S. Wahle, T. Magedanz, Design aspects for a reference M2M communication platform for smart cities, Innovations in Information Technology (IIT), 2013 9th International conference on, Abu Dhabi-UAE, pp. 204–209, 17-19 March 2013, http://dx.doi.org/10.1109/Innovations.2013.6544419.

[21] IEEE 802.16's M2M Task Group, Available from: http://wirelessman.org/m2m/index.html (accessed 16 April 2014).

[22] 3GPP-TS22.368, Service requirements for machine-type communications (MTC); Stage 1 (Release 10) (2011). Available from: www.3gpp.org.

[23] 3GPP-TS23.682, Architecture enhancements to facilitate communications with packet data networks and applications (Release 11) (2013). Available from: www.3gpp.org.

[24] ETSI TS 102 690 v1.1.1, Machine-to-machine communications (M2M); Functional architecture (2011). Available from: http://www.etsi.org/standards.

[25] OneM2M, Available from: http://onem2m.org/.

[26] L. Lange, T. Magedanz, D. Nehls, D. Vingarzan, Evolutionary future internet service platforms enabling seamless cross layer interoperability, in: Internet Communications (BCFIC Riga), 2011 Baltic Congress on Future, 2011, pp. 1–6.

[27] 3GPP-TS29.214, Policy and charging control over Rx reference point (2013). Available from: http://www.3gpp.org/DynaReport/29214.htm (accessed 30 April 2014).

[28] The Open Networking Foundation, OpenFlow Switch Specification Version 1.4.0 (2013). Available from: https://www.opennetworking.org/images/stories/downloads/sdn-resources/onf-specifications/openflow/openflow-spec-v1.4.0.pdf (accessed 30 April 2014).

[29] J. Kempf, B. Johansson, S. Pettersson, H. Luning, T. Nilsson, Moving the mobile evolved packet core to the cloud, in: 2012 IEEE 8th International Conference on Wireless and Mobile Computing, Networking and Communications (WiMob), 2012, pp. 784–791.

[30] L.E. Li, Z.M. Mao, J. Rexford, Toward software-defined cellular networks, in: 2012 European Workshop on Software Defined Networking, 2012, pp. 7–12.

[31] T. Koponen, M. Casado, N. Gude, J. Stribling, L. Poutievski, M. Zhu, R. Ramanathan, Y. Iwata, H. Inoue, T. Hama, S. Shenker, Onix: a distributed control platform for large-scale production networks, in: 9th USENIX OSDI Conference, 2010, pp. 1–6.

[32] A. Tootoonchian, Y. Ganjali, HyperFlow: a distributed control plane for OpenFlow, in: Proceedings of the 2010 Internet Network Management Conference on Research on Enterprise Networking, USENIX Association, Berkeley, CA, 2010.

[33] M. Yu, J. Rexford, M.J. Freedman, J. Wang, Scalable flow-based networking with DIFANE, ACM SIGCOMM Comput. Commun. Rev. 40 (4) (2010) 351–362, http://dx. doi.org/10.1145/1851275.1851224.

[34] A.R. Curtis, J.C. Mogul, J. Tourrilhes, P. Yalagandula, P. Sharma, S. Banerjee, DevoFlow: scaling flow management for high-performance networks. ACM SIGCOMM Comput. Commun. Rev. 41 (4) (2011) 254–265, http://dx.doi.org/10.1145/2043164.2018466.

[35] S. Hassas Yeganeh, Y. Ganjali, Kandoo: a framework for efficient and scalable offloading of control applications, in: Proceedings of the First Workshop on Hot topics in Software Defined Networks—HotSDN '12, 13 August 2012, New York, USA, pp. 19–24, http://dx. doi.org/10.1145/2342441.2342446.

[36] OMA-TS-DM_Protocol-V2, OMA device management protocol V2.0, Open Mobile Alliance (OMA), 2013. Available from: http://technical.openmobilealliance.org/Technical/ technical-information/release-program/current-releases/oma-device-management-v2-0.

[37] OMA-TS-LightweightM2M-V1, Lightweight machine to machine technical specification, Open Mobile Alliance (OMA), 2013. Available from: http://technical. openmobilealliance.org/Technical/technical-information/release-program/current-releases/oma-lightweightm2m-v1-0.

[38] Z. Shelby, K. Hartke, C. Bormann, Constrained application protocol (CoAP), RFC 7252, June 2014. Available from: http://tools.ietf.org/html/rfc7252.

[39] A. Galis, A. Gavras (Eds.), The Future Internet: Future Internet Assembly 2013: Validated Results and New Horizons, Springer, Berlin, Heidelberg, 2013, ISBN: 978-3-642-38081-5.

[40] N. Kushalnagar, G. Montenegro, D.E. Culler, J.W. Hui, RFC 4944: Transmission of IPv6 Packets over IEEE 802.15.4 Networks, 2007. Available from: http://tools.ietf.org/html/ rfc4944.

[41] L.M.L. Oliveira, A.F. de Sousa, J.J.P.C. Rodrigues, Routing and mobility approaches in IPv6 over LoWPAN mesh networks, Int. J. Commun. Syst. 24 (11) (2011) 1445–1466.

[42] V. Devarapalli, K. Chowdhury, S. Gundavelli, B. Patil, K. Leung, RFC 5213: Proxy Mobile IPv6 (2008).

[43] R. Koodli, H. Yokota, B. Patil, K. Chowdhury, F. Xia, RFC 5949: fast Handovers for Proxy Mobile IPv6 (2010).

[44] P. Nikander, A. Gurtov, T.R. Henderson, Host Identity Protocol (HIP): connectivity, mobility, multi-homing, security, and privacy over IPv4 and IPv6 networks, IEEE Commun. Surv. Tut. 12 (2) (2010) 186–204.

[45] B.R. Chandavarkar, G.R.M. Reddy, Survey paper: mobility management in heterogeneous wireless networks, Proc. Eng. 30 (January) (2012) 113–123.

[46] A. Georgakopoulos, K. Tsagkaris, D. Karvounas, P. Vlacheas, P. Demestichas, Cognitive networks for future Internet: status and emerging challenges, IEEE Veh. Technol. Mag. 7 (3) (2012) 48–56, http://dx.doi.org/10.1109/MVT.2012.2204548.

[47] E. Tragos, V. Angelakis, Cognitive radio inspired M2M communications invited paper, in: Wireless Personal Multimedia Communications (WPMC), 2013 16th International Symposium, 2013, pp. 1–5.

[48] Y. Liu, S. Xie, Cognitive machine-to-machine communications: visions and potentials for the smart grid, IEEE Netw. 26 (2012) 6–13.

[49] K. Scott, S. Burleigh, RFC 5050: bundle protocol specification (2007).

[50] W.-B. Pöttner, J. Morgenroth, S. Schildt, L. Wolf, Performance comparison of DTN bundle protocol implementations, in: Proceedings of the 6th ACM Workshop on Challenged networks—CHANTS '11, 2011, pp. 61–63.

[51] R.C. Shah, S. Roy, S. Jain, W. Brunette, Data MULEs: modeling a three-tier architecture for sparse sensor networks, Proc. First IEEE Int. Work. Sens. Netw. Protoc. Appl. 36 (2003) 30–41.

[52] T. Spyropoulos, S. Member, K. Psounis, Efficient routing in intermittently connected mobile networks: the single-copy case, IEEE/ACM Trans. Netw. 16 (1) (2008) 63–76.

[53] A. Balasubramanian, B.N. Levine, A. Venkataramani, Replication routing in DTNs: a resource allocation approach, IEEE/ACM Trans. Netw. 18 (2) (2010) 596–609.

Advanced security taxonomy for machine-to-machine (M2M) communications in 5G capillary networks

12

A. Bartoli[1], M. Dohler[2,3], A. Kountouris[4], D. Barthel[4]
[1]Universitat Politecnica de Catalunya (UPC), Barcelona, Spain; [2]King's College London (KCL), London, UK; [3]Worldsensing, London, UK; [4]Orange, Paris, France

12.1 Introduction

Taking into account today's concept of communication, socialization, and how information is exchanged, it is clear that there is a very powerful trend based on using innovative advanced technologies to improve the efficiency and effectiveness of industrial sectors [1]. Machine-to-machine (M2M) [2] communication networks are an integral part of this paradigm change by allowing the development of automated self-organized networks. In general, self-organized networks are those whose communication systems are able to run without human intervention.

Harbor Research [3] has described the effects of M2M networking as "Pervasive Internet," allowing businesses to make immediate decisions based on accurate, real-time data from near and far portions of critical infrastructures. For this reason, numerous market fields are including the use of wireless means to interconnect specialized smart devices in a variety of new businesses. Indeed, M2M is increasingly used for remote monitoring of environmental conditions (e.g., at landfills); industrial monitoring of chemical containers, pipeline status, and capacity management; alerting operators to dangerous conditions; and reducing maintenance costs and increasing the reliability of systems and service delivery.

At the same time, we observe (Q1 2014) a strong support within industry and academia to design the next-generation wireless communication systems, also known as 5th-generation (5G) systems. M2M will be an integral part of this infrastructure, where one of the key design requirements will be to accommodate a very large amount of wireless M2M (or MTC, i.e., machine-type communications) devices. Numerous challenges lay ahead, such as call admission control for M2M and radio resource management; arguably, the biggest challenge however is to ensure security for these devices as they often monitor and control important industrial processes.

Security threats and resulting challenges are, generally speaking, common to every innovation: novel communications technologies are able to provide many opportunities, but, at the same time, they can open the door to a wide new range of security

Machine-to-machine (M2M) Communications. http://dx.doi.org/10.1016/B978-1-78242-102-3.00012-5

threats that can jeopardize communications. These threats mainly depend on the intrinsic characteristics of each technology. Focusing on typical features of M2M networks, adequate security solutions have to be defined by taking into account the following issues:

- The *wireless medium* makes the physical layer very accessible for an attacker, which can jam, inject, or modify link layer packets without difficulty and which can easily compromise and spoof a smart device.
- Nodes in M2M networks are often "low-end" devices with *constrained resources*, and hence, the use of well-known but expensive security algorithms (e.g., asymmetrical cryptography) is often not feasible and may even be questionable.
- Typical M2M application may implement thousands of devices that are left *unattended* for years of operation without the possibility of human intervention. For this reason, security issues must be guaranteed at the design stage with self-healing mechanisms.
- M2M networks should *support multicast, groupcast, anycast,* and highly directional traffic, and thus, also, security must support several types of communication.
- M2M devices must be able to react when environment changes have been identified. In order to efficiently use the resources and to avoid waste, *self-management* solutions are key improvement for M2M applications.
- In order to connect rural or remote location to the principle backbone network, a capillary system of nodes must be deployed. Since centralized architectures are not always recommended, *distributed secure protocols* ought to be defined. Both centralized architectures and the use of a *cluster node* are not always recommended because if an attacker is able to block a few transmissions from the gateway or a cluster node, the entire network, or a part of it, may be isolated. In some cases, distributing responsibilities among the simple nodes can guarantee better performance from the security point of view (e.g., implement key agreement instead of key distribution).

The above mentioned issues require special attention during the security system design process. Despite security being the center of numerous investigations on both academic and industrial sides, there is no unique and universally accepted taxonomy available to date. We thus aimed to compile a coherent taxonomy composed of the most important elements needed for security studies on embedded systems such as M2M networks. The said taxonomy is first presented in Figure 12.1 and then treated in greater details in the subsequent sections.

Visually summarized in Figure 12.1, the proposed security taxonomy is detailed in Section 12.2 with the availability of system architecture, which generally can be centralized, clustered, or flat. This architecture is reliant on different assets, which are presented in Section 12.3, such as data (sensed data, control data, etc.), logical infrastructure (routing paths, network topology, etc.), and physical devices (nodes, power supply, etc). These assets in turn can be jeopardized by possible threats, described in Section 12.4, typically embodied by denial of service (DoS), falsification of service, leak of service, and time of service. The malicious attacks can be carried out by attackers, which generally enjoy different degrees of attacking capabilities and can be active or passive. This classification is presented in Section 12.5. In Section 12.6, we analyze how attacks are typically perceived and counteracted at different OSI layers (abbreviated as "Lx" where x is the layer under consideration),

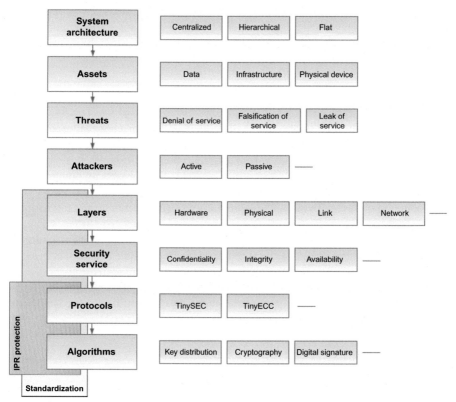

Figure 12.1 Proposed security taxonomy that includes layers, assets, algorithms, and protocols – as discussed in greater details throughout this chapter.

such as L0 subsystem/hardware, L1 physical, L2 link, and L3 network. This therefore requires suitable security services to be offered, such as confidentiality, integrity, and availability. These services are analyzed in Section 12.7. Finally, suitable security protocols and algorithms able to provide the security services, such as cryptography techniques, hash, and digital signatures, are explored in Section 12.8. The conclusion will close this exhaustive security analysis.

12.2   System architecture

The starting point of such security taxonomy is the suitable system architecture of choice. Typically, a design engineer has the choice between the following architectures:

*   Centralized
*   Hierarchical
*   Flat

A specific choice will heavily impact key performance metrics, such as support of different types of control and data flow, end-to-end performance, and security. Each of the above architectures has its advantages and disadvantages from the security point of view. These aspects are briefly discussed below.

12.2.1   Centralized architecture

A centralized architecture implies the availability of a single or a few entities that have control over the entire network. Note that a centralized approach typically means one-hop connectivity to all network members but, in the context of short-range embedded systems, is typically realized via a multihop network. From a security point of view, this single centralized entity needs to monitor the safety of the entire network. The utmost important task with this architecture is, if needed, to generate and to distribute security keys to all the members via a pairwise secure channel established with each member. This requirement might generally be too stringent for embedded multihop systems since a central key server must be continuously available and present in every possible subset of a group in order to support continued operation in the event of arbitrary network partitions. Continuous availability can be addressed by using fault-tolerant and replication techniques; unfortunately, the omnipresence issue is difficult to solve in a scalable and efficient manner.

12.2.2   Hierarchical architecture

A hierarchical architecture implies the availability of clusters, which is controlled by a cluster head that communicates to its associated node members. A simple node is typically but not necessarily associated to a single cluster head. Communication between node and cluster head is typically but necessarily done in a one hop. From a security point of view, the load of key management is now distributed among cluster heads. Typically, each cluster head would now generate and distribute keys only to its associated members or groups. The obvious advantage is that no single point of failure is present and the problem of omnipresence as well as scalability hence diminished. The disadvantage is that more points of attack and failure are created since cluster heads are easier compromised than some centralized security entity. When a cluster head is under attack, the entire cluster can suffer the consequences and thus be isolated from the rest of the network.

12.2.3   Flat architecture

A flat architecture implies that all nodes of the network are equal from a networking point of view and also typically but not necessarily have the same processing capabilities. Short-range embedded systems typically require the presence of multiple hops over such a flat architecture until the sink node or gateway is reached. From a security point of view, the flat peer-to-peer architecture requires that every member contributes an equal share to the common group secret, computed as a function of all members' contributions. This is particularly appropriate for dynamic networks since it avoids the problems with the single point(s) of trust and failure.

12.2.4   Security analysis

Depending on the characteristics of the M2M network communications, different system architectures are recommended. In the following, we discuss several issues:

- *Control traffic*. From a performance point of view, it is thus advantageous to keep the control centralized as the control communication overhead as small as possible; optimized multihop networks with central sink gateway are possible only when continuous unicast control traffic flows are not used. In this sense, self-organized protocols must be defined and implemented. Only broadcast messages should be allowed from the central gateway to every end point.
- *Data traffic*. From a performance point of view, the converge-cast traffic naturally passes in both uplink and downlink through the central entity. The emerging P2P traffic, however, can be channeled via a flat architecture (while the terminals involved in this transmission are instructed by the centralized control signals).
- *Security mechanisms*. A purely centralized approach would yield the aforementioned omnipresence problem. We will thus concentrate on some hierarchical approaches to find a suitable trade-off. Since embedded networks typically cover medium/large area using capillary devices, a centralized presence is not a suitable solution considering security. Self-organized communication protocols should be implemented to provide security in every corner of the network and for the entire network's lifetime.

In summary, a viable embedded system design is likely to require different architectural approaches for control, data, and security mechanisms. Control traffic should be centralized or partially centralized. Data traffic should go through the whole multihop network, from the end points to reach the final gateway. Security mechanism should be self-organized or partially self-organized. This means that depending on which metric is the most important for an M2M network administrator, a specific system architecture is recommended. When several broadcast control messages are necessary, a centralized approach is the best solution. Moreover, when security is a key issue, a flat architecture can guarantee a reliable communication system. Finally, a hierarchical architecture system should be adopted when hybrid solutions are necessary.

12.3   System assets

The architectures described in Section 12.2 are reliant on the following assets that deserve protection:

- Binary data
- Logical infrastructure
- Device components

Each of the above has its peculiarities, which are briefly discussed below.

12.3.1   Binary data

The data circulating in the embedded network are arguably the most valuable asset as it essentially delivers the required data and actuation instructions, as well as controls the data flow in the nodes and network. It is thus composed of uplink sensed data, downlink actuation data, (typically) downlink control data, and node-internal software binaries.

12.3.2   Logical infrastructure

The infrastructure supporting M2M networks is physically composed of the device components discussed below and carrying the data discussed above. Without claim for an exhaustive list, it is generally composed of the following elements: communication channel, network topology, neighborhood, routing paths, and key element (such as duty cycle).

12.3.3   Device components

The embedded network is essentially supported by two types of physical devices, that is, the sensor nodes and the gateway node(s): device and final gateway. The sensor/device nodes are assembled from various components, such as memory and power supply. Since these components in embedded sensor/nodes are generally severely constrained, in these networks, a reliable security system has to be based on energy-efficient mechanisms, such as optimized communication protocols like data aggregation. The gateway, while assembled from similar components as the nodes, is generally not so heavily constrained. However, the fact that it acts essentially as the data bottleneck between embedded network and the core network implies that this component has to work as secure fire wall, respectively, between unsecured area and secured area.

12.3.4   Security analysis

The major assets to protect in M2M networks are the binary data, notably the data and control information, and the infrastructure at large, notably the network topology and routing paths. In the following, we present standards and recommended security solutions for securing the exchanged data in M2M network.

12.4   Security threats

M2M networks are innovative communication systems that suffer several new attacks due to their specific characteristics. For example, wireless communication medium offers opportunity for efficient services; however, it also opens the door to a range of novel attacks from intruders. Typical threats in M2M networks, which are discussed subsequently, can be classified into four categories depending on their specific objectives in M2M networks; "denial-of-service" attacks are the most dangerous and difficult threats from which to defend (Section 12.4.5).

12.4.1 Leak of service

A leak of service essentially implies the exposure of data and control services to the attacker. This does not prevent the central server and each actuator receiving data or control signals, but it leads to a leak of information. It mainly jeopardizes confidential information, and typical attacks in this category are as follows:

- *Eavesdropping.* By listening to the data, the adversary could easily discover the communication contents. Network traffic is also susceptible to monitoring and eavesdropping. This should be no cause for concern given a robust security protocol, but monitoring could lead to attacks. It could also lead to wormhole or black hole attacks.
- *Traffic analysis.* Traffic analysis is typically combined with monitoring and eavesdropping. An increase in the number of transmitted packets between certain nodes could signal that a specific sensor has registered activity. Through the analysis of the traffic, some sensors with special roles or activities can be effectively identified and possibly attacked.
- *Tampering.* Nodes are vulnerable to physical access, such as tampering, which allows the attacker to gain access to the node and thus network.

12.4.2 Falsification of service

Falsification of service essentially implies that data and control services are falsified by the attack. This does not prevent the central server and each actuator receiving meaningful data or control signals, but it may falsify them. This mainly jeopardizes the integrity of the data, and typical attacks in this category are as follows:

- *Event modification.* An adversary may simply alter the event being monitored. For instance, if sensors monitor the outbreak of a fire, the attacker may simply get spatially close to a given sensor or set of sensors and put a lighter close to the sensor.
- *Hello flood attack.* A malicious node can send or replay Hello messages with high transmission power. It creates an illusion of being a neighbor to many nodes in the networks and can confuse the network routing badly, thus the data not reaching its destination or the network being depleted prematurely.
- *Wormhole attack.* In this attack, an adversary tunnels messages received in one part of the network over a low-latency link and replays them in a different part. An adversary situated close to the final gateway may be able to completely disrupt routing by creating a well-placed wormhole.
- *Sybil (multiple identities) attack.* A Sybil attack is defined as a "malicious device illegitimately taking on multiple identities." Using the Sybil attack, an adversary can "be in more than one place at once" as a single node presents multiple identities to other nodes in the network that can significantly reduce the effectiveness of fault-tolerant schemes such as distributed storage, dispersity, and multipath.
- *Node replication (duplication).* Often referred to as impersonation, an attacker here seeks to add a node to an existing sensor network by copying (replicating) the node ID of an existing sensor node.

12.4.3   Denial of service

Denial of service essentially implies that any data and control services are rendered useless by the attack. It prevents the central server and each actuator receiving meaningful data or control signals. It mainly jeopardizes service availability, and typical attacks in this category are as follows:

- *Jamming.* Jamming of a node or set of nodes is typically achieved by transmitting a radio signal that interferes with the radio frequencies being used by the sensor network. This process is able to isolate a node or to disturb their communications.
- *Exhaustion.* The life span of the end devices in a wireless low-power network is limited by the power of the battery. When the power is exhausted, the nodes cannot operate further. For example, the attacker can fake a message asking the devices to continuously retransmit messages to exhaust its energy and eventually cause an out-of-service situation.

12.4.4   Time of service

Time of service attacks mainly imply that the data exchanged through the network are recent. This prevents the central server and each actuator receiving old messages and thus avoids the receipt of fake alarm and/or old valid data content. This can, for example, force some kinds of security activities that are not really necessary, for example, rekeying methods. In a few words, it essentially jeopardizes the "actuality" of the messages, and a typical attack in this category is as follows:

- *Replay attacks.* As the medium is wireless, the attacker can intercept the message flows easily and replays those to start a new session. Such attacks can also be related to DoS as useless receptions can jeopardize the limited resources of end devices.

12.4.5   Security analysis

The majority of outsider attacks can be prevented by simple link layer encryption and authentication using a globally shared key (master or network key). Major classes of attacks not countered by link layer encryption and authentication mechanisms are tampering, jamming, exhaustion, and complex routing attacks, such as Hello and wormhole attacks. Jamming and exhaustion attacks need strong PHY layer solutions [4]. Taking into account Hello and wormhole attacks, although an adversary is prevented from joining the network, nothing prevents it from using a wormhole to tunnel packets sent by legitimate nodes in one part of the network to legitimate nodes in another part to convince them they are neighbors or by amplifying an overheard broadcast packet with sufficient power to be received by every node in the network. The simplest defense against Hello flood attacks is to verify the bidirectionality of a link before taking meaningful action based on a message received over that link [5]. Wormhole and sinkhole attacks are very difficult to defend against, especially when the two are used in combination. Wormholes are hard to detect because they use a private, out-of-band channel invisible to the underlying device. A technique for detecting wormhole attacks is presented in Ref. [6], but it requires extremely tight time synchronization and is thus infeasible for most embedded networks. Because it is extremely difficult to retrofit existing protocols with defenses against these attacks, the best solution is to carefully design routing protocols in which wormholes and sinkholes are meaningless.

12.5 Types of attacks

In this challenging scenario, we now present how attackers can be classified in M2M networks. This analysis has to be used to identify the danger of a particular and thus propose adequate novel security solutions. Attackers can be classified into three different categories: ability, activity, and class.

12.5.1 Ability

The attackers' *ability* is used to describe the *complexity* of a specific attack. In this category, we distinguish three main metrics (cost, skills, and traces):

- *Cost*: It relates to the cost an attacker is required to spend in terms of equipment to carry out an attack successfully. This can range from extremely cheap, where only a soldering iron and some cables are required, to prohibitively high, where top-of-the-line semiconductor test equipment is needed.
- *Skills*: It generally relates to the skills and knowledge that an attacker has to possess for a successful attack. Some attacks might be carried out by a kid after proper instruction, while others might require extensive knowledge of the particular application of the network or a person trained in the use of special equipment. (This property can also be modeled as cost.)
- *Traces*: This relates to the traces left behind by the attack. If after the attack the node is left in the same state as before the attack, including unaltered memory contents, then this is harder to notice than an attack that causes physical destruction of the node.

12.5.2 Activity

An attackers' *activity* is related to the *exposure* of the attackers: they can be passive or active:

- *Passive*: They extract information from the device or from the communication by observing the physical properties of the devices or the network's evolution.
- *Active*: They involve the manipulation (tampering) of the device and/or of the communication.

12.5.3 Class

Finally, *classes* of attackers are related to their competences and/or *capacities* (class I, class II, and class III):

- *Class I* (*clever outsiders*): They are often very intelligent but may have insufficient knowledge of the system. They may have access to only moderately sophisticated equipment. They often try to take advantage of an existing weakness in the system, rather than try to create one.

- *Class II* (*knowledgeable insiders*): They have substantial specialized technical education and experience. They have varying degrees of understanding of parts of the system but potential access to most of it. They often have highly sophisticated tools and instruments for analysis.
- *Class III* (*funded organizations*): They are able to assemble teams of specialists with related and complementary skills backed by great funding resources. They are capable of in-depth analysis of the system, designing sophisticated attacks and using the most advanced analysis tools. They may use class II adversaries as part of the attack team.

12.5.4 Security analysis

The ability and class of typical attacks vary depending on their danger. There exists a relationship between the cost, skills, and traces and the class of attackers. Class I attackers use cheap equipment, have low–medium skills, and leave some traces. On the other hand, class III attackers typically use expensive equipment, have high skills, and leave a low number of traces. Finally, the difference between passive and active attacks is very clear. Typical passive attacks are traffic analysis and eavesdropping. Active attacks are normally all others.

12.6 Layers under attack

Capitalizing on the availability of the above security threats, the adversaries' attacks have a different impact at different OSI communication layers, where we will restrict ourselves to the following:

- L0 subsystem/hardware
- L1 physical layer
- L2 link layer
- L3 network layer

Using the above-described threats, we will now briefly summarize on a per-layer basis the possible attacks and typical countermeasures taken.

12.6.1 L0 hardware

Typical attacks carried out at the subsystem level and respective countermeasures are as follows:

- *Destruction.* This typically implies that nodes or devices are physically destroyed, removed, or stolen. Within the price limit of a sensor node, one suitable countermeasure is to design a very robust case that protects the entire node and also allows mounting it to some unmovable structure; note however that, for example, the typical glass casing used with road sensors costs around €80 (Q1 2014). Another option is to have a dense enough topology so that some fallout can be tolerated. Note that another way to destroy a node or even large fields of sensor nodes is to use Taser-like devices that create a highly directive EM beam rendering the circuitry useless.

- *Tampering*. This implies that physical access to the node is established without destroying it. It allows for various personification attacks, such as sibyl and wormhole. To counteract this, a similar measure as with destruction can be taken. In addition, self-destructing mechanisms could be deployed once physical intrusion is detected. Furthermore, to protect the binary code contents present on the node, encryption with changing keys is recommended. To counter a physical attack on the microcontroller, code attestation is recommended. Furthermore, in the case of a physical attack on the external memory (EEPROM, electrically erasable programmable read-only memory), code obfuscation can be used. If the actual sensor is attacked, various trust management schemes can be of use.

12.6.2   L1 PHY layer

Typical attacks carried out at the L1 PHY layer and respective countermeasures are as follows:

- *Jamming*. At the PHY layer, this implies that a point-to-point link is disturbed by an adversary transmitting at high power and the same frequency band used by the nodes. Several countermeasures are available. First, an ultra-narrowband emergency channel can be maintained, which usually costs little extra bandwidth and little extra hardware requirements; narrowband channels are known to have a significantly large susceptibility that potentially allows the nodes to communicate "through" the interference. Second, an ultra-wideband (UWB) radio can be used for communications, which is usually resistant to interference of less bandwidth. Notwithstanding this, if an adversary has a powerful wideband jammer, then UWB will not help. Third, for embedded systems currently on the market, such as IEEE 802.15.4, "frequency hopping" or "surfing channel" techniques may help as long as not all hopping bands are jammed and/or interfered. Fourth, a very strong link layer channel code can be used that, together with suitable link layer retransmission schemes, may just be enough to facilitate communication.
- *Eavesdropping*. Since the wireless medium is essentially a broadcast medium, any adversary may just be able to eavesdrop on ongoing transmissions, given that he or she can decipher its contents. An interesting and emerging countermeasure currently being investigated is the use of physical layer security. Here, a time division duplexing link is presumed allowing a legitimate transmitter and receiver to exchange the channel state information (CSI). This CSI is unique and only pertinent to the established physical layer link. It can thus be used to encrypt the contents of the physical layer data stream, making it difficult for an adversary to decipher the stream already at physical layer. Another interesting countermeasure is to use the physical layer to estimate some node inherent properties, such as position and distance (which can be obtained from the time and/or angle of arrivals), which allows excluding nodes that are not within predefined limits.
- *Exhaustion*. Various higher layer attacks may lead to an exhaustion of the battery. A typical attack launched at the PHY layer is to exhaust the radio's power supply. Exhaustion typically happens due to collisions, that is, transmitting a malicious packet with the aim to make it collide with a useful packet; overhearing, that is, force nodes to listen to packets that are advertised as being of interest but actually being malicious; idle listening, that is, force nodes to wait for a packet that was promised to be transmitted but which is not; retransmissions, that is, force nodes to retransmit a packet continuously even though it has been well received; and interrogation, that is, force nodes to issue clear-to-send messages by continuously broadcasting a request-to-send message. A typical countermeasure would be to design a suitable link layer, which prevents most of above exhaustion mechanisms. Additive measures are

node authorization (e.g., by extra protection of network ID), node authentication, message verification Cyclic Redundancy check Code (CRC), and message encryption. Possible countermeasures are to allow for provision of natural energy resources that recharge quicker, such as solar, vibration, and magnetic induction. Another possible countermeasure is to duty cycle the battery in that power is only provided at predefined times that cannot be overridden by higher layer requirements.

12.6.3   L2 MAC layer

Typical attacks carried out at the L2 MAC layer and respective countermeasures are as follows:

*   *Sybil.* Once an adversary has access to the network, he or she can take multiple identities and thus cause significant harm at link layer. Notably, these identities can occupy the channel and thus prevent legitimate nodes to communicate meaningfully. Second, it can influence data aggregation mechanisms employed at link layer where entirely uncorrelated data are provided, hence only allowing for packet "addition" but no compression. Another attack can be carried out in the context of voting, since, in the presence of a high network connectivity, some voting mechanisms are sometimes used to establish the next-hop forwarder of choice; the false identities thus can influence or even stuff the voting ballot. Again, countermeasures must be taken that essentially prevent false identities, such as rekeying.
*   *Desynchronization.* An embedded multihop network typically relies on strong synchronization between nodes. An adversary may trigger signals, typically embedded into the link layer control information, which cause nodes to desynchronize and thus disconnect the network. An example is the Dust Networks industrial monitoring platform where synchronization signals are inserted by the transmitter in the data packet and by the receiver piggybacked in the ACK packet. If the classical IEEE 802.15.4 link layer was to be used, where the ACKs are not secured, an adversary node may simply modify this piggybacked synchronization signal. In general, a suitable MAC design with some basic security mechanisms may prevent this, such as not using all and not always the same nodes to synchronize.
*   *Traffic analysis.* An adversary could monitor the activity of the network by simply analyzing the occupancy of the channel. While the information gained from this is minimal, a possible countermeasure is to dummy packets in quieter hours of the system.
*   *Eavesdropping.* Once access is gained to the network at link level, an adversary may simply eavesdrop on the ongoing transmissions and extract required information. To counteract this attack, suitable encryption schemes need to be used.

12.6.4   L3 NTW layer

Typical attacks carried out at the L3 NTW layer and respective countermeasures are as follows:

*   *Node replication/event modification.* A prerequisite to countering this type of attacks is to ensure that the communicating nodes are authenticated prior to data encryption applied in the networking operation. Authentication ensures that the nodes are who they claim to be even though it does not provide an indication of whether the node has been compromised.
*   *Sinkhole attack/selective forwarding/wormhole attack.* These can typically be countered by performing some regular monitoring of the network using source routing protocols. Furthermore, if permissible, some physical monitoring of the field devices can be of advantage.

- *Traffic analysis/sniffing attacks.* An adversary could monitor the activity of links and conclude on the choice of routes and thus networking topology. Countermeasure could pertain to the insertion of dummy packets into unused routes.

12.6.5 Security analysis

L0 subsystem layer attacks can be challenged from hardware and specific software solutions that work on the hardware components. In the case of tampering, code attestation and obfuscation are important solutions [7,8]. Recommended L1 PHY layer solutions use frequency hopping [9], CSI security [10], and authentication preamble (Section 12.4.5) for securing the M2M networks, respectively, against jamming, eavesdropping, and exhaustion attacks. L2 MAC layer and L3 NTW layer solutions have to implement standards security solutions for offering recommended security services. These services are presented in the following section.

12.7 Security services

In order to face the threats described above per layer, recommended solutions for M2M networks are typically archived by implementing the CIA [11] (confidentiality, integrity, and availability) security model. This model provides an adequate level of security. The security requirements recommended for M2M networks are discussed subsequently.

12.7.1 Confidentiality

Confidentiality essentially means keeping information secret from unauthorized parties. An embedded network should not leak data readings to neighboring networks or adversaries. The confidentiality objective is required in the sensors' environment to protect information within the nodes and traveling between the nodes of the network or between the devices and the final gateway from disclosure, since an adversary having the appropriate equipment may eavesdrop on the communication. By eavesdropping, the adversary could overhear critical information such as sensitive data and routing information.

12.7.2 Integrity

Integrity means that the data produced and consumed by the network must not be maliciously altered. Unlike confidentiality, integrity is, in most cases, a mandatory property. The wireless channel can be accessed by anyone; thus, any peer (outsiders and insiders) can manipulate the contents of the messages that traverse the network. Even more, data loss or damage may occur due to the harsh communication environment, and in the worst case, the network will accept corrupted data. As the main objective of an embedded network is to provide services to its users, the network will fail in its purpose if the reliability of those services cannot be assured due to inconsistencies in the information.

12.7.3 Availability

Availability implies that the users of an embedded network must be capable of accessing its services when they need them. The importance of availability of nodes when they are needed cannot be ignored. For example, when the network is used for monitoring purpose in manufacturing system, unavailability of nodes may fail to detect possible accidents. Availability ensures that sensor nodes are active in the network to fulfill the functionality of the network. It should be ensured that security mechanisms imposed for data confidentiality and authentication are allowing the authorized nodes to participate in the processing of data or communication when their services are needed. As nodes have limited battery power, unnecessary computations may exhaust them before their normal lifetime and make them unavailable. Sometimes, deployed security protocols or mechanisms in embedded networks are exploited by the adversaries to exhaust the nodes by its resources and make them unavailable for the network.

In addition to the above security requirements, we identify *freshness* as an important additional security service.

12.7.4 Freshness

Freshness means the data should be recent in order to allow protection with old messages. This is an important security requirement to ensure that no message has been replayed. It means that the messages are in a specific order and they cannot be reused. This prevents the adversaries from confusing the network by replaying the captured messages exchanged between sensor nodes. To achieve freshness, security protocols must be designed in such a way that they can identify duplicate packets and discard them.

12.8 Security protocols and algorithms

Confidentiality, integrity, availability, and freshness are important security services for protecting M2M networks from external attackers. In the following, we present typical protocols and algorithms that provide these security services.

12.8.1 Security services

The following security services are of importance:

- *Confidentiality*. Symmetrical and asymmetrical key cryptography primitives are able to offer the necessary mechanisms for protecting the confidentiality of the information flow in a communication channel. For achieving this confidentiality level, in the former case, it's necessary that both the origin and the destination share the same security credential (i.e., secret key), which is utilized for both encryption and decryption. In the latter case, two keys are implemented as security credential: a key called secret key, which has to be kept private, and another key named public key, which is publicly known. Any operation done with the private key can only be reversed with the public key, and vice versa. In general, symmetrical key methods are implemented in typical M2M networks since they use less energy. Examples of symmetrical primitives used in low-power networks are Advanced Encryption Standard (AES), Twofish, RC5, etc. Following NIST recommendations [12], these algorithms have to be used together with secure modes of operation: CBC (cipher block chaining), CCM (Counter with MAC) [13], GCM (Galois/Counter Mode), OCB (Offset Codebook Mode) [14], etc. On the other hand, novel asymmetrical cryptography solutions related to elliptic curve methods, such as TinyECC [15], which are now providing high security level with low energy consumption, are elliptic curve Diffie–Hellman (ECDH) key agreement scheme, the Elliptic Curve Digital Signature Algorithm (ECDSA), and the Elliptic Curve Integrated Encryption Scheme (ECIES) [16]. Notably, ECDH is a variant of the Diffie–Hellman key agreement protocol on elliptic curve groups, ECDSA is a variant of the Digital Signature Algorithm that operates on elliptic curve groups, and finally, ECIES is a public-key encryption scheme that provides semantic security against an adversary who is allowed to use chosen-plaintext and chosen-ciphertext attacks. ECIES is also known as the Elliptic Curve Augmented Encryption Scheme or simply the Elliptic Curve Encryption Scheme. Finally, since these last techniques are most of the time protected from own patents and royalty, symmetrical techniques are still preferred for M2M purposes.

- *Integrity*. Hash primitives are utilized in order to compress a set of data of variable length into a set of bits of fixed length. The result is a "digital fingerprint" of the data, guaranteeing integrity, identified as a hash value. A cryptographic hash function must satisfy two properties: (i) Given a hash value "h," it should be hard to find a message m such that $\text{hash}(m) = $ "h" and (ii) it should be hard to find two different messages m_1 and m_2 such that $\text{hash}(m_1) = \text{hash}(m_2)$. Recommended hash functions for M2M network are SHA-1, SHA-2 [17], etc.

- *Availability*. Generally speaking, availability relies on the proper operation of the network nodes and their communication links. Therefore, the said availability is mainly jeopardized by DoS attacks, which can be classified into (i) attacks damaging network nodes; (ii) attacks disturbing the communication links, for example, by means of jamming techniques; and (iii) attacks exhausting the network nodes, for example, by engaging them into meaningless packet exchanges that consume their precious batteries and thus significantly shorten their lifetime. Protection of the network nodes is mainly related with physical security that should allow keeping nodes out of reach of attackers. Regarding communications, several security measures have been proposed in order to prevent or mitigate jamming-based attacks, most of them based on frequency hopping and channel surfing techniques (Section 12.6.5). However, to the best of our knowledge, authentication preamble and cross layer link layer techniques [18] are the only energy-efficient solutions able to mitigate the effects of exhaustion attacks (Section 12.4.5).

- *Freshness*. Simple techniques, such as a timestamp and a counter, are able to guarantee freshness for the exchanged data [19]. Its length will be related to the amount of sent/received packets per time interval. This value can be encrypted or just used as input data for the hash function.

12.8.2   Security analysis

Since typical M2M networks are complex multihop communication systems, security services have to be provided in a smart system design. For such a reason, an appropriate security system has to provide end-to-end (between end nodes and the central server or the cluster node) and hop-by-hop (within every link) security services, as well as key management solutions. Finally, also typical implemented energy-saving mechanisms, such as data aggregation, have to include security methods.

End-to-end and hop-by-hop security services. End-to-end security is achieved by means of a shared secret between every node and the central server; hop-by-hop security is done at link layer by means of pairwise secrets between every network node and its one-hop neighbors:

- *End-to-end*: The aim of end-to-end security is to protect the data from unauthorized eavesdropping (confidentiality), to allow the destination to check the integrity of the received data and its freshness, and to unequivocally identify the source of such data (authentication). End-to-end security is important for protecting the data from the source of the message to the final destination. Since this security control is made at the final server, end-to-end security is typically provided at application layer.
- *Hop-by-hop*: End-to-end security is checked at the final destination; however, before reaching the gateway or Internet destination, the packets must go through one or more wireless links that are by nature exposed to attackers. As a result, if no security is provided in order to restrict the access to the media, only the destination point will be able to detect altered, dropped, or fake packets. This fact exposes the network to exhaustion attacks since those packets will waste precious energy at the intermediate nodes (routers). Consequently, hop-by-hop integrity, authentication, and freshness should also be provided at link layer.

Key management. Cryptographic mechanisms rely on the secure management of the necessary keying material. Keys are analogous to the combination of a safe. If the combination becomes known to an adversary, the strongest safe provides no security against penetration [20]. For such reason, an important aspect of M2M networks is to define a reliable rekeying mechanism for allowing a key change when this process is needed. In this specific case, this method also provides forward and backward secrecy. With forward secrecy, a device should not be able to read any future message exchange in the network after it leaves this system. On the other hand, with backward secrecy, a joining entity should not be able to read any previously transmitted message in the network. These properties may not be important in certain scenarios, for example, when there is no need to hide the contents of the network from old nodes and new nodes are authorized to perform the same tasks as their partners. This may happen when master long-term, or static, keys are implemented to compute short-term, or ephemeral, keys. In this case, the long-term keys do not have to provide both forward secrecy and backward secrecy. However, there exist M2M applications where these properties must be taken into account, such as in networks where only authorized devices are allowed to perform certain tasks.

Interesting rekeying methods used in commercial M2M network are now presented. The Wi-Fi Protected Access (WPA) [21] security protocol has become the industry standard for securing 802.11 networks. Using a preshared encryption key or digital certificates, the WPA algorithm Temporal Key Integrity Protocol (TKIP) securely encrypts data and provides authentication to said networks. TKIP was designed to be a transition between old hardware and new encryption models. The IEEE 802.11i protocol improved upon the WPA algorithm (TKIP) to the new that uses a better encryption algorithm: AES. As a major step forward, the protocol also specifies more advanced key distribution techniques, which result in better session security to prevent eavesdropping. In addition, ZigBee PRO high security mode [22] presents an innovative rekeying method. The Trust Center maintains a list of devices, link keys, and network keys that it needs to control and enforce the policies of network key updates and network admittance. In addition, this security mode provides also the use of master keys: these optional keys are not used to encrypt frames while they are used as an initial shared secret between two devices when they perform the key establishment procedure (SKKE, Symmetric-Key Key Exchange) to generate link keys. Thus, with these kinds of keys, a couple of nodes are able to deal with a secret key (link key) without the supervision of a Trust Center, consequently saving energy. Furthermore, protections for device authentication and key management and distribution, including the use of the SKKE and PKKE (Public-Key Key Exchange), are provided, among other security features.

Secure data aggregation mechanisms. In M2M networks, the deployed devices use low-power components; thus, efficient protocols have to be implemented. In most of the case, in order to not overload the network with unnecessary messages, and thus not waste energy and facilitate wireless communications, many systems perform in-network aggregation. Most existing aggregation algorithms and systems do not include any provisions for security, and consequently, these systems are vulnerable to a wide variety of attacks. In particular, compromised nodes can be used to inject false data that lead to incorrect aggregates being computed at the final gateway. In the following, we present recommended data aggregation protocols that not only are able to provide efficient energy-saving solutions but also guarantee security services.

In general, the main objective of data aggregation techniques is to increase the network lifetime and to improve the effectiveness of wireless channel communications. Network lifetime is increased by reducing the energy consumption of the entities involved in the communication system by decreasing the number of bytes for sending/receiving, while the effectiveness of the wireless communications can be improved by decreasing the bandwidth usage; less messages exchanged are thus more likely to find the wireless channel free. However, increasing network lifetime, data aggregation protocols may degrade important quality of service metrics in wireless networks, such as data accuracy, latency, fault tolerance, and security. Therefore, the design of an efficient data aggregation protocol is an inherently challenging task because the protocol designer must trade between energy efficiency, data accuracy, latency, fault tolerance, and security. In order to achieve this trade-off, data aggregation techniques are tightly coupled with how packets are routed through the network. Redundant packets should be avoided and useless overheard should be decreased.

SDA is the first secure data aggregation scheme that was proposed by Hu and Evans [23]. This protocol was proposed to study the problem of data aggregation once one node is compromised. This protocol was designed to ensure that a single compromised node can only mislead the network about its own reading; resilience against a compromised node is archived by implementing a μ-TESLA protocol for verifying the authenticity of the aggregation processes performed within the network. In order to improve SDA scheme, ESA was proposed by Jadia and Mathuria [24]. ESA scheme, instead of using μ-TESLA to authenticate the base stations, uses one-hop pairwise keys to encrypt data between a node and its parent and two-hop pairwise keys to encrypt data between a node and its grandparent. Castelluccia et al. in Ref. [25] proposed a new homomorphic encryption scheme that allows intermediate sensors to aggregate encrypted data of their children without having to decrypt. As a result, even if an aggregator is compromised, it cannot learn the data of its children, resulting in much stronger privacy than a simple aggregation scheme using hop-by-hop encryption. Papadopoulos et al. in Ref. [26] presented a novel secure data aggregation that provides both integrity and confidentiality covering a wide range of aggregates and returning exact results. The authors' contribution focuses on providing secure aggregation for M2M networks where (i) devices are deployed in open and unsafe environments and (ii) the aggregation process is outsourced to an untrustworthy service. Finally, Bartoli et al. in Ref. [27] presented an efficient data aggregation scheme that uses end-to-end and hop-by-hop security services for providing a high-security suite for low-power networks. This solution is particularly recommended for those applications, such as metering services, where energy-efficient solutions have to both introduce energy-saving and safeguard each individual sensor reading (lossless aggregation).

12.9   Concluding remarks

Embedded devices in M2M networks aim to infer the sensed data and deliver this information reliably and securely to the final gateway that in general represents the service provider center or network actuator. In order to complete this process, the sensed data, the infrastructure, and the communication protocols should be secured against numerous threats.

In this chapter, we have presented several approaches to efficiently secure M2M network components and to protect against typical energy-constrained networks attacks. These approaches provide confidentiality, integrity, and availability during the entire network's lifetime (freshness). Notably, this taxonomy is important because it suits the needs of wireless M2M networks regarding their specific security issues.

In this work, the analysis of the most-common architectures, the valuable assets, the typical threats, the counteracting services, the protocols carrying out these services, and the algorithms that these protocols has given rise to the following general own assessments:

- Security is a truly global issue spanning through all protocol layers and across all network elements.
- Preventing attacks at a given OSI layer essentially eliminates threats at higher layers (given that an attacker does not gain physical access to a node).

With that said, we strongly believe that some basic security mechanisms at link and physical layer can be tremendously useful when protecting higher layers, notably the networking and application layers.

An interesting line of research related to M2M (MTC) in 5G networks pertains to security work spanning across different systems, such as 3GPP and Wi-Fi systems. Centralized or globally trusted security authentication might be the right approach for these types of emerging systems. Furthermore, security-by-design approaches in L3—involving future Internet design approaches—are a promising approach where security is guaranteed end to end. Overall, the community has understood that security cannot be offered as an add-on after the main engineering work of a system, but has to be part of the design process from the very beginning.

References

[1] IBM, Smarter planet initiative, 2010.

[2] Carbon War Room analysis, Machine to Machine Technologies: Unlocking the Potential of a $1 Trillion Industry editors. T. Lee, H. MCMaHon. February 2013. http://www. grahampeacedesignmail.com/cwr/cwr_m2m_down_singles.pdf.

[3] Harbor Research study: "Pervasive Internet & Smart Services Market Forecast", 2009.

[4] A. Bartoli, J. Hernandez-Serrano, M. Soriano, M. Dohler, A. Kountouris, D. Barthel, Optimizing energy-efficiency of PHY-layer authentication in machine-to-machine networks, in: Globecom Workshops (GC Wkshps), 2012 IEEE, 3–7 December 2012, pp. 1663–1668, http://dx.doi.org/10.1109/GLOCOMW.2012.6477835.

[5] H.K. Kalita, A. Kar, Wireless sensor network security analysis, Int. J. Next-Gener. Networks 0975-7252, 1 (1) (December 2009), ISSN: 0975-7023.

[6] N. Song, L. Qian, X. Li, Wormhole attacks detection in wireless ad hoc networks: a statistical analysis approach, in: Proceedings of the 19th IEEE International Parallel and Distributed Processing Symposium (IPDPS), Workshop 17, Volume 18, Washington, DC, USA, 2005, p. 289.

[7] T. AbuHmed, N. Nyamaa, D.H. Nyang, Software-based remote code attestation in wireless sensor network, in: Global Telecommunications Conference (GLOBECOM), 30 November 4 December 2009, IEEE Computer Society, Washington, DC, USA, 2009, pp. 1–8, http://dx.doi.org/10.1109/GLOCOM.2009.5425280.

[8] S.M. Darwish, S.K. Guirguis, M.S. Zalat, Stealthy code obfuscation technique for software security, in: International Conference on Computer Engineering and Systems (ICCES), 30 November 2 December 2010, pp. 93–99, http://dx.doi.org/10.1109/ICCES.2010.5674830.

[9] I. Mansour, G. Chalhoub, A. Quilliot, Security architecture for wireless sensor networks using frequency hopping and public key management, in: IEEE International Conference on Networking, Sensing and Control (ICNSC), 11–13 April 2011, pp. 526–531, http://dx.doi.org/10.1109/ICNSC.2011.5874890.

[10] M. Tahir, S.P.W. Jarot, M.U. Siddiqi, Wireless physical layer security using channel state information, in: International Conference on Computer and Communication Engineering (ICCCE), 11–12 May 2010, pp. 1–5, http://dx.doi.org/10.1109/ICCCE.2010.5556862.

[11] CTIA—The Wireless Association, Mobile Cybersecurity and the Internet of Things Empowering M2M Communication. http://www.ctia.org/docs/default-source/default-document-library/ctia-iot-white-paper.pdf, 2013.

[12] NIST, National Institute of Standards and Technology (NIST), 2011 (accessed August 2011).

[13] NIST, Recommendation for Block Cipher Modes of Operation: The CCM Mode for Authentication and Confidentiality. NIST Special Publication 800-38C, May 2004. http://csrc.nist.gov/publications/nistpubs/800-38C/SP800-38C_updated-July20_2007.pdf.

[14] NIST, Recommendation for Block Cipher Modes of Operation. http://csrc.nist.gov/publi cations/nistpubs/800-38a/sp800-38a.pdf, 2001.

[15] A. Liu, P. Ning, TinyECC: a configurable library for elliptic curve cryptography in wireless sensor networks, in: International Conference on Information Processing in Sensor Networks (IPSN), 22–24 April 2008, pp. 245–256.

[16] M.J. Dubai, T.R. Mahesh, P.A. Ghosh, Design of new security algorithm: using hybrid Cryptography architecture, in: 3rd International Conference on Electronics Computer Technology (ICECT) (Volume 5), 8–10 April 2011, IEEE Computer Society, Washington, DC, USA, 2011, pp. 99–101, http://dx.doi.org/10.1109/ICECTECH.2011.5941965.

[17] A.A.L. Selvakumar, C.S. Ganadhas, The evaluation report of SHA-256 crypt analysis hash function, in: International Conference on Communication Software and Networks (ICCSN), 27–28 February 2009, pp. 588–592, http://dx.doi.org/10.1109/ICCSN.2009.50.

[18] C.-T. Hsueh, C.-Y. Wen, Y.-C. Ouyang, A secure scheme for power exhausting attacks in wireless sensor networks, in: Third International Conference on Ubiquitous and Future Networks (ICUFN), 15–17 June 2011, pp. 258–263, http://dx.doi.org/10.1109/ICUFN.2011.5949172.

[19] X. Wu, M. Zhang, X. Yang, Time-stamp based mutual authentication protocol for mobile RFID system, in: 22nd Wireless and Optical Communication Conference (WOCC), 16–18 May 2013, pp. 702–706, http://dx.doi.org/10.1109/WOCC.2013.6676465.

[20] M. Pattaranantakul, A. Janthong, K. Sanguannam, P. Sangwongngam, K. Sripimanwat, Secure and efficient key management technique in quantum cryptography network, in: Fourth International Conference on Ubiquitous and Future Networks (ICUFN), 4–6 July 2012, pp. 280–285, http://dx.doi.org/10.1109/ICUFN.2012.6261711.

[21] A.H. Lashkari, M.M.S. Danesh, B. Samadi, A survey on wireless security protocols (WEP, WPA and WPA2/802.11i), in: 2nd IEEE International Conference on Computer Science and Information Technology (ICCSIT), 8–11 August 2009, pp. 48–52, http://dx.doi.org/10.1109/ICCSIT.2009.5234856.

[22] ZigBee website: https://www.zigbee.org/.

[23] L. Hu, D. Evans, Secure aggregation for wireless networks, in: Proceedings of the 2003 Symposium on Applications and the Internet Workshops (SAINT '03 Workshops), 27–31 January 2003, pp. 384–391, http://dx.doi.org/10.1109/SAINTW.2003.1210191.

[24] P. Jadia, A. Mathuria, Efficient secure aggregation in sensor networks, in: Proceedings of the 11th international conference on High Performance Computing, 19–22 December 2004, Bangalore, India, 2005, pp. 40–49, http://dx.doi.org/10.1007/978-3-540-30474-6_10.

[25] C. Castelluccia, A.C.-F. Chan, E. Mykletun, G. Tsudik, Efficient and provably secure aggregation of encrypted data in wireless sensor networks, ACM Trans. Sens. Netw. 5 (3) (2009) 20, http://dx.doi.org/10.1145/1525856.1525858.

[26] S. Papadopoulos, A. Kiayias, D. Papadias, Exact in-network aggregation with integrity and confidentiality, IEEE Trans. Knowl. Data Eng. 24 (10) (2012) 1760–1773.

[27] A. Bartoli, J. Hernandez-Serrano, M. Soriano, M. Dohler, A. Kountouris, D. Barthel, Secure lossless aggregation over fading and shadowing channels for smart grid M2M networks, IEEE Trans. Smart Grid 2 (4) (December 2011) 844–864, http://dx.doi.org/10.1109/TSG.2011.2162431.

Establishing security in machine-to-machine (M2M) communication devices and services

13

F. Ennesser, H. Ganem
Gemalto, Meudon, France

13.1 Introduction

Future security challenges in the Internet of Things (IoT) can be well assessed from those encountered on the Internet today and should not be underestimated. In fact, the main security threats from the Internet world will become even more critical in the IoT due to specific vulnerabilities that are inherent to machine-to-machine (M2M) environments:

- Like individual computers today, M2M devices will become targets of attacks. While by nature these devices generally operate without human attendance, they may be part of systems whose proper operation is critical (e.g., safety hazards caused by improper operation of a power grid).
- A main source of threats on the Internet comes from unanticipated security breaches in software that are only discovered after deployment. As M2M computing environments often have limited capabilities due to low-energy consumption and cost constraints, their operating systems are often weaker than on mainframe computers. In addition, bugs are harder to fix as devices may not be always online, are located in unfriendly locations, or are subject to bandwidth limitations.
- Billions of M2M devices connected to communication infrastructures will create potentially weak entry points into networks and infrastructures for attackers, which may compromise valuable assets and further increases the risks arising from seamless access to infrastructures.
- Privacy breaches are already a frequent problem on the Internet, though the exposure of individual currently results mainly from their intended, controllable actions. But in tomorrow's IoT, the M2M devices around us are likely to expose privacy-sensitive information about our lives without our awareness. Furthermore, the classification of the information manipulated by devices may evolve according to context.

This chapter investigates how M2M-specific constraints affect security and privacy and describes principles and solutions that may be deployed to alleviate the associated risks.

Machine-to-machine (M2M) Communications. http://dx.doi.org/10.1016/B978-1-78242-102-3.00013-7

13.2   Requirements and constraints for establishing security in M2M communications

While the experience acquired in telecommunication systems security and cybersecurity can be reused in M2M communications, several specificities require security measures that deserve particular attention, as outlined below.

13.2.1   Unattended devices

By nature, M2M devices do not enjoy babysitting by human users, which makes security more tricky to address. M2M devices may not have human interface and need to be configured with minimum human intervention. And the association between an M2M device and an application or service may not be predictable before deployment (e.g., for consumer devices acquired off the shelf) or may need to change dynamically during the lifetime without physically visiting all deployed devices. A solution is to rely on a specific remote infrastructure entity (comparable to the Authentication, Authorization and Accounting (AAA) servers used in cellular networks) to support automated security bootstrapping (i.e., personalization of security keys) and remote configuration of the M2M devices, as well as reconfiguration and other security update operations (e.g., security updates affecting security algorithms). The nature of M2M devices creates dependencies on a remote management infrastructure that needs to be resilient. This further encourages leveraging on other devices or gateways, even belonging to other stakeholders, to improve device reachability and communication reliability at minimum additional cost. The remote management infrastructure should provide end-to-end security association between end devices while relieving intermediate nodes from the decryption/re-encryption operations resulting from traditional hop-by-hop communication and security management approach. And interdomain security mechanisms involving multiple networks in M2M communications will be needed.

13.2.2   Impact of multitude

The issues resulting from multitudes of connected devices, such as their geographic dispersion (static or dynamic), their reachability, and their lifetime expectations, result in cost and energy constraints. All these issues have an impact on security.

As multitudes of devices are expected to become connected, the service cost per device must be optimized. Furthermore, in many applications, the devices, especially mobile ones, have no connection to external power sources, which require optimization of their energy consumption. The trend toward cost- and energy-"constrained" devices leads to minimization of the communication and computation overheads. This makes solving the security challenges even more difficult.

13.2.3   Communication overhead

The first effects are on communication capabilities: M2M devices cannot be expected to be constantly reachable online and require wake-up mechanisms. They typically support limited wireless communication range, which makes capillary routing through other devices so attractive. In turn, communication constraints have security implications. M2M services that require only occasional data access on wireless networks put strong constraints on the cost of security solutions such as network authentication requirements. Security mechanisms must be adapted to the sporadic nature of communications, and specific mechanisms must be implemented to enable applications to control the security of transmitted data end to end, over wide area network (WAN) through gateways, and into capillary networks.

13.2.4 Computation overhead

From a security standpoint, a sensitive aspect in terms of computation overhead is the ability to generate random (unpredictable) values. The combination with the communication overhead also becomes particularly sensitive as many applications require the confidentiality of very short messages. This makes the size of the cipher text critical, which is a disadvantage for public-key encryption. Lightweight algorithms for block ciphers, stream ciphers, and hash functions are still in development. As Internet Protocol (IP) packet headers also need to be compressed to minimize overhead, data and header compression need to take place prior to encryption, all at a very low layer of the protocol stack. Obviously, overheads from multiple layers of confidentiality/ integrity protection must be avoided where possible.

13.2.5 Spatial considerations

The following spatial constraints also need to be taken into account:

- Some devices are hardly accessible because of hazardous locations (e.g., in nuclear reactor) or directly embedded in the raw material and infrastructure (e.g., humidity sensor in a wall). As deploying devices in such environments is difficult and costly, such applications generally also involve long lifetime expectancy. In such cases, reliable means for remote security administration are required, possibly including capabilities to change service subscription (e.g., switching to a different carrier), to upgrade security software, or to detect possible attacks.
- On the contrary, other devices are exposed in public locations where they are physically accessible to adverse individuals (e.g., traffic lights). There, the secrets that root the security in the devices require hardware protection to prevent discovery through tampering by physical attackers. This protection must be commensurate with the costs incurred in case the secrets are compromised.
- Other applications, such as intelligent vehicle traffic optimization, involve multitudes of mobile devices, which constantly impacts network topology. This impacts mainly device management infrastructures and results, for example, in a need to manage groups of devices dynamically and to support broadcast and multicast communication modes. For example, mobile devices that are likely to require reauthentication or authorization in bulk, such as passengers in a bus, should be grouped dynamically to avoid overloading the infrastructure. Some of the security management tasks can advantageously be delegated to intermediate devices or gateways to off-load the core infrastructure, especially for frequent operations such as reauthentication and authorization in mobile capillary networks. But solutions to manage group security and to protect broadcast and multicast communications are not current practice today.

13.2.6 Many-to-many communications

As M2M evolves from client–server applications (one to many) to the IoT (many to many), the need to share information between multiple actors (e.g., public transport, healthcare services, and municipality all making use of traffic information in a smart city infrastructure) without breaking the security chain implies dissociating the data advertising and dissemination service from the information security chain that needs to be trusted by all stakeholders. The latter responsibility may again involve an independent party if required for proper trust establishment between all actors. The exchange of information with external entities (devices/servers belonging to other domains) requires a remote authorization infrastructure enabling deployment of shared credentials, to support security services for delivery and data access (e.g., confidentiality, integrity, nonrepudiation, and audit). The IoT implies that devices affiliated to one service domain can seamlessly communicate with devices within other domains, in the same way as subscribers affiliated with one telecommunication operator seamlessly reaching subscribers affiliated to other telecommunication operators. Basic devices in the IoT, e.g., sensors and actuators, act as generators or recipients of data streams and can announce their capabilities to other entities to enable a flexible subscription/notification system supported by a trusted authorization mechanism.

13.2.7 Ecosystem liability considerations

Some M2M data are security-critical, as expressed by the attention paid to the protection of infrastructures that are vital to the life of human communities such as energy or healthcare systems. In these contexts, operators of M2M applications cannot take the risks to expose data to the third parties involved in M2M services and need to encrypt transmitted data end-to-end, making them opaque to the intermediate actors. This can be supported without implementing redundant encryption layers, by delegating the distribution of the secrets used to secure the data streams to a third-party trusted by both the M2M service provider and the operator of the M2M application, as described in Section 13.3.3. This solution is also of interest for service providers who are not willing to assume liabilities resulting from security breaches in critical applications.

13.2.8 Privacy aspects

Finally, the fact that many devices may expose data about the personal life of individuals adds an important privacy protection dimension in many applications such as healthcare, transportation, or house control. Indeed, Article 9 of the 1948 Universal Declaration of Human Rights entitles individuals to protect their privacy. Yet many M2M devices such as Global Positioning System (GPS) trackers or utility meters transmit data exposing the personal life of individuals (e.g., are they at home, taking a shower, and how many people are in their house on Christmas evening), without warning. Therefore, means for privacy preservation must be provided in M2M systems through local device configuration or remote management interfaces. This system must also be future-proof, considering that data classification may evolve due to evolution of the society, regulations, and algorithmic computation capabilities such as data correlation.

13.3   Trust models in M2M ecosystems

In order to meet the security requirements for M2M applications, particular attention should be devoted to the security architecture. It should enable system level confidentiality, authenticity, and privacy for all involved entities to encourage adoption of an M2M solution on a large scale. Data confidentiality, privacy, and trust are three key issues spanning the data transport, service, and application levels in an open M2M ecosystem in which different business actors may be involved in a given application scenario.

The question of who distributes and manages the credentials is probably one that will have the most significant impact over the security architecture. This determines the trust models. The importance of this issue is such that it is worth identifying the different trust models. For M2M communications, three such models may be identified:

* Ad hoc application level security.
* Security managed by the M2M service provider.
* Security managed by an independent trust manager.

Let's examine the details of those three models.

13.3.1   Ad hoc application level security

The ad hoc application level security model involves credentials that are essentially managed at the application level, by the business entity operating the M2M application. This type of security management has been and is still widely used with stand-alone client–server applications. The main advantage of this model lies in the flexibility of the credential management. Also, the selection of security protocols and algorithms may be perfectly tailored to the risks associated with the application, as their choice can be entirely controlled by the application developer. This model is also quite suitable to achieve end-to-end security, because traditional M2M applications involve communication end points that are usually part of the same application. This will however no longer be the case in the IoT model.

The weak point of this security model lies in the lack of interoperability and scalability, resulting in difficulties to achieve secure communication between entities that are not part of the same application and have not been planned in advance to communicate together. Hence, different trust models are required to support the IoT "many-to-many" requirement.

13.3.2   M2M service provider-managed model

In this model, security is considered as an enabling service for M2M communication and the credentials are managed and distributed by an M2M service provider. The management of the security by a dedicated business entity leads to a better scalability

(ability to achieve communications with a growing number of entities). Another advantage of this model is a greater interoperability between applications developed and operated by independent parties, thanks to the centralized management of the credentials under the responsibility of the M2M service provider.

However, communicating data sources and destinations may be controlled by owners affiliated to different M2M service providers. In this case, the traditional way to secure the communications is to implement hop-by-hop security, where each hop of the communication from source to destination is secured using distinct credentials managed by distinct parties as illustrated in Figure 13.1. This is different from end-to-end security where data are protected from source to destination using a single set of credentials.

The hop-by-hop model is the security model used to secure telephony communications: Each provider secures the communication with its subscribers, and different providers secure the communications between themselves.

An underlying implicit assumption is that there is a circle of trust between the different service providers. While this assumption has proved to be reasonable within the restricted group of mobile telecommunication operators with strong service commitment, it may be more difficult to ensure with a large number of M2M service providers, addressing multiple business segments.

Also, the hop-by-hop security model involves rekeying operations at the level of the M2M service providers, and hence, all transmitted data are available in clear at the level of the M2M service platform.

But M2M service providers are not always involved in dealing with the semantics of the data they distribute. For example, an M2M service provider may forward health data originating from a body sensor to a healthcare institution without being involved in this patient health management and may not have the *trust* level required to handle such sensitive data; or the service provider may not be aware of the data sensitivity, if the risk assessment was not communicated by the customer. In some other cases, the M2M service provider, in order to be compliant with regulations, may have to put in place data security measures that will carry a significant price tag and be unwilling to assume the *liability* or the *cost* associated with protecting customer privacy and would therefore prefer disseminating data that remain opaque to them.

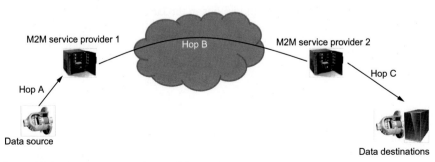

Figure 13.1 Hop-by-hop security model.

End-to-end security, involving the use of a single set of credentials from source to destination, is a solution to this problem but will require the deployment of an adequate credential management and distribution system.

13.3.3 Trust manager security model

The role and functions offered by the M2M service provider could be split in two categories:

- The functions related to the "discovery" and "dissemination" of data originating from devices to all possible interested parties (the service functions).
- The functions related to the management of trust and security (the security functions).

Traditionally, service and security functions have been endorsed by the same business entity. However, there is no absolute necessity for it. The assumption that the service and security functions may be offered by two independent business parties leads to a third security model: the trust manager security model. The corresponding architecture is summarized in Figure 13.2.

In this model, the M2M trust provider, a business entity possibly distinct from the M2M service provider, manages access rights to the devices. Different M2M trust providers may coexist and compete for business on the market. Device owners select the trust provider of their choice.

The role of the trust manager is to:

- authenticate data sources and data destination entities,
- manage the authorizations for access to M2M data, and
- enable end-to-end protection of data using group keys.

Enabling end-to-end protection of data using group keys involves the setup of a global distribution of group credentials to data sources and destinations owned by parties possibly affiliated to different M2M service providers. A group key is delivered to

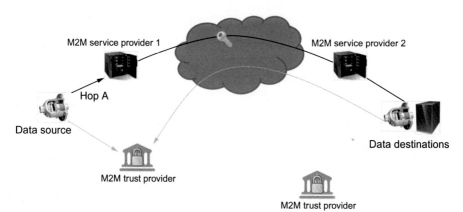

Figure 13.2 Trust manager security model.

a data source device and is being used to cipher transmitted data. It is also confidentially delivered to all authorized destinations.

To make sure that all parties interested and authorized to communicate with an M2M device may do so, the M2M service provider is managing the data discovery and dissemination aspects. To that end, the M2M service provider enforces the authorization decisions made by the M2M trust manager, who bears responsibility for protecting information.

13.4 Protecting credentials through their lifetime in M2M systems

13.4.1 Security fundamentals

Security always relies on secrets that must be protected against discovery at a level that is commensurate with the liabilities incurred in case these credentials are compromised. The required security level varies according to the M2M applications and needs to be estimated by a risk assessment process. Since security is a chain whose strength is always conditioned by the weakest link, the most challenging aspect is to maintain a consistent security level not only through an entire system but also throughout the entire lifetime of the credentials, from initial generation to deployment and operation to final revocation.

13.4.2 Protecting secrets in exposed equipment

The choice of a security solution to protect secrets in deployed assets shall be evaluated by running a proper risk assessment process. The level of exposure of an asset to attackers (e.g., asset exposed only to remote attacks vs. asset physically accessible to attackers) is an important aspect in the choice of a security solution, as hardware protection measures must be implemented to protect secrets in assets subject to physical attacks. Basically, the following security solutions, from the weakest to the strongest, are available for protecting security credentials inside equipment during operation:

1. The protection of secrets relies only on software running on a general-purpose processing unit: protective measures are software-only. Sensitive data intended to be kept as "secret" are stored in standard memory in general-purpose processing hardware, and sensitive functions are executed on the general-purpose processing hardware of the supporting equipment. In such cases, little effort is needed for an attacker with physical access to obtain the secrets, and even remote attackers may find ways to tweak the software to expose the intended secrets. Basically, this solution should be reserved for equipment in safe premises that are not even exposed to remote attackers.

2. The protection of secrets relies on a trusted execution environment (TEE): there, a hardware root of trust is used to protect the boot sequence of a general-purpose processing hardware, and an enforced isolation enables sensitive code, data, and resources to be separated from an unprotected operating environment, software, and memory. The code running in the protected environment is cryptographically verified for integrity assurance. Such a system is

more secure than software-only approach against remote attacks, but attackers having physical access to the equipment are still able to obtain secrets with reasonable effort. Such solutions are therefore mostly suitable for equipment that is not physically exposed to potential attackers.

3. The protection of secrets is provided by a tamper-resistant security module: sensitive data and functions are performed by a dedicated hardware security module that implements strong protective measures against physical attacks, such as the detection of abnormal operating conditions and scrambling plus hardware masking of the memory and side-channel analysis (e.g., electromagnetic emissions) of operations involving sensitive data. This hardware security module is independent from the general-purpose processing unit of the equipment and provides its services through dedicated and standardized interfaces. This solution is required to protect unattended equipment in exposed environments.

Note that a single hardware element can securely host several cryptographically isolated security environments that can each be remotely administered through a secure link by their respective service providers. The GlobalPlatform specifications in particular, already deployed in some industries, provide a flexible means for the owner of a security element to delegate dedicated protected environments to a third party, enabling each actor to provision its services independently in a single secure element. For example, a smart card embedded in a device can be used to provide not only cellular telecommunication service ((Global System for Mobile Communications) GSM/ 3rd Generation Partnership Project (3GPP) subscriber identity module (SIM) applications) but also M2M dissemination services handled by a specialized service provider (as standardized in European Telecommunications Standards Institute Technical Specification (ETSI TS) 102 921) and a couple of application security domains supporting the functionalities served by an M2M device. GlobalPlatform further enables stakeholders to remotely download and manage their own security applets independently in a standardized manner, ensuring interoperability throughout secure elements. This capability relies on a virtual machine (typically Java Card) featuring application programming interfaces supported by standardized secure elements such as A type of Integrated Circuit Card specified in the ETSI TS 102 221 specification (UICCs) standardized by ETSI.

13.4.3 Efficient security approach

A widely successful security solution, adopted in major industries where high security requirements combine with massive deployment needs (e.g., banking cards and cellular telecommunication), relies on preprovisioning credentials on dedicated security element (e.g., smart cards) that interface with their hosting devices in a standardized manner. In this model, the protection of credentials from their creation to their distribution in devices and infrastructures is handled by a specialized actor proficient with the management of secure facilities and logistic chains and independent from the device manufacturer: Thus, the first actor can focus solely on the individual customization and security constraints while the latter focuses on mass manufacturing to provide economies of scale. The deployment of credentials in independent hardware security element also minimizes the cost of the security certification process (e.g., Common Criteria for Information Technology Security Evaluation as per ISO

15408), which is required to enforce secure implementations of sensitive applications, like those affecting critical infrastructures (energy, payments, healthcare, etc.).

13.4.4 Emerging alternative: physically unclonable functions

An emerging alternative that may reduce the cost of secure credential deployments relies on hardware fingerprint: This is the principle of physically unclonable functions (PUFs). PUFs are innovative primitives to derive secrets from complex physical characteristics of integrated circuits (ICs) rather than storing keys in digital memory. Another advantage of PUFs is that they do not require any specific personalization process or programming and testing steps. The drawback is that this approach implies the existence of a central authority able to uniquely identify each device, which may prevent anonymity.

The principle of a PUF is based on a "one-way function," making it possible to compute an image for every input but hard to invert when given the image of a random input. When a physical stimulus is applied to the hardware, it reacts in an unpredictable way due to the complex interactions between the stimulus and the physical microstructure of the device itself. The microstructure of the device depends on physical factors introduced during manufacturing process and is unpredictable. The applied stimulus is called the challenge, and the reaction of the PUF is called the response. A precise challenge together with its response forms a challenge–response pair. The PUF device identity is determined by the properties of a microstructure.

The advantage here is that the device is not revealed by the challenge–response procedure, so the PUF device is resistant to spoofing attacks. The hardware cost is limited and dependent on the number of challenge and response bits. However, the outputs when using PUF are likely to be slightly different on each evaluation, even on the same IC for the same challenge (due to noise). Therefore, a PUF can generate volatile secret keys used for cryptographic operation in two steps.

First, an error correlation process, consisting of initialization and regeneration, ensures that the PUF can produce the same output even when there are environmental changes such as voltage and temperature fluctuations.

The second step is the generation process in which PUF outputs are converted into cryptographic keys.

While traditional secure elements involve the generation and storage of cryptographic keys in nonvolatile memory, PUFs offer the possibility to perform key derivation upon power-up and only store the resulting cryptographic secrets in volatile memory, which cannot be accessed once the device power is turned off. This characteristic of the so-called "hardware-intrinsic security" techniques significantly increases the difficulty of compromising the device.

Pairing of PUFs requires creating a unique fingerprint characterizing the electronic hardware in an initial bootstrapping phase, which can then be used to uniquely identify the device. The fingerprint provides a hardware-intrinsic key that can then be used to share with a server either an asymmetrical key pair or a symmetrical key derived from this hardware-intrinsic key generated during power-up. Thanks to their unique fingerprint, PUFs may provide an economic way to authenticate devices in the IoT, by relying on methods to perform a security bootstrap between the device and a server using their hardware-intrinsic key, as described in Section 13.5.

For example, at power-up, the device generates a private key using its hardware-intrinsic key and uses this key to authenticate successfully with the server. A variant of this is to use the hardware-intrinsic key to cipher the private key prior to storing it in the nonvolatile (flash) memory of the device. At power-up, the hardware-intrinsic key is used to decipher the private key, which is itself used to authenticate the device and secure its communication with the server.

13.5 Security bootstrap in the M2M system

13.5.1 Preprovisioning of credentials and physical binding

The presence of a secure element inside a device provides the possibility to store data securely and perform confidential execution of code. The security approach described above (see Section 13.4.4) easily accommodates preprovisioning of credentials in secure facilities prior to deployment, including the cases where several stakeholders (e.g., access network operator, M2M service provider, and M2M application operator) need to deploy their own credentials independently.

An important question to be dealt with relates to the way to establish a security association between the device/secure element and an infrastructure server, to be able to authenticate devices and provide further security services such as encryption and integrity protection. However, the ability to authenticate a device based on secrets in a secure environment will not be the same when this secure environment is bound to the device and when it is removable, in which case some sort of logical binding may need to be established.

While the traditional approach to distribute personalized secrets in deployed devices relies on removable secure elements, there are several use cases in M2M communications that require the secure elements to be physically and/or logically bound to their hosting device. This is the case, for example, for SIM cards included in traffic lights, which may be stolen and inserted in handsets in order to place voice calls, unless the associated subscription is restricted to a specific usage only.

SIM cards are a perfect example of removable secure element. As they are easily transportable between communication modules, their removability makes them better suited to authenticate subscription to services (e.g., communication service that the user may use for different devices) than to authenticate M2M devices themselves. Therefore, removable secure elements such as smart cards are mostly suitable for the M2M consumer market. The specificities in this market include miniaturized form factors and ruggedness to withstand specific environmental constraints. Furthermore, distribution and operating costs need to be optimized to remain commensurate with the average revenue expected from the M2M application. This cost can however be shared between several stakeholders, such as when the M2M service subscription and the M2M application service are provided as third-party applications on the SIM card used to provide cellular connectivity to a device: modern SIM cards comply with the UICC specifications from ETSI Technical Committee (TC) Smart Card Platform (SCP), which generally supports such possibilities. Note that ETSI TS 102 671 provides some means to physically or logically bind a UICC platform to a hosting device.

On the other hand, nonremovable secure elements are tightly embedded in their hosting device. They cannot be transported from device to device and therefore are well suited to authenticate the devices themselves, as well as associated subscription to services.

13.5.2 Need for late-stage personalization, dynamic provisioning, and security administration

For usage related to M2M, we will focus on "nonremovable" secure elements because of their ability to help in authenticating a device.

In many circumstances, the secure element is deeply embedded within a complex device or needs to be physically bound to it because of environmental constraints (e.g., vibrations or heat): In such cases, the device/secure element needs to be initially provisioned during manufacturing or deployment, without a direct physical connection to the embedded secure element.

In several cases, the long expected lifetime of deployed devices as well as regulatory constraints additionally requires the possibility to change a device affiliation within its lifetime (e.g., in the case of a car changing owner, as in secondhand), which results in the need to reinitialize or provision a new secret (not known from the former owner/provider) in a device that was already deployed or to administrate existing credentials dynamically during device lifetime. And in several other cases, a product can only be associated with a stakeholder after deployment, as such association often depends on user choice or geographic location (e.g., buying the device in another country).

All the above-used cases require the possibility of changing credentials during an equipment's lifetime. Relevant scenarios are considered in the following sections.

13.5.3 Remote bootstrapping of prepersonalized secure elements: late-stage personalization

The possibilities for remote security bootstrapping depend on the initial provisioning scenarios: a secret, controlled by either the device manufacturer or a service provider, may be already present or not.

The secure element manufacturer may personalize devices at manufacturing time with initial secrets that will be used as initial seeds. Typically, a secure element is purchased by a device manufacturer that embeds it in a nonremovable way inside an M2M device. The device is later purchased by a device owner.

These secrets may be considered as the seed that will enable differentiation between two identical devices manufactured in the same conditions. This seed may be used by a third-party trusted provider (trusted service manager, TSM) to remotely diversify the secrets and bootstrap the security between the device and the server. This process is illustrated in Figure 13.3.

The secure element manufacturer will personalize secure elements at manufacturing time with initial secrets that will be used as initial seeds.

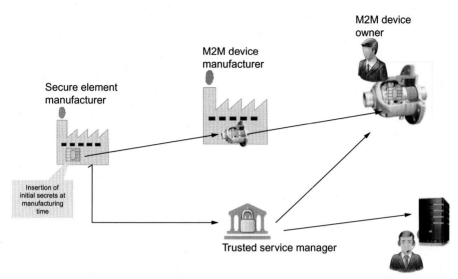

Figure 13.3 Secure element management via trusted service manager.

The secure element is purchased by a device manufacturer that will embed it in a nonremovable way inside an M2M device.

The TSM will retrieve the seed inserted in the device from the secure element manufacturer and will use that seed to establish a remote secure access to the secure element embedded inside the device. It will remotely execute a key diversification process in the secure element.

The TSM will then create a secure connection with the service provider and will provide the newly generated key; at this point, the service provider (server) and the device share a common secret and may use it to authenticate and secure their communications.

A business model similar to the one described in Figure 13.3 has already been implemented for the deployment of near-field communication services in the cellular telecommunication world. Process and communication protocols between the multiple parties involved have already been specified by the GlobalPlatform organization. Those protocols adapt easily to the case of M2M communications.

A special case of this scenario is when the device is owned by the service provider and has already been deployed in the field. For example, devices may be electricity meters deployed in households and the service provider may be an electricity company. Resorting to a TSM is an attractive solution when there is a need to provision and administer a large number of devices without human intervention or when additional legal liability or privacy concerns come into play.

13.5.4 Dynamic bootstrapping

Another advantage of resorting to a trusted third party is the fact that the security bootstrapping process may be executed again, should the ownership of the device change. In this case, the TSM is able to remotely replace the keys stored in the device so that the new owner gets complete control over the device while the previous owner looses this control. This is particularly important when a large number of devices have already been deployed on the field and device replacement is not an option (the example of the electricity meter provides again a good illustration of this situation).

13.5.5   Bootstrapping by derivation from pre-existing credentials

Another frequent scenario is when the M2M device needs to be affiliated with one service provider, such as a telecommunication service provider, prior to operation. For example, the automatic emergency call systems deployed in modern vehicles in case of accidents may be provisioned with a subscription to a mobile network operator according to a contract negotiated by the manufacturer. The device has therefore to be provisioned with an initial seed to support this service. For example, 3GPP devices are provisioned with 3GPP credentials stored securely on a SIM card (UICC) embedded within the device. When there is a need for a third-party provider to provision new secrets on the SIM/secure element for a service other than the 3GPP communication (in our case, the M2M service), one can think of reusing previously distributed credentials to perform the security bootstrap, assuming proper trust agreement between the service provider and the network operator. This is the purpose of the Generic Bootstrapping Architecture defined by 3GPP: the 3GPP telecommunication operator offers to use the 3GPP credentials in order to bootstrap the definition of a new shared secret between the device and the third-party service provider (in our case, the M2M service provider).

13.5.6   Out-of-band-assisted bootstrapping

The last scenario deals with a device with embedded security that has never been provisioned with secrets prior to its acquisition by its current owner. The fact that the secure element does not share any secret with anyone at the time of purchase precludes the use of the remote provisioning methods described above. However, the security pairing techniques described in the next section, involving the generation of keys on the secure device that can be transmitted to the server protected by a secret communicated to the server by an out-of-band channel, can be applied.

Those methods can support a change of device ownership: the new owner just runs again the security bootstrapping process in order to define new shared secrets with the server. However, the methods requiring human intervention for implementation of the out-of-band channel are not suited to the management of a large population of normally unattended devices, such as an electricity meter fleet.

13.5.7   Key pair generation in communicating objects

A first method aims at generating a pair of asymmetrical keys from a hardware-intrinsic key. The device will retain the private key and will send the public key to the server protected by a passphrase. The sequence diagram when using this scheme for pairing a PUF with a server is represented in Figure 13.4. The passphrase is communicated

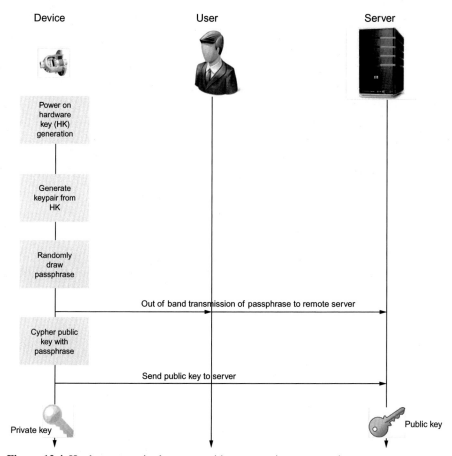

Figure 13.4 Hardware security bootstrap with asymmetric cryptography.

by the device to the server by an out-of-band channel. The implementation of this out-of-band channel may involve actions from the user. In this case, the transmission of the passphrase from the device to the remote server is achieved in two steps:

Step 1: Transmission of the passphrase from the device to the user. When the device cannot display such passphrase (due to the lack of user interface), Light-Emitting Diode (LED) lights or speaker sounds can be used, or the passphrase could come embedded in the device package.

Step 2: The user communicates the passphrase to the server. Possible implementations are as follows:

- Via a web interface (after a form of device registration).
- Via a telephony interface (and touch tones). In this case, there may be a need to provide a device identifier as well. With this method, a device including a speaker could generate directly the appropriate key tones: the user just has to place the device close to the handset.
- Via SMS on a cellular phone network.

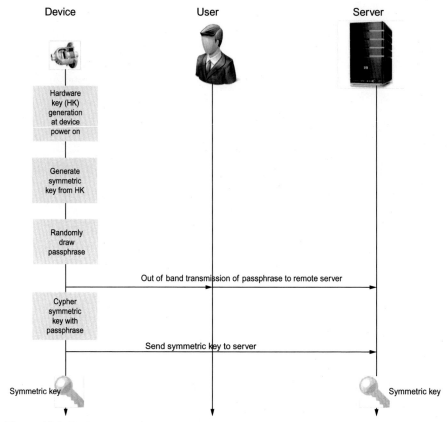

Figure 13.5 Hardware security bootstrap with symmetric cryptrography.

The server, after receiving the passphrase, is able to recover the public key. At this point, the server and the device may protect their communication by using the key pair.

The passphrase transmitted from the device to the server via the out-of-band transmission methods may also be used to protect the transmission of a symmetrical key. At the end of this session, the device and the server share the same symmetrical key and may use it for authenticating and securing their communications. The sequence of operation associated with this method when pairing a PUF is shown in Figure 13.5.

13.6   Bridging M2M security to the last mile: from WAN to LAN

M2M applications often involve reaching devices located within a local area network (LAN) from the WAN. This generally involves the use of a device, gateway, or router located at the border of the LAN, with one network interface on the LAN and one on

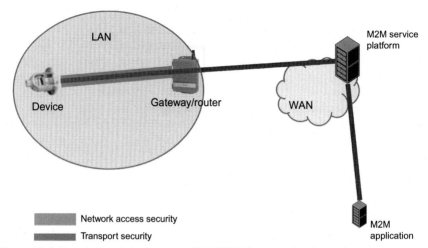

Figure 13.6 Security layers in composite LAN/WAN communication path.

the WAN. This section discusses the security mechanisms involved in protecting data communications between a LAN device and a communication end point on the WAN.

Figure 13.6 shows a typical communication architecture: a device in the LAN communicates with an M2M application via an M2M service platform located in the cloud. Different security schemes may be involved to address different communication layers, as outlined in the figure.

On the LAN side, MAC level security, also called *radio layer security* (in dark grey in the figure), is a common solution to protect the privacy of local area wireless communications. Most wireless communication protocols (Wi-Fi, Institute of Electrical and Electronics Engineers (IEEE) 802.15.4, and Bluetooth) specify such radio layer protection. This security scheme will protect the whole data payload including the IP headers carrying end point communication addresses. However, the data protection with this scheme stops at the border of the LAN.

Radioprotection may also exist on the WAN side when a radiocommunication infrastructure (e.g., 3GPP communications) is involved. To keep things simple, this scenario is not considered here.

As the data move beyond the border device, transport layer security (TLS) is commonly used (in red in the figure) to protect communications between two end points. secure socket layers and its successor TLS are the typical solutions used to protect Transmission Control Protocol (TCP) communication, and Datagram Transport Layer Security (DTLS), a protocol derived from TLS, is addressing the security of User Datagram Protocol (UDP) communication. TLS will leave IP headers and the request URL in clear when used along with HTTP.

TLS usually relies upon long-term credentials to mutually (or not) authenticate the client and the server. An ephemeral session key is usually derived during a handshake exchange, and this session key is used to actually protect the data exchanged between the two points during the session.

A third type of security, *the application security*, is sometimes used to protect data from end to end. It involves a ciphering of the application payload.

It is also possible to use TLS to achieve end-to-end security.

However, M2M applications commonly involve "one to many" communication schemes. In such cases, a pairwise negotiated ephemeral key is not an adequate solution and needs to be replaced by a group key negotiation. Technical solutions for this do exist, but are not widely deployed and need to be matured.

A specific application security layer will be implemented if there is a good reason for the entity owning one or the two communication end points to keep the control of credential management. The discussion below will deal only with the combination of radio and transport level security.

13.6.1   Gateway security models

In Figure 13.6, the LAN part of the communication path shows two candidate protocols for security:

- The radio layer security provides an exhaustive protection for the data transmitted within the LAN, but not operant beyond the LAN. The credentials of this security layer are typically managed by the LAN owner.
- The TLS covers both the LAN and WAN parts of the transmission, but leaving IP headers in clear in the LAN wireless transmission and therefore imperfectly protecting communication privacy. The credentials of this security layer are typically managed by the service or application provider. In the case shown in Figure 13.6, they are managed by the M2M service provider.

Note: Another drawback of using the TLS alone is that an attack carried in the LAN, such as the injection of fake connection packets, may have consequences on the WAN side (e.g., the fake packets will travel across the LAN before action may be taken).

The way to handle the two security protocols will depend on the type of the border device. Three typical cases will be considered:

- The border device is a gateway, ending the IP communication path.
- The border device is a gateway using NAT (network address translation).
- The border device is a border router (typically used in IPV6 configuration).

13.6.1.1   Security model with a gateway ending the IP communication path

In this scenario, the gateway is the only point of contact from the WAN, hiding the presence of individual devices in the LAN. The TLS may be terminated at the gateway, which may forward the packets to end nodes within the LAN using only radio data protection.

When dealing with energy or computing power-constrained end devices, this configuration avoids the implementation of multiple security layers inside the constrained device, thus saving energy and improving the battery duration.

Its main drawback is that device addressing is not dealt with at the IP level and must be addressed at the application level.

13.6.1.2   Security model with a gateway using NAT

In this case, the gateway is still the only point of contact between WAN and LAN. However, the transport layer protection extends up to the LAN individual nodes, and each communication node may be transport layer-protected using a distinct session key. The advantage lies in the possibility of each node to protect its communication not only from outsiders but also from other nodes in the LAN.

This solution however has two drawbacks:

•   In order to achieve good privacy, radioprotection should be used in addition to transport layer protection on the LAN wireless transmission, requiring the implementation of two security layers within devices. This is not a very suitable solution for energy-constrained devices.

•   The NAT scheme generally requires that all communication sessions be device-initiated, making scenarios involving incoming asynchronous requests from the WAN to LAN nodes more complex, or sometimes impossible to implement.

13.6.1.3   Security model with a border router device

This solution is suitable in an IPV6 LAN where individual devices may be attributed globally addressable IP addresses. In this case, each LAN node can be addressed individually from the WAN, and the border router simply routes data from the WAN to the LAN nodes. This enables the handling of incoming asynchronous requests from the WAN to the LAN nodes.

On the security aspect, the transport security extends up to each LAN node. As in the previous case, the need for good privacy protection will generally require to superimpose radio and transport level security layers on the LAN side. There are however specific scenarios where this may be avoided. One such scenario is described below.

13.6.2   Specific case: multihop capillary networks

Multihop LANs involve transmitting data from a device through adjacent nodes in order to reach a final destination node, as illustrated in Figure 13.7.

However, when using multihop networks, many applications will not require individual security up to the leaf nodes, but rather global security within the network. In other words, individual nodes belonging to the LAN need to protect their data from outsiders, but not necessarily from other nodes within the LAN.

In this case, transport layer protection may stop at the entry point of the LAN and be replaced within the LAN by radio layer protection, as shown in Figure 13.7.

This configuration enables addressing each node of the LAN independently from the WAN. Thus, incoming asynchronous requests from the WAN directed toward LAN nodes may be handled.

The communication path is protected by TLS on the WAN part and by the radio layer security on the LAN part, offering a good privacy protection on the LAN side.

Energy-constrained devices located in the LAN only need to implement radio layer security, thereby minimizing the energy overhead linked to security.

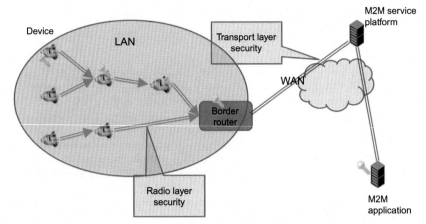

Figure 13.7 Security of the communications in multihop LAN networks.

13.6.3   Bridging different security layers

Two types of security layers have been described in the previous section:

* Radio layer security.
* Transport layer security.

These two security layers differ not only by the layers of communication addressed but also by the entities managing the associated credentials.

In the scenario outlined in Figure 13.7, the M2M service provider may manage the TLS credentials, while the LAN owner would be responsible for the management of the radio layer security credentials.

Bridging the two security layers implies linking the generation of their credentials to reduce the number of actors involved in credential management.

For example, radio layer credentials may be generated automatically as a secondary set of credentials derived from transport layer credentials. Under such a scenario, the M2M service provider may offer the remote management of both types of credentials as a service, to relieve the LAN owners (not directly involved in credential management) from such responsibility. This bridging operation may occur at different places, although it would be logical to locate it in the border device.

13.7   Conclusion

13.7.1   Further sources of information/advice

Standards developed for various industrial sectors can serve as useful references for securing M2M applications, provided that they satisfy the specific requirements and constraints of the M2M application. Aside from field-specific standards, the most interesting general sources of information in this respect are provided below.

In order to provision and preserve secrets in exposed M2M devices from manufacturing to deployment and throughout the product lifetime, specifications from the ETSI TC SCP (available at http://portal.etsi.org) and from the GlobalPlatform organization (http://www.globalpaltform.org) are the most useful. The ETSI TS 102 221 defines the UICC, a secure hardware security platform widely deployed in mobile telecommunication devices, which support a standardized application programming framework (ETSI TS 102 241) and associated remote file and application management capabilities (ETSI TS 102 225 and TS 102 226). ETSI TS 102 600 defines ruggedized hardware form factors associated with this platform that are suitable for M2M environments. The GlobalPlatform Card Specifications and associated amendments further extend this framework with the capability of sharing the platform securely between multiple stakeholders by means of independent security domains that can each be remotely administered independently. The SIM cards used in mobile telecommunication networks such as the 3GPP (http://www.3gpp.org) or its CDMA equivalent body 3GPP2 (http://www.3gpp2.org) have so far been the most successful standardized applications for the UICC platforms, but the ETSI TC "Smart M2M" has now standardized a related application for M2M services as part of ETSI TS 102 921, which could be further extended for specific industry domain organizations in specifying their own dedicated M2M applications. The ETSI M2M specifications TS 102 689 (architecture level) and TS 102 921 (protocol level) further provide a basic security framework for M2M services that can also accommodate lower-security demands, as enabled by implementations based on, e.g., a TEE specified by GlobalPlatform. But they do not currently address security behind gateways, which can be provided by LAN-specific standards such as the IEEE 802.15 family (http://www.ieee.org) or the ZigBee Alliance (http://www.zigbee.org). We look forward to the oneM2M international standardization partnership (http://www.oneM2M.org) to provide a more complete security framework that will address the security needs of M2M applications, such as the bootstrapping of end-to-end credentials.

13.7.2   Future trends

The previous sections provided an overview of the challenges involved in securing IoT communications. However, there is no "one size fits all" solution regarding security. The threats associated with each application need to be assessed in order to define the appropriate security mechanisms to put in place.

When trying to protect assets of significant value, the motivation to implement those mechanisms will be high. But, when confronted with the need to spend significant resources to implement security in an application where the risk level is badly defined or underestimated, an easy choice could be to forget altogether about security. The problem, as discussed above, is that one aspect of the risk associated with IoT comes from the multitude: a large number of unprotected "low-risk" applications may create a significant threat to our society. The solution to this is to lower the cost and the difficulty to secure M2M applications, simplifying all along the user experience.

Today, interoperability is a major focus in the M2M community. We are witnessing the emergence of M2M interoperable platforms such as the one being proposed by oneM2M. Those platforms do not only promise to enable ad hoc interoperability between objects and applications: by exposing a number of services and Application Programming Interface (APIs) easily accessible to developers, they will simplify the coding of M2M applications and lower their cost. Security is part of this value proposition. The deployment of M2M interoperable platforms will provide a set of standardized security mechanisms to rely upon, simplify the implementation of security in the applications, thanks to the security services exposed by the platform, and ultimately favor wide-scale implementation of security. There is however a need to identify and expand this set of security services to cover a number of requirements that are not or badly addressed today:

- End-to-end security between M2M communication end points enables the protection of the whole communication path with a single set of credentials. End-to-end security often relies upon globally accessible credential distribution mechanisms that need to be easily accessible by applications and need to be supported by M2M service platforms.
- "One to many" communications involve one data source sending data to multiple recipients: this type of data communication is common in IoT applications. Multicast techniques provide in this context an efficient dissemination of the data. However, the supporting techniques to bootstrap and implement security are not always mature and need to be enhanced in order to be widely adopted.
- Securing the communication originating from small affordable objects presents a real challenge today. Such devices are typically limited in memory and computing power and very often energy-constrained. Implementation frequently leads to the impossibility of supporting security due to their limited resources. There is a need to identify energy-friendly cryptographic algorithms and security bootstrap mechanisms well suited for devices without or with very limited human interface capabilities. Furthermore, small objects used in wireless sensor networks may organize dynamically in ad hoc mesh networks presenting specific risks that need to be addressed.

Security in the world of IoT also implies the capability of M2M devices and networks to be resilient to cyber attacks. The same type of large-scale data monitoring and control solutions that appear today to protect Internet nodes against denial of service and other types of malware may need to be implemented to protect M2M networks against unpredictable aggressions.

Part Three

Network optimization for M2M communications

Group-based optimization of large groups of devices in machine-to-machine (M2M) communications networks

14

T. Norp
TNO, Delft, The Netherlands

14.1 Introduction

Many machine-to-machine (M2M) applications involve large groups of M2M devices. A typical electricity company that uses M2M for smart metering may employ millions of M2M devices. A car company that equips its cars with a remote diagnostics application may sell a few hundred thousand M2M-enabled cars each year. There are M2M applications with only a single device or a few devices. However, a network operator will generally not consider a customer with less than 1000 devices to be an M2M customer.

Often, all devices in an M2M application are owned by the same company or organization that also owns the application. The application owner may use the M2M application for internal processes. For example, a water management organization may use M2M to measure water levels. All devices and the server these M2M devices talk to are controlled by the same organization. Another common business model for M2M is that the devices are owned by end customers but that the application owner controls the M2M application on the devices. The application owner may use the M2M application to provide services to its customers. An example is the navigation device company TomTom™, which includes mobile connectivity in navigation devices to provide real-time traffic updates to its customers in their vehicles.

Figure 14.1 shows an example of a business model for an e-book reader company. The e-book reader company sells e-book readers that come with a mobile subscription. Through the mobile subscription, end users can download new e-books via a mobile network. However, it is not the end user but the e-book reader company that owns the subscriptions with the mobile operator. The e-book reader company sells e-book readers with preinstalled SIM cards. The cost for the subscriptions is included in the price of the device and the price of the e-book downloads. The e-book reader company will not use a different mobile operator for every single country the e-book reader is sold in, but use only a few mobile operators worldwide. This implies that if the e-book reader company sells millions of these e-book readers with mobile communication capabilities, it will also have millions of M2M subscriptions. For the mobile operators, the e-book reader company will be a very large customer.

Machine-to-machine (M2M) Communications. http://dx.doi.org/10.1016/B978-1-78242-102-3.00014-9

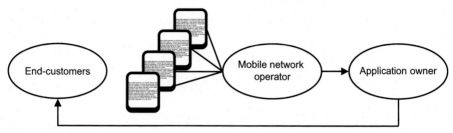

Figure 14.1 Business model example with e-book readers.

Mobile operators have special provisions for corporate customers that need mobile subscriptions for a few hundred or thousand employees. There are special interfaces for the corporate customers to manage subscriptions, and the mobile operator will provide combined billing for all the subscriptions. It makes sense to also provide similar facilities to M2M customers with even much larger number of subscriptions. However, the requirements of M2M customers are different, and therefore, mobile operators have created M2M-specific services for large M2M accounts. Also, 3GPP standardization has addressed mobile network optimizations for group-based M2M applications.

This chapter will explain what type of optimizations mobile networks can provide for group-based M2M applications. The focus is on optimizations of the M2M data communication provided by the mobile network, that is, where the mobile network operator is acting as connectivity provider. First, an overview will be provided of the different mobile network optimizations for groups. In the subsequent sections, each of these different group optimizations will be explained in more detail.

14.2   Mobile network optimizations for groups of M2M devices

Current mobile network technologies such as long-term evolution (LTE) were not designed with M2M communications in mind. Optimizations are needed to adapt mobile networks to the requirements of M2M devices that may only send small amounts of data, may have severe power constraints, and almost always require low-cost solutions. Chapter 3 of this book addresses what type of optimizations can be done to make public mobile networks better suited for M2M. Also, Ref. [1] discusses mobile network optimizations for M2M. 3GPP standard specifications of mobile network improvements of machine-type communications (MTC) can be found in 3GPP Technical Specification 22.368 [2].

M2M improvements for mobile networks include optimizations for handling large groups of M2M devices. The common subscription model in mobile networks is that a single customer has a single device with a single subscription. In contrast, with M2M communications, a single M2M customer often has up to millions of devices and subscriptions. This aspect of M2M communications makes it clearly of interest to look at how mobile operators can provide enhancements for M2M customers with large groups of devices.

Billing and provisioning are the first aspects to look at when providing solutions for customers that can have up to millions of subscriptions. Clearly, a large M2M customer will not want to receive individual bills for every subscription. Also, provisioning interfaces need to be better than what is available for individual subscribers. Billing and provisioning interfaces are generally not standardized and differ between different operators. Similarly for M2M communications, operators compete on providing optimal support for handling large amounts of M2M subscriptions.

Though there is no formal standard for M2M billing and provisioning interfaces, there are a number of alliances of mobile operators that provide M2M solutions according to a common proprietary standard. It is easier for M2M customers to work with multiple operators from an alliance as the information technology interfaces for billing and provisioning are the same. Different alliances compete on providing the best solutions, and switching between alliances is far less easy. Examples of such alliances are the *Global M2M Association* (with TeliaSonera, Orange, Deutsche Telekom, Telecom Italia, Bell Canada, and SoftBank Mobile) and the *M2M World Alliance* (with Rogers, Telefónica, KPN, VimpelCom, Etisalat, SingTel, NTT DOCOMO, and Telstra). The exact functionality that these alliances provide, such as interface descriptions, is not public information. Therefore, we cannot describe the optimizations provided by these alliances in detail.

3GPP standardization has also addressed group optimizations. There have been a number of different work items that address group optimizations. Table 14.1 shows an overview of the different functions that are under consideration in 3GPP.

Unfortunately, the status of the work on group optimizations within 3GPP is less clear (as of Q1 2014). The stage 1 service requirements for group optimizations were first defined together with most other M2M optimizations in 3GPP Technical Specification 22.368 [2] within Release 10. However, the stage 2 architecture work on M2M in Release 10 focused on the more urgent M2M congestion and overload control work. In Release 11, stage 2 work produced a basic M2M architecture that can be used for much of the group optimizations. However, the group optimizations themselves

Table 14.1 Overview of group optimizations in 3GPP

Group-based addressing	Triggering or sending messages to large numbers of devices in a particular geographic area through local broadcast
Group-based policing	Policing the aggregated bit rate (or data volume) of a group of devices
Group-based charging	Generating call detail records per group of devices instead of per individual device
Group identifiers	MTC groups need to be identifiable throughout the 3GPP network

were not yet specified. Finally, in Release 12, a work item was started dedicated to group-based M2M optimizations. However, later on, within the Release 12, this work item was deprioritized in favor of work to support public safety services in mobile networks. In Release 12, there will be no stage 2 specifications on group optimizations. The related service requirements have been brought forward to Release 13, and the work on group optimizations has been taken up again in Release 13. Table 14.2 provides an overview of how the work on group optimizations is distributed over the different 3GPP Releases.

In 3GPP Release 12, there has been a lot of focus on enabling public safety services. One of the Release 12 work items related to public safety services is also related to groups. Group call service enablement aims to support group calls in critical communications, for example, push-to-talk. The functionality that is defined within this work item may also be very useful for M2M group optimizations.

Release 12 also saw a change to the group charging requirements. The original requirements implied that it should be possible to generate charging detail records (CDRs) per group of devices instead of per individual M2M subscription. It was felt however that the requirements to minimize the generation of CDRs were more relevant to wholesale charging instead of charging for groups of M2M devices. Consequently, the group charging requirements in 3GPP Technical Specification 22.368 were removed, and general bulk charging requirements were added to the general 3GPP charging requirements in 3GPP Technical Specification 22.115 [3].

Overall, it is still unclear what will be the outcome of 3GPP standardized optimizations for M2M groups or when they will be finalized. What follow in the subsequent sections are insights in what has been achieved so far and what other possible optimizations are possible.

14.3   Managing large groups of M2M subscriptions

An M2M application owner has to manage a large amount of subscriptions. It is not practical to do this with the support operators give to ordinary subscribers; M2M application owners need a lot more support. Large corporate customers that need mobile communications for a few hundred employees have similar requirements. Specific functionality is needed to easily add or remove subscriptions or change different subscription options. With M2M, the numbers are even greater; where a large corporate customer has a few thousand subscriptions, a large M2M application owner may have a few million subscriptions.

M2M applications additionally have the problem that it generally is very difficult to get access to the devices. If a reset is needed of the communications module in a smart meter, a service engineer has to visit the house in which the smart meter is installed. That makes even a simple reset very expensive. For M2M application owners, it is very important that management of the devices can be done remotely. There are several remote device management frameworks that can be used for remote management of M2M devices; for example, the Broadband Forum has specified the TR-069 [4]

Table 14.2 Group optimizations in the different 3GPP releases

	Release 10	Release 11	Release 12	Release 13
Stage 1	• Group-based addressing • Group-based policing • Group charging • Group identifiers	–	• Group call service enablement requirements • Removal of group charging	–
Stage 2	–	• Basic MTC architecture	• Work item on group-based MTC features started • De-prioritization of group-based MTC features for Release 12 • Group call service enablement	• Continuation of group-based addressing • Continuation of group-based policing

management protocol and the Open Mobile Alliance (OMA) has specified OMA device management.

Often, M2M devices are manufactured with embedded SIM cards. This has a specific impact on subscription management. The manufacturer of the M2M devices will get a large amount of SIM cards from the operator. The operator will preprovision these SIM cards in its system. However, the SIM cards are only activated when the devices are sold and activated. Only from that moment onwards, the application owner will pay subscription fees. Specific agreements are made such that the manufacturer can test the devices with the SIM cards in the factory—often in a different country from where the devices will be used—without generating large roaming charges. With remote subscription management, it is even possible to fit the devices with a generic SIM card and remotely change the subscription of the SIM card to an appropriate operator. This way, it is possible to use local operators wherever the devices are used, while keeping the manufacturing and logistics processes for the devices simple.

Remotely setting and retrieving parameters on an M2M device assume that the mobile network connection with the M2M device is functioning. But with mobile communications, you cannot always assume that a connection is available, the device may be out of mobile network coverage, it may be out of power, there may be a technical issue in the device, or there may be a network issue with the connection. To find out whether there is a mobile connection to the M2M device or whether there is something wrong, application owners want to be able to troubleshoot M2M connections remotely. Information on the status of M2M device can be provided to M2M application owners on the basis of information that is available in the mobile network. For example, if an M2M device is out of coverage, the network will no longer receive the regular location update messages from the device. By using available network information, the M2M application owner can monitor the status of the M2M device without constantly sending small messages to check if the connection is still working. 3GPP [2] has specified requirements for remote MTC monitoring. MTC monitoring is not strictly seen in 3GPP as a group-based feature as it works on individual devices and subscriptions as well. However, MTC monitoring only becomes important when managing large numbers of MTC devices.

Billing is another aspect of managing groups of M2M devices. An application owner does not want a large amount of individual bills for individual devices. On the other hand, the application owner wants to be able to monitor the behavior of individual devices. Itemized billing for each individual device is needed to prevent fraud and troubleshooting. If one device is using much more data than other devices, there may be something wrong with the device. A higher than normal data usage for an M2M device could indicate fraud. A case of M2M fraud happened in South Africa [5], where SIMs were stolen from traffic lights and then used to make phone calls. There may also be a technical fault that results in a particular device sending much more data than normal. An example is a device that requests a firmware upgrade but does not manage to install it properly and then repeatedly keeps requesting more upgrades. Monitoring the usage data of individual devices is important to flag problems, to repair faults, or to stop fraud as soon as possible (see example in Box 14.1).

Box 14.1 Use case example 1

An energy company has one million metering devices deployed. These metering devices are connected to the mobile network of Mobile Operator A. All metering devices have subscriptions with exact settings and options in the subscription profile.

The energy company wants to update part of the subscription profile (e.g., a new default Access Point Name—APN setting) of all the one million devices.

Group optimizations are also possible for the handling of subscription profiles within the mobile network [1]. A large part of the subscription profile (e.g., the max bit rates) will be the same for all subscriptions in a group of subscriptions. All these common elements can be stored in a group subscription profile. The individual subscription profiles can then refer to the group subscription instead of storing these common parameters. In case an application owner needs to make changes to the common parameters, only a change to one group subscription profile is needed instead of potentially changing millions of individual subscriptions.

In mobility management procedures (e.g., the attach procedure), the subscription profile is downloaded to serving network (in either the mobility management entity (MME) or the serving GPRS support node (SGSN)). Here too, it is possible to use group-based subscriptions. When the first device from a group performs an attach procedure at a particular MME/SGSN, then both the individual subscription profile and the group-based subscription that it refers to have to be downloaded. However, when a subsequent device from the same group attaches, there is no need to download the group subscription profile again. Also, if there is an update of the group service profile, the group profile needs to be updated to the MME/SGSN only once.

14.4   Group-based messaging

Group-based messaging is about sending messages to groups of M2M devices efficiently. These messages can be used to send instructions or to trigger M2M devices to report to a network server. To efficiently send data to a large group of devices, broadcast technology can be used. By using broadcast, a message does not have to be sent to each M2M device individually (see example in Box 14.2).

Using broadcast also implies that there is a specific geographic area in which messages are sent. This implies that group-based messaging is not really suitable for applications with a very wide geographic scope, such as fleet management for trucks. Group-based messaging is much more suited for smart city-related applications as these generally are confined to the geographic area of a city.

Geographic area restrictions can also be used as a benefit. For example, in an application to control streetlights, the streetlights can be switched on in a particular

---

Box 14.2 Use case example 2

A city has installed underground garbage containers where citizens can deposit their waste. Each container is fitted with an M2M-sensor that reports to the garbage collection agency when the container is full and needs to be emptied.

When a lot of garbage is expected, for example, around the December holiday season, the garbage collection agency wants advance information about containers that are already more than half full and are likely to fill up over the holidays. Therefore, a single broadcast message is sent to all containers to trigger them to report their status.

---

geographic area by targeting the broadcast to that geographic area. All streetlights will belong to one group, and the geographic area is used to indicate which lights need to be switched on or off. If all lights need to be switched on or off, a message is broadcast in the full service area. However, when checking for broken lights during maintenance, it is possible to switch on lights only in a specific area.

3GPP has been working on group-based messaging as an extension of the basic MTC architecture [6]. This basic architecture, as specified in Release 11, is depicted in Figure 14.2. Via a new interface Tsp, the service capability server (SCS) can send a

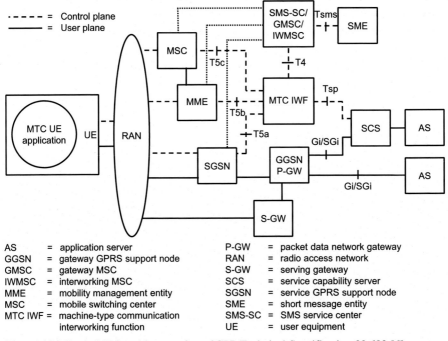

AS	= application server	P-GW	= packet data network gateway
GGSN	= gateway GPRS support node	RAN	= radio access network
GMSC	= gateway MSC	S-GW	= serving gateway
IWMSC	= interworking MSC	SCS	= service capability server
MME	= mobility management entity	SGSN	= service GPRS support node
MSC	= mobile switching center	SME	= short message entity
MTC IWF	= machine-type communication	SMS-SC	= SMS service center
	interworking function	UE	= user equipment

Figure 14.2 Basic MTC architecture from 3GPP Technical Specification 23.682 [6].

trigger message to be sent over the mobile network. The MTC interworking function (IWF) decides which way trigger message is delivered, either via SMS using the T4 interface or via signaling using the T5 interface.[1] The MTC IWF acts as the gateway between the mobile network and the M2M service capabilities, which can be outside the mobile operator domain.

The basic architecture from Figure 14.2 can also be used for group-based messaging. The same Tsp interface can be used to send broadcast triggers from the SCS to the mobile network. To broadcast in mobile network, two different 3GPP standardized technologies can be used: cell broadcast service (CBS) [7] and multimedia broadcast/multicast service (MBMS) [8].

CBS allows to send unacknowledged messages to all devices within a particular region. CBS messages are broadcast to defined geographic areas known as cell broadcast areas. The broadcast area can vary from a single cell to all cells in a mobile network.

A cell broadcast message page is composed of 82 octets. Up to 15 of these pages may be concatenated to form a cell broadcast message. Messages are sent over a dedicated cell broadcast channel (CBCH). This implies that radio resources have to be reserved for cell broadcast.

Cell broadcast messages can be repeated multiple times to ensure that most devices have received the message. The repetition period between repeated messages and the number of repetitions can be set per broadcast message. A sequence number in the messages ensures that user equipments (UEs) can distinguish new from repeated messages.

Multiple "channels" of cell broadcast messages can be supported using message identifiers. UEs are configured to only receive CBS messages with specific message identifiers. This allows different services to be provided in parallel (e.g., weather information and traffic information).

Cell broadcast is also used for distribution of public warning service messages [9]. This implies that operators that for regulatory reasons have to implement public warning service will have to implement CBS. As the public warning service is unlikely to require a lot of capacity, there will be remaining CBS capacity that can be used for other services such as group-based messaging.

Another broadcast technology to use for group-based messaging is eMBMS (evolved multimedia broadcast/multicast service). eMBMS is the Evolved Packet System version of MBMS. eMBMS supports both LTE and universal mobile telecommunications system (UMTS) radio. Where the previous version of MBMS encompassed both broadcast and multicast, the eMBMS only supports broadcast.

Within 3GPP Release 12, group-based messaging solutions have been studied for both CBS-based group messaging and MBMS-based group messaging [10]. The architecture in Figure 14.3 is based on the architecture from Figure 14.2. Broadcast messages are sent via Tsp. The MTC IWF can decide which delivery method to use. If the MTC IWF chooses to use cell broadcast, the group message is sent toward the cell broadcast center (CBC). From the CBC, the message is sent as a normal CBS message, either via GSM, using the base station controller (BSC) and base transceiver station (BTS); via

---

[1]Note that T5-based procedures have not yet been defined in 3GPP Release 12.

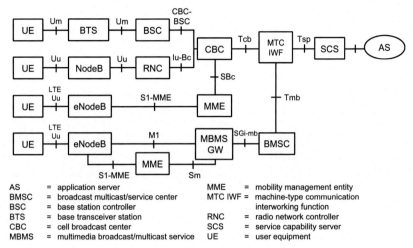

Figure 14.3 Group-based messaging architecture.

UMTS, using the radio network controller (RNC) and NodeB; or via LTE, using the MME and eNodeB. In case the MTC IWF decides to use MBMS as the delivery mechanism, the broadcast message is sent to the broadcast multicast/service center (BMSC). The BMSC uses MBMS technology to deliver the group message to the devices.

In Release 12, 3GPP has also defined support for group communication for critical communications [11,12]. This solution too is based on eMBMS. The group communication functionality defined for critical communication matches the requirements for M2M group messaging as well. The difference is that group communication for public safety does not use the MTC architecture shown in Figure 14.2.

The interface on Tsp will need some adaptations to handle group-based messaging. One of the aspects is that for group-based messaging, the group identifier and the geographic area in which the message has to be distributed are important. This information will have to be passed over the Tsp interface. Furthermore, the SCS should be able to control aspects like repetition period and number of repetitions. This also will need to be sent over Tsp.

When broadcast messages are sent over the radio interface, they may also be intercepted by devices that are not part of the group. As this may be a problem, the content of the broadcast messages needs to be encrypted. An application-specific encryption can be used to encrypt the data before it is delivered over the Tsp. The devices within the group have to be provisioned (e.g., using OMA Device Management) with the right decryption key for the group.

Though the concept behind the MTC IWF was that the MTC IWF function decides which delivery mechanism to use, in case of group messaging, the SCS should be able to decide what delivery mechanism to use. There is quite a difference between using CBS, which can only send short trigger messages, and using eMBMS, which can be used to distribute much larger messages (e.g., sending a complete firmware upgrade). The Tsp interface should also enable SCS to select the delivery mechanism.

One of the risks of sending a trigger message to a large group of devices is that they may all respond at the same time. It is important to ensure that these responses are spread out over time. A mechanism to do this is to include a back-off time value in the group-based message. Before responding the UE will apply a random delay based on this back-off time value.

The area in which a group message is broadcast has an impact on capacity requirements. A focused area for broadcast implies that broadcast capacity is used only in a limited area. On the other hand, wide-area broadcast uses up a lot of capacity and should be discouraged. The operator can define a pricing scheme for group-based messaging that includes the geographic area in which the broadcast takes place. This ensures that the wide-area broadcast is only used when needed.

Devices can scan the network for cell broadcast transmissions even when they are not attached. If devices can be kept detached while still being able to trigger them, this enables a reduction of the number of simultaneous attached users. The number of simultaneous attached users is an important measure in determining the capacity of mobile network. In pay-as-you-grow agreements between mobile network operators and network vendors, the number of simultaneous attached users is often one of the parameters that determine price. Therefore, there is an incentive for mobile operators to keep down the number of simultaneous attached users. Broadcast can also be used to trigger individual nonattached devices when location is known sufficiently accurately or when value is high (e.g., triggering a car that is reported stolen to report its position).

14.5 Policy control for groups of M2M devices

Another group optimization that has been addressed by 3GPP is policy control for groups of M2M devices. The concept behind policy control for groups of devices is to police the aggregate bit rates of all devices in a group, rather than the individual bit rates of each device (see example in Box 14.3).

Box 14.3 Use case example 3

A company that provides managed printing services has 10,000 printers deployed in office buildings. All these printers are remotely managed via a mobile network connection. Under normal circumstances, the printers generate only a small amount of data to report usage or faults. However, sometimes, more extensive remote maintenance or a firmware upgrade generates a lot of data communication for a single printer.

The printer company has arranged with the operator a maximum aggregated bit rate that suffices for day-to-day operation of all printers together with a firmware upgrade for a couple of printers, but not for firmware upgrades of all printers at the same time. The printer company gets a discount from the mobile operator for being "network-friendly" and limiting to a maximum aggregated network load.

Group-based policy control works best for large groups of devices. Because all devices together can generate a significant network load, the operator and application owner can agree on scheduling the devices to communicate during the nonbusy hours of the day. It then becomes important to ensure that not all devices communicate at the same time in the agreed time window but to spread out the traffic over the agreed time window. Without such arrangements, an application is likely to schedule coUse case communication for all devices at the start of the agreed time window, generating a peak in data usage that may even be higher than the average busy hour traffic. For the operator, it is important to spread out traffic and cap the maximum bit rate for the total application. The application owner on the other hand wants to keep maximum flexibility when to schedule traffic. Therefore, it is better not to distribute a maximum bit rate evenly over all devices, but to agree on a maximum aggregated bit rate for all devices in the group. That leaves the application owner with the freedom to decide how the traffic is distributed over the devices. For example, if there are many devices that only report small amount of data, some of the devices can do a firmware download. But an agreed maximum aggregated bit rate will not allow a firmware download for all devices at the same time.

Figure 14.4 shows the advantage of policing aggregated bit rates versus simply policing the average bit rate. The average bit rate is the agreed maximum total bit rate for the complete group, divided by the number of devices. If the mobile network polices traffic based on the average bit rates, the bit rate for each of the devices is capped individually. It is clear that with average bit rate policing, the total bit rate for the group will never exceed the maximum group bit rate. With aggregate bit rate policing, the aggregate bit rate will not exceed the maximum group bit rate either. However, the bit rate of an individual device (e.g., device 2 in Figure 14.4) can be larger than the average bit rate, as long as this is compensated by a lower than average bit rate for other devices.

To police the aggregate bit rate for a group of devices, the first step is to meter the bit rate for the group of devices. The next step is to enforce that the metered bit rate will not be higher than the agreed maximum. For both steps, the 3GPP the policy and

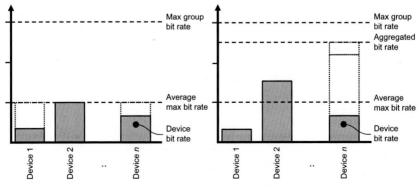

Figure 14.4 Policing aggregated bit rates versus average bit rates.

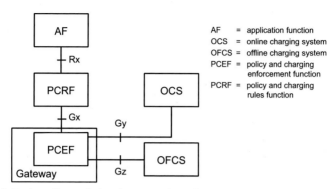

Figure 14.5 Basic policy and charging control architecture.

charging control architecture can be used [13]. A basic version of the 3GPP policy and charging control architecture is shown in Figure 14.5.

Within the architecture in Figure 14.5, the policy and charging rules function (PCRF) is responsible for the policies. These policies generally have the form of a filter (e.g., based on originating and terminating IP addresses, originating and terminating port numbers, or access point names (APNs)) together with a rule that should apply for the data that are identified by that filter. Such a rule could be to meter the traffic and discard all data that are over a maximum bit rate. The enforcement of these policies is done in the policy and charging enforcement function (PCEF). The PCEF is located in a gateway (e.g., the gateway GPRS support node (GGSN) for 2G and 3G packet architectures or a packet data network gateway (P-GW) for the Evolved Packet Core).

The architecture also includes an online charging system (OCS) for online charging (e.g., to support prepaid charging) and an offline charging system for offline charging (e.g., to support postpaid charging).

With group aggregate bit rates, all data from a group of devices should be combined. Similar aggregate bit rates are already supported in 3GPP policy and charging control; the APN-AMBR (aggregate maximum bit rate) aggregates all nonguaranteed bit rate data from a specific UE to a specific APN, and the UE-AMBR aggregates all nonguaranteed bit rate data to a specific UE over all APNs. For group aggregate bit rates, an aggregate has to be made of all nonguaranteed bit rate data for all UEs in a specific group. How the PCEF can determine which data connections are related to the UEs in a specific group has not yet been addressed in 3GPP.

When policing a group AMBR in the downlink, it is sufficient to discard packets that exceed the maximum bit rate. However, discarding data packets in uplink may not help. Uplink packets that are discarded in the gateway/PCEF have already been sent over the radio interface, and it is the radio interface capacity that operators want to protect with policing. TCP/IP rate control may help to reduce the traffic generated by the devices if there is a significant amount of data per IP session. However, if each IP session contains only a small amount of data, for example, an application with a large amounts of devices that each send only a single IP packet at a time, TCP/IP rate control does not work. A discarded packet would then only result in the device resending the same packet, increasing the radio resource usage.

To enforce the group AMBR in the uplink, a better option may be to reduce the individual maximum bit rates of data connections. This can be done by the PCRF if it is informed that the group AMBR is exceeded. If the PCRF informs the gateway that the individual maximum bit rates of some of the devices within the group have to be reduced, the adaptation of the maximum bit rate for the bearer connection is signaled to the UE. The UE will then its own policing. A guaranteed way to reduce bit rate is to inform all devices that the maximum bit rate for their data connection is reduced to the average bit rate. However, this implies that all data connections need to be updated. A simpler way may be to no longer accept new connection requests and to start to zero rate random connections from within the group until the bit rate has dropped under the maximum group bit rate. When the bit rate has significantly dropped below the maximum group bit rate, policy control can again start to accept new connections or reinstate the old maximum bit rates for the connections that were zero-rated.

A major issue with policing group-aggregated bit rates is that generally, the data connections for different devices within the group would be routed via different gateways. However, if multiple gateways are involved, the different gateways would have to keep each other up to date in real time about the amount of data metered on each gateway. To keep things simple, 3GPP decided in 23.769 [10] to only use a single APN and a single gateway for the group. This alleviates the problem of metering data exchange between gateways. But it also implies that applications for which group-aggregated policing applies will have one gateway as a single point of failure. Most large M2M application owners prefer to be connected to multiple gateways for redundancy.

An alternative solution for policing group-aggregated bit rates makes use of the OCS to collect the metering information from multiple gateways and to enforce the group policy. For every data connection within the group, online charging mechanisms are used to inform the OCS of how much data is consumed for that device. The gateway sends a credit control request to the OCS, for each data connection. The OCS responds with a credit control answer in which it gives a grant for an amount of data that can be sent together with a time limit. The gateway allows data transport for the data connection until the data amount is almost consumed or the time limit is almost expired. When the gateway does a new credit control request, it also indicates how much of the data budget was consumed from the previous grant. If the grants are not made too large, the OCS can get a fairly accurate view of how much data was sent in a particular time window by adding up the credit control requests for all the devices in the group. Dividing this amount of data by the length of the time window gives the aggregate bit rate averages over that time window. Inevitably, this method will result in some averaging of the measured aggregate bit rate. However, if the main interest from the operator is to ensure spreading of data communication over the day, a time window of 1 min would be sufficiently short.

Which particular method of policing group aggregate bit rates will be specified within 3GPP is not yet (Q1 2014) clear. Target is that a solution for group aggregate bit rate will be specified within the Release 13 timeframe.

14.6 Groups and group identifiers

When an application owner wants to apply a specific group optimization to a group of devices, it is necessary to identify which group these optimizations need to apply to. A straightforward method to identify a group is with a group identifier. The group identifier identifies a list of devices or subscriptions that all belong to the same group.

More complicated to answer is whether a single group should apply for multiple group optimizations at the same time or whether it should be specific for specific group optimizations. Related to this is the question whether devices can be in multiple groups or always just in a single group (see example in Box 14.4).

If a single group is used for multiple group optimizations, groups can become fragmented. It is quite possible that an application owner would prefer a set of groups for group-based addressing that is different from the preferred set of groups for group-based policing. With combined groups, all features have to be the same for a group, which implies that the groups have to be smaller than would be optimal from a service point of view.

Whether a device can belong to multiple groups depends on the group optimization to which this group is applied. With group-based addressing, a single device can belong to multiple groups. The device will react to messages sent to each of the groups the device belongs to. However, if the group identifier is associated with a particular group subscription profile, a subscription can only belong to one group at a time. If a subscription belongs to two different group subscription profiles, this will lead to conflicting service settings.

Box 14.4 Use case example 4

A city has deployed several thousand environmental sensors throughout the city to measure air pollution, smog, etc. As the measurement network was established gradually, there are different sensors with different capabilities; not all sensors can measure the same. Furthermore, over the years, sensors have been sourced from different suppliers.

The city wants to be able to easily trigger all sensors with the same measurement capability and therefore has defined group addressing groups for each measurement capability. As sensors can have multiple capabilities, they can be in multiple groups.

Next to these groups, a different set of groups is defined to trigger firmware upgrades. These groups are dependent on the make and version of the sensor and not on the capability of the sensor.

Finally, the mobile network subscriptions are grouped for group-based subscription management. Here, devices with similar mobile network requirements are grouped together.

3GPP is not very clear yet on the relation between groups and group-based identifiers. On the one hand, 3GPP Technical Specification 22.368 states that "An MTC Device can belong to more than one MTC Group." On the other hand, it states that "The system shall provide a mechanism to associate an MTC Device to a single MTC Group." The assumption behind the requirements in 3GPP Technical Specification 22.368 seems to be a group subscription profile as it states that "each Group Based MTC Feature is applicable to all the members of the MTC Group."

There are also different requirements on the format of a group identifier. For a group-based subscription profile, the group identifier has to be globally unique as the group identifier identifies the group service profile that is downloaded to the visited network (see Section 1.3). For group-based addressing, the requirement is that the group identifier is not too long. The group identifier will have to be broadcast, which implies that a long group identifier is wasteful on broadcast resources. The Tsp interface may have other requirements on group identifiers. As a SCS may be connected to different operators, the group identifier may have to indicate the operator that hosts the group.

Conclusion is that there are several aspects to consider on how to organize groups and group identifiers. It seems that 3GPP has not really tackled this aspect. In the 3GPP architecture work on group features, identifiers have been considered for each optimization individually. There is not a unified concept of group identifier for multiple group-based features.

14.7 Conclusions

M2M communication often involves group of devices. Therefore, it makes sense to optimize mobile network communications for groups. Some group optimizations are defined as a proprietary solution by a mobile operator (e.g., group-based provisioning and charging). Also, 3GPP has been working on group optimizations (e.g., group-based messaging, group-based policing, group-based subscriptions, and group identifiers). A first outlook can be given on which group optimizations 3GPP will standardize and how they will work. However, 3GPP standardization of group-based optimization is not yet (Q4 2014) finalized.

References

[1] T. Norp, B. Landais, M2M optimizations in public mobile networks, in: M2M Communications: A Systems Approach, John Wiley & Sons, Chichester, England, 2012.
[2] 3GPP Technical Specification 22.368, Service requirements for machine-type communications (MTC); Stage 1. Available from: www.3gpp.org.
[3] 3GPP Technical Specification 22.115, Service aspects; Charging and billing. Available from: www.3gpp.org.
[4] The Broadband Forum Technical Report TR-069, CPE WAN Management Protocol.

[5] The Guardian, 6 January 2011. Available from www.theguardian.com/world/2011/jan/06/
     johannesburg-traffic-light-thieves-sim.
[6] 3GPP Technical Specification 23.682, Architecture enhancements to facilitate communi-
     cations with packet data networks and applications. Available from: www.3gpp.org.
[7] 3GPP Technical Specification 23.041, Technical realization of cell broadcast service
     (CBS). Available from: www.3gpp.org.
[8] 3GPP Technical Specification 23.246, Multimedia broadcast/multicast service (MBMS);
     Architecture and functional description. Available from: www.3gpp.org.
[9] 3GPP Technical Specification 22.268, Public warning system (PWS) requirements. Avail-
     able from: www.3gpp.org.
[10] 3GPP Technical Report 23.769, Group based enhancements. Available from: www.3gpp.
     org.
[11] 3GPP Technical Specification 23.468, Group communication system enablers for LTE;
     Stage 2. Available from: www.3gpp.org.
[12] 3GPP Technical Report 23.768, Study on architecture enhancements to support group
     communication system enablers for LTE. Available from: www.3gpp.org.
[13] 3GPP Technical Specification 23.203, Policy and charging control architecture. Available
     from: www.3gpp.org.

Optimizing power saving in cellular networks for machine-to-machine (M2M) communications

H. Chao, J. Wu
Research & Innovation Center, Bell Laboratories China, Alcatel-Lucent
Shanghai Bell Co. Ltd, Shanghai, China

15.1 Introduction

Power management is one of the most important issues to be dealt with for machine-to-machine (M2M) communications in cellular networks. In some M2M use cases, the power source of M2M devices, i.e., batteries, cannot be replaced or recharged as easily as those of human's mobile phones [1]. If it is not possible to charge or replace the battery, the lifetime of the battery may even determine the lifetime of the device. Even for scenarios where user equipment (UE) may consume power from an external power supply, it may be desirable to consume less power for energy efficiency purposes [2]. Therefore, cellular networks should provide M2M services with the least possible energy required in order to extend the lifetime of M2M devices.

Power control issues for wireless sensor networks have been extensively investigated. A thorough survey on relevant power control approaches was classified and presented under different contexts by Pantazis and Vergados [3]. However, cellular networks cannot simply borrow the approaches discussed in Ref. [3] for wireless sensor networks due to the different system and network characteristics. Unlike cellular networks with a hierarchical architecture, wireless sensor networks are dynamic network topologies, which can vary from simple star networks to advanced multihop wireless mesh networks, while cellular networks are typically with hierarchical architectures. Further, although the schemes in Ref. [3] may work well for certain scenarios, they not necessarily perform the best for all possible situations. Thus, it is highly likely that dedicated solutions will be found for cellular networks.

The conventional existing cellular networks are mainly designed for human-to-human (H2H) communications and thus may not work efficiently for M2M services. In recent years, cellular operators are expecting system solutions to deploy M2M services on their systems with minimal deployment costs. Since 2005, 3GPP (the 3rd Generation Partnership Project) has initiated the study item of M2M communications in GSM (Global System of Mobile communication) and UMTS (Universal Mobile Telecommunications System) systems to investigate potential requirements to facilitate improvements and the more efficient use of radio and network resources [4]. Since September 2008, 3GPP has started the study on LTE (long-term evolution) and LTE-Advanced (LTE-A) network improvements of machine-type communications

Machine-to-machine (M2M) Communications. http://dx.doi.org/10.1016/B978-1-78242-102-3.00015-0

(MTC) as one of the new features of Release 10. Requirement analysis [1], improvements for core network (CN) [5], and improvements for radio access network (RAN) [6] have been outlined and are discussed in different work groups of 3GPP. In recent years, the normative work on Release 12 has initiated a dedicated building block "UE Power Consumption Optimizations (UEPCOP)" to find candidate power-saving solutions [2]. Several mechanisms are proposed, which however require further evaluations and RAN impact analysis. In the remainder of this chapter, the term MTC is applied to emphasize discussions under the 3GPP infrastructure, and those UEs for MTC are called as MTC UEs.

It is well-known that high-level functionalities of cellular networks are grouped into access stratum (AS) and non-access stratum (NAS). In general, candidate mechanisms in AS and NAS can be divided into two categories to reduce power consumption for MTC UEs (direct method and indirect method):

1. In the direct method, the removal of unnecessary M2M activities will be beneficial to conserve the power of MTC devices. 3GPP has provided mechanisms in AS and NAS, respectively, to reduce power consumption for MTC devices in terms of reducing the frequency of certain activities.
2. In the indirect method, improvements in network operations and optimized signaling flow for both AS and NAS may help to reduce the power consumption of MTC UEs.

Therefore, it is worthwhile discussing power-saving improvements in both the MTC UE side and the network side. In the rest of this chapter, we discuss direct solutions and indirect solutions, respectively, with clear classification of their AS or NAS functionalities.

The chapter is organized as follows. In Section 15.2, we first discuss the direct solutions in the MTC UE side, within which both AS and NAS improvements for extended idle mode are explained. Then, we discuss indirect solutions in Sections 15.3 and 15.4. Our proposed power-saving approaches in RAN are presented in Section 15.3, mainly addressing the paging issue. The CN-side optimization for M2M is presented in Section 15.4. In each section, we discuss the performance results for either 3GPP solutions or our proposed approaches.

15.2   Extended idle mode for M2M devices

15.2.1   Requirements

Unlike human to human terminals, a lot of M2M devices related to specific applications such as metering and vendor machines are not expected to move randomly and even stay in fixed locations. 3GPP defines this new feature for MTC UEs as low mobility [5]. As a consequence, the Mobility Management (MM) for M2M in cellular systems can be optimized to capture the new feature. Especially for idle-mode UEs, they are not actually "idle". Periodic tasks or event-triggered tasks consume their power, e.g., system information reading, measurement, cell/Radio Access Technology (RAT)/Public Land Mobile Network (PLMN) selection/reselection and paging detecting, and periodic transition to connected mode for periodic MM procedures. It is not necessary for low-mobility MTC UEs to perform the above tasks as frequently as normal UEs. In this section, optimizations to idle-mode MTC UEs are proposed in both AS and NAS.

15.2.2 Optimizations in the non-access stratum

15.2.2.1 Extended periodic updating timers

For idle-mode terminals, one of the NAS-level power consumption tasks is the periodic mobility management procedure. This mechanism is created to track whether the terminals are available for paging or not. For M2M devices, especially for those low-mobility devices, the mobility management signaling is expected to reduce the network resource usage. It also helps to reduce the power consumption for the M2M devices in a direct way. Therefore, we start the section with the periodic updating timers controlling the periodic mobility management procedure.

For normal terminals, the periodic location area update (LAU) procedure is controlled in the terminal by a broadcast NAS timer T3212 for circuit-switched (CS) systems. Similarly, the periodic routing area update (RAU) procedure (in UMTS) and tracking area update (TAU) procedure (in LTE/LTE-A) are controlled by NAS timer T3312 and NAS timer T3412, respectively. In contrast to LAU, the value of T3312 and T3412 is provided by the network operator during the attach procedure and can be changed during the RAU procedure and TAU procedure, respectively. The network activates the periodic updating timer when allocating a nonzero value to the terminal. If the allocated value is zero, the periodic updating timer is deactivated. Then, the terminal does not need to perform the periodic LAU/RAU/TAU procedures.

In order to reduce periodic LAU/RAU/TAU signaling, the extended NAS timer is introduced to reduce the frequency of the periodic LAU procedure for CS systems and the RAU procedure or TAU procedure for PS systems [5]. Extended NAS timer can only be allocated to terminals supporting the extended periodic updating timer. If one mobile station (MS) supports extended periodic timer T3212, it includes the "MS network feature support IE" in the attach request or LAU request messages. If one UE supports extended periodic timer T3312/T3412, it includes the "MS network feature support IE" in the attach request or RAU/TAU request messages.

To lengthen the timer T3212, the network allocates to the MS with a periodic LAU timer, i.e., "Per MS T3212," that may be different to the broadcast value, only in the Location Updating Accept message. However, the network can allocate an extended value of T3312 and T3412 to the UE in Attach Accept or RAU/TAU Accept messages. Taking LTE/LTE-A UE as an example, the maximum value of T3412 is 192 min [7,8], while the specified default value of T3412 is 54 min [7]. The maximum value of T3412 extended value is 320 h [7,8]. With such a value, the power consumption for MTC UE to perform periodic TAU procedures is saved. The procedure of the extended periodic updating timer allocation is shown in Figure 15.1.

It should be noted that the extended periodic updating timer can be applied for signaling overload control for the CN node and for power-saving purposes.

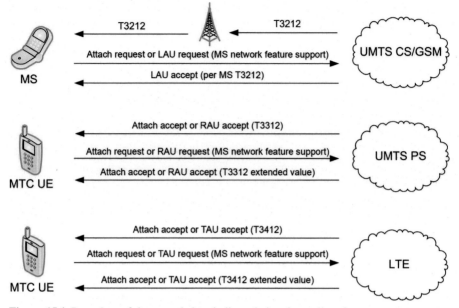

Figure 15.1 Procedure of the extended periodic updating timer allocation.

The CN node is SGSN (Serving GPRS (General Packet Radio Service) Support Node) for UMTS or MME (Mobility Management Element) for LTE/LTE-A. If the CN node is experiencing an overload situation, the CN node may decide to set periodic updating timers to higher values or decide to deactivate them for MTC UEs indicating "low-priority access" in previous Attach or LAU/RAU/TAU procedures. From the network side, reducing the periodic LAU/RAU/TAU signaling for those MTC UEs may introduce the least adverse effects on the whole system.

15.2.2.2   Power-saving state

A solution named as "allowed time period after RAU/TAU" is proposed for Release 10 MTC [5]. The basic idea is that downlink data transfer for MTC can be limited in an allowed time after MTC UE performed the TAU/RAU procedure. Then, MTC UEs may power off after the allowed time period. However, RAU/TAU is not the only case the MTC UE initiates the uplink transmission. It is more reasonable to provide an opportunity to the downlink data transmission each time the UE is connected to the network. Moreover, power saving in terms of power-off may not be a good choice as the UE has to perform the full Attach procedure next time it needs to initiate uplink transmission. We believe that is one of the reasons that an improved power-saving state solution has been proposed for REL12 in Ref. [2].

In the proposed power-saving state solution, an MTC UE may enter the power-saving mode (PSM) to reduce its power consumption [8,9] via shortening its reachable time in idle mode [2]. MTC UEs can be configured to change to PSM from the idle mode after an active timer expires. The active timer is configured by the network and starts each time the MTC UE enters into the idle mode from the connected mode. When the PSM is activated, the MTC UE stops all idle-mode AS-level tasks such as system information reading, measurement, cell/RAT/PLMN selection/reselection and paging detecting, and NAS MM procedures. However, NAS timers still take effect. While the MTC UE remains attached during the power-saving state, it can return to normal state more quickly than a UE using the "allowed time period after RAU/TAU" solution.

In order to illustrate the impact of the power-saving state to the standardization, we simplify the EPS Mobility Management model with simplified states and show their relative power consumption in Figure 15.2. In the connected state, the MTC UE communicates with the network. In the idle state, the MTC UE does not have a connection with the core network. The consumed power of connected state is higher than that of idle state. The introduction of power-saving state reduces the power consumptions further as UE stops all AS tasks during the period. The periodic updating timer controls the length the UE camped in the power-saving state. The actual length or duration of the active timer determines the timer period that the PSM-activated MTC UE.

One may be inspired by Figure 15.2 that the power-saving state solution seems extending the discontinuous reception (DRX) cycle, which equals to the periodic updating timer. Are there any differences between the power-saving state solution and the extended DRX solution? The answer is YES. And we will discuss the extended DRX cycle solution in the section on "Extended DRX in idle mode". These two solutions are different and the power-saving state solution has its own benefits as below [10]:

1. The active timer can be activated and deactivated. Then, the power-saving state can be timely changed due to UEs' requirements or the application situation, which is more flexible.
2. The period of active timer could include one or more DRX cycles, which is enough for the UEs to detect downlink data.
3. In the previous section, we discussed that the default value of periodic updating timer in LTE/LTE-A is 54 min and the extended periodic updating timer for MTC could be hours. However, no matter what value will be specified for an extended DRX cycle, it is probably not of the same order of magnitudes.

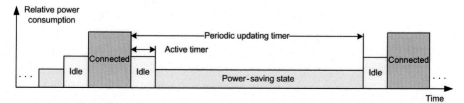

Figure 15.2 Relative power consumption for the power-saving state solution.

15.2.3 Optimizations in the access stratum

15.2.3.1 Extended DRX in idle mode

The extended DRX in idle-mode solution is an improvement in the access stratum. It is proposed to reduce the frequency for MTC UEs to wake up and monitor the paging channel. With this solution, the maximum DRX cycle in idle mode is extended with longer value. Consequently, the paging transmission period is also extended based on the extended DRX cycle applied to the UE. The solution is required backward compatible to make sure normal UEs, i.e., those UEs not requiring low power consumption, are not impacted. This means that the current DRX parameter setting method cannot be reused. Let us review the DRX parameter setting in LTE/LTE-A system at first and then summarize how extended DRX parameter can be configured.

In LTE/LTE-A systems, the default value of DRX cycle is broadcast by the eNB in the air interface. The default value is informed to the MME via the S1 setup procedure. If a UE has its own specific DRX cycle preference, it informs the MME about the value of the specific DRX cycle via NAS signaling, e.g., attach request or TAU request. Each time when the MME initiates a paging message to the eNB, it includes the specific DRX for the indicated UE in the paging message if the value is smaller than the default one. Then, the eNB knows which value, i.e., the default or the specific one, shall be applied to transmit the paging message for the indicated UE.

Similarly, there are three problems below to be solved using the configuration of the extended DRX cycle:

1. The core network and the MTC UEs need to know whether the connected Evolved Universal Terrestrial Radio Access Network (E-UTRAN) supports the extended DRX cycle or not.
2. The eNB needs to know which value for the paging message for a certain UE shall be applied.
3. The MTC UE needs to negotiate with the network for its preferred extended DRX value.

In this solution, the support of extended DRX cycle is assumed for all the eNBs within the same tracking area. Based on the assumption, two options exist to solve the above problems [2]. The main difference between the above two options is whether or not the default value of the extended DRX cycle shall be broadcasted in the air interface. In option A, the answer is YES. This is contributed to the consideration that the extended DRX cycle has potential impact for the extension of buffers for pending paging messages. The default value indicates the maximum the RAN wants to allow. Therefore, MTC UE cannot propose its preferred specific extended DRX cycle without limitations. In option B, the answer is NO. The underlying understanding is that, since the MME knows the value of the default extended DRX cycle, it can judge whether or not the specific value proposed by the MTC UE is reasonable. If not, MME transmits the default value to the MTC UE as the allowed extended DRX cycle.

- Option A: This option is called as extended DRX broadcast way. A similar framework as the normal DRX cycle is applied, i.e., there exist differences between the default extended DRX cycle and the specific DRX cycle; E-UTRAN informs its support of the extended DRX cycle in the system information broadcasting in the air interface; the default extended DRX value is informed to the MME the same way as the default normal DRX value. However, MTC UE supporting the extended DRX cycle shall inform the MME about its support in the attach request or TAU request message since not all MTC UEs in the TA require the low power consumption. If UE supporting the extended DRX cycle does not receive the extended DRX cycle in the system information, it knows that the cell/TA does not support it and does not need to inform MME about its supporting for extended DRX. The MTC UE can only use the normal DRX cycle. MTC UE can provide its preferred specific extended DRX cycle to the MME in NAS signaling, e.g., attach request or TAU request. In the paging message, MME tells the eNB whether or not the paged MTC UE supports the extended DRX and optionally includes the smaller value of the default extended DRX cycle and the specific extended DRX cycle if MME obtains the specific extended DRX cycle from the MTC UE.

- Option B: This option is an NAS only way, which means there is no default extended DRX cycle broadcast in the air interface. The value of the default extended DRX cycle applied in the RAN is transmitted to the MME. MTC UE should inform the network its desire using of extended DRX by including its preferred extended DRX cycle value. Allowed extended DRX cycle is always informed by the MME to the MTC UE via the NAS signaling, e.g., Attach Accept. If MTC UE supporting the extended DRX cycle does not receive the extended DRX cycle in the Attach Accept message, it can only use the normal DRX cycle. The eNB always obtains the applied extended DRX cycle for a certain MTC UE from the MME in a paging message.

Besides the default value of the extended DRX cycle, the extended DRX cycle has more impacts on all network nodes, i.e., MTC UE, system, and performance requirements. In fact, the most important impact of the extended DRX cycle on the RAN depends on the maximum value of the extended DRX cycle. In Table 15.1, we summarize the possible standard impacts [2,11,12].

15.2.3.2 Power-saving state

As described in Section 15.2.2.2 on the "Power-saving state" in the non-access stratum, while the PSM-activated MTC UE, the MTC UE does not perform any idle-mode AS tasks. It indicates that the MTC UE activated PSM does not execute functions associated with RRC Idle (Radio Resource Control Idle) mode. Different states of RRC idle mode have been specified in Refs. [13,14], in which different tasks are allocated to different states. During the discussion period in 3GPP on the "applicability" of the power-saving solution in RAN, there is an opinion that the UE does not necessarily perform the measurement and cell reselection before it returns back to the normal idle state from the power-saving state. Hence, this power-saving state can be equivalent to a "new" state of RRC idle mode.

For MTC UEs with an activated active timer allocated by the MME, an improved RRC_IDLE states and state transitions are proposed in Ref. [10] as shown in Figure 15.3. The basic idea is to introduce a new dormant state in idle mode, in which unnecessary activities of low-mobility MTC UEs are removed. It should be emphasized that the proposed new dormant state, labeled "camped dormant," is different from the standalone "sleep mode" of H2H communications in IEEE 802.16 [15]. In fact, the H2H mobile stations (MSs) in "sleep mode" of IEEE 802.16 are switched between unavailability intervals and availability intervals. This power-saving manner is similar to the DRX defined for legacy H2H terminals in RRC idle mode of 3GPP. MS in "idle mode" of IEEE 802.16 always works in availability, while 3GPP does not define a

Table 15.1 Possible standard impact analysis (© 3GPP)

Factors	Impact descriptions
Impacts to radio protocols	- If the DRX cycle is extended longer than the maximum system frame number, i.e., beyond 1024 radio frames in LTE or 4096 radio frames in HSPA, potential modifications to paging are expected. Modifications could include an updated way to calculate paging frame and paging occasion, UE_ID extension, and reliability of paging reception - If the DRX cycle is extended to the extent that the current allowed modification period of system information is not an integer multiple of the DRX cycle, the MTC UE cannot have valid system information blocks (SIBs) all the time as today. Then, UE is expected to check and acquire the latest system information before every paging occasion and uplink transmission - Updates to RRM requirements may be necessary (e.g., measurement and radio link monitoring) - UE and eNB capability support
Impact on mobility	- Mobility is supported; however, cell reselection may be delayed and take longer due to possibly reduced frequency of measurements - Stationary devices may be less impacted by this issue
Impacts to S1/Iu signaling	- Extended DRX cycle capability support - (Option A) CN tells RAN whether the paged MTC UE supports the extended DRX or not and optionally includes the smaller value of the default extended DRX cycle and the specific extended DRX cycle if CN gets the specific extended DRX cycle from the MTC UE - (Option B) CN tells RAN about the applied extended DRX cycle for the paged MTC UE in the paging message
Impact to network implementation	- Support of extended DRX values and potential paging enhancement - Potential extension of buffers for pending paging messages - (Option A) Broadcast the default extended DRX cycle in the air interface - (Option B) Support of protocol extensions to enable negotiation of capability of extended DRX cycles with MTC UE
Impact to UE implementation	- Support of extended DRX values and potential paging enhancement - (Option A) Indication of own support of extended DRX value to the core network. Support of extended UE-specific DRX values no longer than the default one - (Option B) Support of protocol extensions to enable negotiation of capability of extended DRX cycles with the core network
Impact on UE performance	- Longer access delays for mobile terminated services - If the MTC UE is required to check and acquire the latest system information before every paging occasion and uplink transmission, the power consumption gain could be reduced dramatically

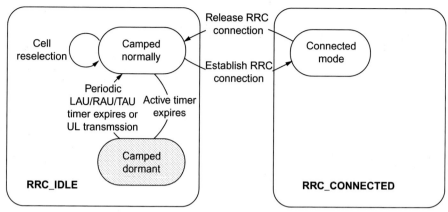

Figure 15.3 States and state transition in RRC idle mode.

corresponding mode. The RRC_CONNECTED state in Figure 15.3 indicates the CELL_FACH or CELL_DCH of UMTS, or RRC_CONNECTED state of LTE/LTE-A.

Although the power-saving state solution will probably be applicable to UEs in low mobility, the UE could move during the time period in the camped dormant state. When the UE finds its location is changed from the previous one when it was in the camped normally state, it is proposed that the UE performs the necessary measurement and cell reselection before the UE initiates the periodic LAU/RAU/TAU or UL transmission to avoid itself coming back to the camped normally state in a wrong cell or location area (LA)/routing area (RA)/tracking area (TA) for a certain period of time.

As illustrated in Figure 15.3, new tasks for RRC_IDLE mode UEs are defined as below.

- Camped normally state
    - When transiting from connected state, UE stays in the "camped normally" state for a time period limited by an active timer allocated by the core network. Upon the expire of the active timer, it enters into the camped dormant state;
    - Each time the UE comes back to the "camped normally" state from the "camped dormant" state.
        - Before further transition to the connected mode for UL data transmission or periodic LAU/RAU/TAU procedure as required by the upper layer, based on possible location change, it could
            - read system information,
            - perform measurements for the cell reselection evaluation procedure,
            - perform cell reselection procedure.
- Camped dormant state
    - UE in "camped dormant" state stops all access stratum tasks. State transfer from "camped dormant" to "camped normally" could happen
        - before the expire of periodic LAU/ RAU/TAU timer,
        - due to UL data transmission required by the upper layer.

To simplify the impact of PSM to the AS specification, 3GPP very recently has finally decided not to introduce new substate of RRC into Refs. [13,14]. Corresponding changes of PSM are described as, when PSM starts, indicated by the NAS, the AS configuration is kept, all running timers continue to run, but the UE need not perform any idle-mode tasks [13,14].

15.2.4   Performances

In Section 15.2.1, we have already established that idle-mode UEs do not actually "idle" due to periodic tasks and event-triggered tasks. With the power-saving state solution introduced in Sections 15.2.2.2 and 15.2.3.2, there is no doubt that tasks saved in idle mode are dramatically reduced. In this section, power savings for idle-mode MTC UEs are evaluated in terms of some amount of activities in LTE/LTE-A system. We will see the extent of the saved tasks for PSM-activated MTC UE comparing to those for normal UEs. For a simple expression, PSM-activated MTC UE is called as power-saving UE in the rest of this section.

To simplify the problem, here, we do not consider measurements of inter-RAT (radio access technology) cell as in Ref. [16]. Thus, activities for normal idle-mode UEs include the following seven types [16]: system information reading, paging detecting, serving cell measurement and evaluation (including measurement and evaluation selection criterion S for serving cell), intrafrequency cell measurement and evaluation (including intrafrequency cell detection, measurement and evaluation), interfrequency cell measurement and evaluation (including interfrequency cell detection, measurement and evaluation), periodic MM procedure, and UL data transmission. Except for UL data transmission, the frequencies of above activities are defined in Ref. [17]. In Table 15.2, other applied evaluation parameters are given, among which we apply the default value of the periodic TAU timer defined in Ref. [7].

It is shown in Figure 15.4 that the more tasks for power-saving UEs, the longer the active timer. For the same value of active timer, the longer DRX cycle brings more tasks for power-saving UEs. The percentage amount of tasks for power-saving UEs compared to normal UEs is illustrated in Figure 15.5. Up to 98.32%, tasks can be avoided in the case that the DRX cycle is 0.64 s.

Table 15.2 **Parameters for idle-mode activities**

Parameters	Value
Evaluation time (min)	60
Number of interfrequency carriers	1
Number of intrafrequency cells	2
Periodic TAU timer (min)	54
Active timer (min)	1, 2, 5
DRX cycle for H2H (s)	0.32, 0.64, 1.28, 2.56
Frequency of MTC UE transport UL data	Four times per hour

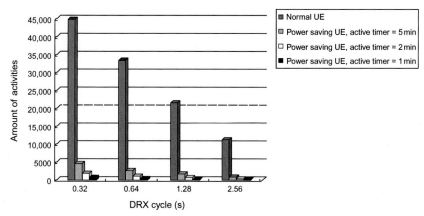

Figure 15.4 Task comparison of normal UE and power-saving UE.

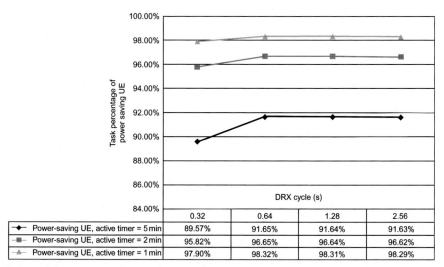

	0.32	0.64	1.28	2.56
Power-saving UE, active timer = 5 min	89.57%	91.65%	91.64%	91.63%
Power-saving UE, active timer = 2 min	95.82%	96.65%	96.64%	96.62%
Power-saving UE, active timer = 1 min	97.90%	98.32%	98.31%	98.29%

Figure 15.5 Task percentage of power-saving UE.

15.3 Paging idle-mode M2M device in a power-efficient manner

15.3.1 Challenges

15.3.1.1 Requirements for paging M2M devices in a group manner

M2M communications involve a very large number of terminals [4,18,19]. Terminal types could include metering devices, vendor machines, cars, home appliances, various monitoring machines, and other types of equipment. Allowing such a large number of terminals to access the network brings significant challenges to the networks due to potential signaling congestions and overloads. We know that paging messages lead to the establishment of the RRC connection in the AS and further transmission of NAS messages such as service request. Paging a group of terminals simultaneously shall avoid surges of uplink signaling.

From the efficiency point of view, currently, MTC applications typically involve more than 1000 subscriptions for a single customer [2]. There are some benefits in optimizing handling groups of MTC devices for both customer and operator. Group-based paging can efficiently distribute the paging message to those members of an MTC group that are located in a particular geographic area. This requirement indicates two aspects. One is that the distribution of group-based paging shall be efficient. The other is that the group-based paging shall be able to facilitate MTC customers to require MTC data from their MTC terminals with consideration of location information.

15.3.1.2 Limitation of current H2H paging mechanism

To receive paging messages, H2H terminals should monitor paging indication periodically in a physical channel at certain paging occasions (POs) in a predefined paging frame (PF) [20]. The network determines the radio frames (i.e., the PF) and the subframes within that PF (i.e., the PO). In LTE/LTE-A, subframes 0, 4, 5, and 9 in a radio frame are candidate POs for H2H paging. The configured number of paging subframes within one radio frame used for H2H paging is transmitted via system information block type 2 (SIB2).

In order to increase paging capacity, multiple terminals are allocated with the same PF/PO. Then, woken-up terminals have to dig into the paging message to detect their identifier, which finally indicates the paging target the paging message is addressed to. Although the existing paging mechanism for H2H allows one or more terminals to be paged by a single paging message, generally, this two-step H2H paging mechanism cannot be reused directly for M2M as it cannot meet the requirements discussed in Section 15.3.1.1 above.

At first, the paging area for H2H terminals is based on the mobility management function in the core network. Since the UE is in idle mode, UE's location is only known at the CN level. Therefore, paging is distributed over an LA (for CS system), RA (for UMTS PS system), or TA (for LTE/LTE-A), which means all cells belonging to the LA, RA, or TA have to distribute the paging message in the air interface. It is not possible in today's H2H systems for customized paging used by M2M communications to consider location information. For example, it is impossible that if we want to trigger electronic metering devices in a certain building to report to its MTC server/ MTC user (defined in Ref. [1]).

Second, one H2H paging message can only carry a limited number of UE identifiers. For example, up to 16 UE identifiers can be carried in a single LTE/LTE-A paging message [21]. This limited number is much less than that of MTC applications with thousand subscriptions. Furthermore, the existing H2H paging mechanisms are not efficient for paging a group of terminals. The UE identifier carried by the paging message could be S-Temporary Mobile Subscriber Identity (S-TMSI) or International Mobile Subscriber Identity (IMSI). If it is S-TMSI, 40 bit is necessary for each indicated UE. If it is IMSI, multiple decimal digits (not exceed 15) have to be carried for each indicated UE. Paging a group of terminals will occupy large precious wireless resources. More importantly, H2H paging mechanism cannot perform paging for a certain type or group of M2M devices separating from H2H terminals. New methods need to be pursued.

Finally, if H2H terminals and M2M device are paged together, a larger number of H2H terminals and M2M devices need to wake up at the same PF/PO. This results in both terminal power wasting and a higher false-alarm probability. Although 3GPP has discussed possible solutions using M2M group ID to identify a group of M2M devices, it is still an open issue on how to define the M2M group ID and how to properly utilize it in existing RAN.

15.3.2   Group-based paging for MTC

Motivated by the above observations of the limitations of the H2H paging mechanism, this section proposes a location-based group paging mechanism for MTC in the NAS level and a three-layer efficient paging mechanism for MTC in the AS level, respectively. The whole solution solves the abovementioned problems without introducing negative impacts to legacy H2H paging.

15.3.2.1   Location-based paging in the NAS level

Location-based paging is flexible to support diversity MTC applications. The basic idea of the solution is to add location information into the control plane (CP) signaling in the CN that will further trigger paging message from the CN node to the RAN node. By this way, the CN node can therefore initiate the group paging procedure in the expected paging area. The architecture of proposed location-based paging is shown in Figure 15.6. The architecture is based on the MTC infrastructure defined in Ref. [9]. The MTC application in the external network is typically hosted by an application server (AS) and may make use of a service capability server (SCS) for additional value-added services. The connection between the AS and the SCS is beyond the 3GPP scope and thus not depicted in the current architecture. The SCS can be located outside or inside the network operator's domain, which is out of the

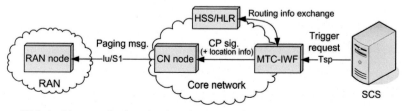

Figure 15.6 Architecture for location-based group paging.

scope of this chapter. The MTC interworking function (MTC-IWF) relays or transmits MTC signaling from SCS over Tsp interface. The CN node is MSC (mobile switching center) for GSM, SGSN for UMTS, or MME for LTE/LTE-A. The RAN node is BTS for GSM, RNC for UMTS, or eNB for LTE/LTE-A.

Taking the LTE/LTE-A as an example, the detailed procedure is shown in Figure 15.7 [22]. One MTC UE may have one or more external group identifiers. Each of them indicates one MTC application or service provided by a certain mobile network operator and accessed via an indicated domain [2,9]. In Ref. [2], more than one option is discussed for possible ways to map the external group identifier into the internal group identifier. In the example shown in Figure 15.7, the external group identifier used by the SCS is interpreted by the MTC-IWF into the internal group identifier applied in the 3GPP network.

1. Trusted SCS sends the group-based device trigger message to MTC-IWF carrying external group identifier, location/area of the indicated MTC group. The group-based device trigger message is used to trigger MTC UEs to initiate communication with the MTC Server [1].
2. The MTC-IWF selects the group-based paging as the delivery method of the group-based device trigger message.
3. The MTC-IWF maps the location/area into the geographic area for the distribution of the group message used in 3GPP.
4. The MTC-IWF interrogates appropriate HSS (Home Subscriber Server)/HLR (Home Location Register) to map the external group identifier into the internal group identifier.
5. The MTC-IWF interrogates appropriate HSS/HLR to select proper MME node(s) to transmit the group-based device trigger message.
6. The MTC-IWF sends a CP signaling such as submit request with the internal group identifier and the geographic area of the indicated MTC group to the selected MME node(s).
7. The selected MME initiates the group-based paging procedure in the indicated geographic area. Internal group identifier instead of UE ID is used in the paging message.

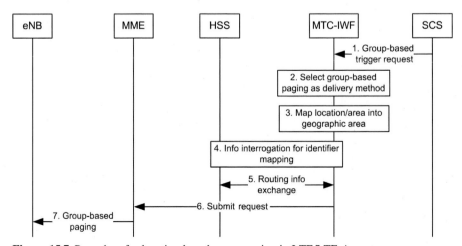

Figure 15.7 Procedure for location-based group paging in LTE/LTE-A.

15.3.2.2 Three-layer paging in the AS level

The architecture of the proposed efficient three-layer paging mechanism for MTC in the AS level is shown in Figure 15.8.

- Layer one: dedicated MTC PO

  The key point of layer one is that dedicated PO for MTC is defined to narrow down the paging range from all camped terminals to MTC UEs alone. It means that in dedicated PO for MTC, H2H terminals do not waste their power to wake up for detecting paging messages.

  Unlike H2H communication, whose PF/PO is calculated by terminal ID, PF/PO of MTC is calculated by MTC internal group identifier mentioned in Section "Location-based paging in NAS level". MTC UEs calculate PF and PO using [16]

$$\text{SFN mod } T = (T/\text{Nf}) \times (\text{Group\_ID mod Nf}) \tag{15.1}$$

$$i\_s = \text{floor}\,(\text{Group\_ID}/\text{Nf})\,\text{mod Ns} \tag{15.2}$$

where SFN is the system frame number identifying the PF; T is the DRX cycle of the UE, which is min(UE-specific DRX cycle, default DRX cycle); Nf is the number of PFs for MTC within one paging cycle; i_s is an index of subframe pointing to a predefined subframe pattern; Ns is the number of paging subframes within one radio frame used for MTC paging; and Group_ID is the MTC internal group identifier modulo operation over Mmod. Mmod is MTC mod coefficient.

To page a certain group of MTC UEs, paging indication is set in corresponding PO in PF. For LTE/LTE-A system, the paging indication is P-RNTI (paging-radio network temporary identifier) carried in PDCCH (physical downlink control channel). Idle-mode MTC UEs wake up in their dedicated MTC paging occasions and detect the paging indication. If an indication is detected, the paging detection procedure moves to the second layer. Otherwise, the procedure is finished.

- Layer two: paging target

  In the second layer, the paging range is further narrowed down to expected group of MTC UEs. All MTC UEs detected paging indication will decode the paging message, in which

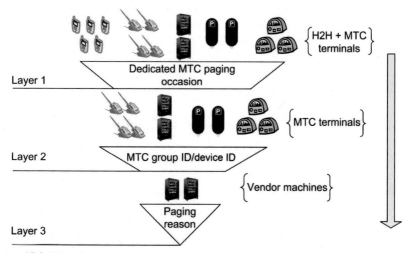

Figure 15.8 Three-layer paging mechanism for MTC.

MTC group ID or MTC UE ID is carried as indicated paging target. If MTC group ID is detected in the paging message, all the MTC UEs belonging to the MTC group are paged and therefore need to respond to the received paging message. If device ID is found in the paging message, only the indicated MTC UE needs to respond. If an MTC UE does not find its own MTC group ID or device ID in the received paging message, its paging detection procedure is finished. It belongs to MTC UEs that are false-alarmed. Otherwise, the procedure enters the third layer.

• Layer three: paging reason
The last layer is to tell MTC UEs why they are paged. In a real system, this can be achieved by different ways, e.g., multiple parallel indications or flags in the paging message or a single information element with different values. Here, they are collectively referred to as paging reason. The detailed message format is out of the scope of this chapter. No matter which way would be finally accepted by the standards, we propose to allow one paging message to be able to page one or more (groups of) MTC UEs with different paging reasons.

Candidate paging reasons are call setup, MTC report, and MTC system information acquisition. If the reason is call setup, it requests the paged device(s) to set up call connection with the network. If the reason is MTC report, the network wants the paged device(s) to respond a report. Regarding the reason of MTC system information acquisition, it means that the network requires the paged device(s) to update latest MTC system information block.

15.3.3 Performance

Power-saving performance is measured in terms of false-alarm probability for paging in simulations, since the larger the probability is, the more power is consumed for all the relevant UEs. Simulation parameters are tabulated in Table 15.3. Smaller mod coefficient is applied for MTC compared to H2H as the number of paging targets and the amount of paging resources within one DRX cycle are reduced with the proposed group-based paging mechanism.

For each paged H2H terminal, those terminals having the same result of "IMSI mod Nmod" wake up in the same PO [14]. IMSI is the International Mobile Subscriber Identity used to identify a H2H terminal. Thus, the false-alarm probability is defined by Equation (15.3) for H2H. In the case that shared PO is used for MTC UEs and H2H terminals, the false-alarm probability for all terminals camped in the system is given by Equation (15.4):

Table 15.3 Parameters for false-alarm probability

Expression	Semantics description	Value
N	Number of H2H terminals	10,000
m	Number of MTC UEs/number of H2H terminals	3, 5, 10, 15, 20, 25, 30
Nm	Number of MTC UEs	$m \times N$
Nmod	mod coefficient of H2H	1024 [14]
Ngroup	Number of MTC UEs in one MTC group	128, 256, 512, 1024
Mmod	mod coefficient of MTC	256

$$P = (\text{floor}\,(N/\text{Nmod}) - 1)/\text{floor}\,(N/\text{Nmod}) \tag{15.3}$$

$$\text{Pm} = (\text{floor}\,((N+\text{Nm})/\text{Nmod}) - 1)/\text{floor}\,((N+\text{Nm})/\text{Nmod}) \tag{15.4}$$

In our proposed group-based paging mechanism, dedicated PO is applied for MTC. In this case, the false-alarm probability for legacy H2H terminals is a constant in that legacy terminals do not need to wake up at MTC PO. The false alarm for MTC UEs occurs when different MTC groups are configured with the same dedicated MTC PO. The false-alarm probability of MTC group is calculated by Equation (15.3):

$$\text{Pm\_d} = \text{floor}\,(\text{Nm}/\text{Ngroup}/\text{Mmod} - 1)/\text{floor}\,(\text{Nm}/\text{Ngroup}/\text{Mmod}) \tag{15.5}$$

Figure 15.9 illustrates the false-alarm probability of the shared PO paging mechanism and that of the proposed group-based paging mechanism, considering the false-alarm probability of H2H UEs as a baseline. When using shared PO for both H2H UEs and MTC UEs, the false-alarm probability for both H2H UEs and MTC UEs is significantly increased. It is shown that the larger number of MTC UEs results in the higher false-alarm probability. With the idea of group-based paging and dedicated PO, the false-alarm probability of different MTC groups is obviously reduced. Smaller false-alarm probability of group-based MTC paging is achieved when the number

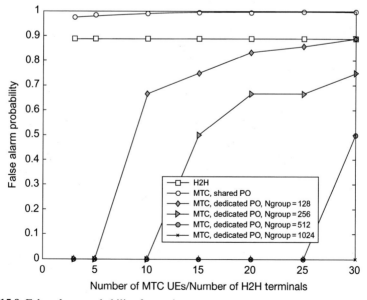

Figure 15.9 False-alarm probability for paging.

of MTC UEs in one MTC group is larger. For a certain number of MTC UEs in one MTC group, the same trend as the shared PO paging mechanism, i.e., the larger number of MTC UEs, results in the higher probability.

Larger Mmod is foreseen to introduce smaller Pm_d. However, larger Mmod in return brings more scattered Pos in the timescale for different MTC groups, which makes the network consume more paging resources to deliver paging indications for all MTC groups. We believe different numbers of Mmod would not impact the trend of the curves. That is the reason that we do not show simulation results for different values of Mmod.

15.4 Power saving for uplink data transmission

15.4.1 Requirements for signaling optimization

15.4.1.1 Analysis on M2M applications

Within the 3GPP scope, MTC scenarios include the communications between devices and the MTC server and the communications between the devices themselves [1]. In the former scenario, traffic mainly appears in the uplink. M2M devices can send data to the network in a periodic way or an event-triggered way. For example, metering devices report their measured values periodically, while vendor machines send report only when they find some goods are almost sold out. However, there is another typical application worth paying attention to, which is especially meaningful for surveillance, tracing, and tracking systems. In some cases, when urgent or sudden events occur, the MTC server/user always needs instant reports from certain machines to analyze what is taking place or what may probably occur in the near future. In this section, we focus on the finalization of this kind of application. We firstly analyze the limitations of the existing method and then present our improved solution.

15.4.1.2 Available application-based method

An MT (mobile-terminating) message can be applied to trigger uplink data transmissions. Based on received MT message, the target MTC UE then can know that it shall respond using an instant report to MTC server in terms of, e.g., SMS (short message service) or MMS (multimedia messaging service). The whole signaling flow is shown in Figure 15.10. The detailed signaling flow is explained as below:

1. SCS initiates the group-based device trigger request message to MTC-IWF including the SMS to MTC UE.
2. MTC server sends one SMS to the SMS service center (SMS-SC). SMS-SC as one of the SMS entities involved in the signaling flow is not shown separately for simplicity.
3. SMS entities interrogate HSS/HLR to select proper CN node(s) to transfer the SMS. CN node HSS/HLR is not shown for simplicity.
4. SMS entities forward the SMS to the right CN node(s). CN node(s) could be MSC for GSM, SGSN for UMTS, or MME for LTE/LTE-A.

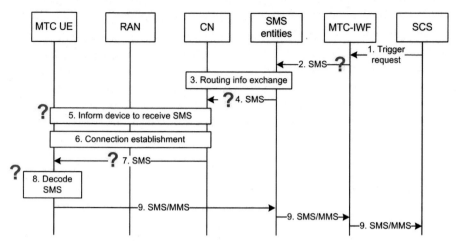

Figure 15.10 Available method to trigger instant UL transmission.

5. Then, the target MTC UE is informed that there is an SMS message to receive typically via a paging message.
6. Each informed MTC UE needs to setup the connection with the network to receive the SMS message.
7. MTC UE downloads the SMS from the connected CN node.
8. MTC UE knows it is required to send MTC data via SMS/MMS to the MTC server upon successful decoding the received SMS/MMS message.
9. Required SMS/MMS is sent from the MTC UE to the MTC server.

This method is an application-level solution, and therefore, it is time-consuming and power wasting. If more than one MTC UEs are involved, it also wastes system resources since separate messages have to be sent to each of the target MTC UE. In the whole procedure, there are optimization rooms for application-level steps (i.e., the 2nd, 4th, 7th, and 8th steps) and the 5th step of paging messages, as shown "?" signs in Figure 15.10 (in new procedure, the steps with "?" signs could be avoided).

15.4.2 On-demand uplink transmission

Motivated by the above analysis of available methods, a method labeled "on-demand uplink transmission" is proposed. The basic idea is to allow the MTC server to require instant MTC report from MTC UEs by triggering the paging message in the access stratum. The proposed method is applicable to various cellular networks. In Figure 15.11, taking 3GPP systems as an example, the whole optimized signaling flow is shown. The power of MTC UEs as well as system resource is thus saved for uplink data transmission required by the network.

In the proposed method (Figure 5.11), dashed lines indicate the CP signaling to trigger the UL transmission. The continuous line indicates the UL transmission. Steps 2–4 utilize the group-based paging mechanism proposed in Section 15.3.2. The detailed signaling flow is explained below:

1. SCS initiates the new instant report requisition message to MTC-IWF with external group identifier, location/area of the indicated one or more MTC groups.
2. MTC-IWF is allowed to initiate the existing routing information exchange procedure to HSS/HLR. With this step, MTC-IWF knows to which CN nodes shall be involved in step 3 for which MTC groups are identified by internal group identifiers. The MTC-IWF also maps the location/area into the geographic area for the distribution of the group message used in 3GPP.
3. MTC-IWF sends a CP signaling with the internal group identifier and the geographic area of the indicated MTC group to the selected CN node(s). Selected CN node(s) is MSC for GSM, SGSN for MTS, and MME for LTE/LTE-A.
4. The CN node(s) interprets the received message into group-based paging messages and sends to relevant RAN node(s) in the indicated geographic area. RAN node(s) distributes the group-based paging message in the indicated geographic area.
5. MTC UE decodes the received paging message. If the reason "M2M report" is detected, the MTC UE establishes a connection with the network for uplink data transmission.
6. Paged MTC UEs encapsulate the MTC report into predefined type, i.e., SMS or MMS. The terminating point of the instant report is preconfigured MTC server, which can be stored in the devices via operation and maintenance.

15.4.3 Performances

It is difficult to calculate how much power is saved due to signaling optimization, and thus, only qualitative analysis is provided here. In summary, both device power and system resources are saved in the proposed on-demand uplink transmission method. Comparing Figure 15.10 with Figure 15.11, the proposed method avoids steps 2, 7, and 8 of available method. In return, the power consumption of devices is avoided for processing those application activities. The triggering for CN operation is changed from SMS (step 4 in Figure 15.10) to a CP signaling (step 3 in Figure 15.11), which avoids larger payload. We have already known the efficiency improvement of group-based paging (step 4 in Figure 15.11) compared to today's UE-specific paging (step 5 in Figure 15.10) in Section 15.3. What's more, comparing step 6 in Figure 15.10 with step 5 in Figure 15.11 with the same function, the former one contains one more

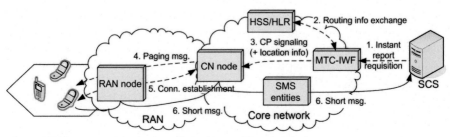

Figure 15.11 Proposed method to trigger instant UL transmission.

signaling from SGSN/MME to MSC to request the SMS data for each triggered MTC UE (not shown in Figure 15.10 for simplicity).

15.5 Conclusions

In this chapter, we focus on the power-saving issue for M2M communications in cellular networks. We discuss direct solutions and indirect solutions, respectively, with clear classification of their AS or NAS functionalities and numerous performance results. Direct solutions discuss both AS and NAS improvements for extended idle mode in the MTC UE side. In the device side, we introduce sleep state with the removal of unnecessary activities in both AS and NAS, through which the power is saved directly for low-mobility M2M devices. Indirect solutions are presented from the RAN side and the CN side, respectively. In the RAN solution, a location-based group paging mechanism for MTC in NAS level and a three-layer paging mechanism for MTC in AS level are proposed to improve the efficiency and reduce the power consumption. In the CN solution, the needs to support instant report application are analyzed at first. An "on-demand uplink transmission" method is proposed to overcome the drawbacks of available solution, with which the power of M2M devices is saved, as well as both radio and core network resources are used more efficiently. Researches on power-saving issues of M2M communications are being further deepened and extended. They are important not only for future system development but also for application deployment. Some future research directions may include power-saving issues under the requirements of large data rates and/or extended coverage, distributed mechanisms for power-saving issues, and joint consideration of power savings and quality of services.

References

[1] 3GPP TS 22.368 v12.2.0, Service requirements for machine-type communications; stage 1, March 2013. Available from: www.3gpp.org.
[2] 3GPP Technical Report 23.887 v1.4.1, Architectural enhancements for machine type and other mobile data applications communications, November 2013. Available from: www. 3gpp.org.
[3] N.A. Pantazis, D.D. Vergados, A survey on power control issues in wireless sensor networks, IEEE Commun. Surv. Tutorials 9, (4) (2007) 86–107.
[4] 3GPP Technical Report 22.868 v8.0.0, Study on facilitating machine to machine communication in 3GPP systems, March 2007. Available from: www.3gpp.org.
[5] 3GPP Technical Report 23.888 v11.0.0, System improvements for machine-type communications, September 2012. Available from: www.3gpp.org.
[6] 3GPP Technical Report 37.868 v11.0.0, Study on RAN improvements for machine-type communications, October 2011. Available from: www.3gpp.org.
[7] 3GPP TS 24.301 v12.4.0, Non-access-stratum (NAS) protocol for evolved packet system (EPS), March 2015. Available from: www.3gpp.org.

 [8] 3GPP TS 24.008, v12.5.0, Mobile radio interface layer 3 specification; core network protocols; stage 3, March 2015. Available from: www.3gpp.org.
 [9] 3GPP TS 23.682, v12.0.0, Architecture enhancements to facilitate communications with packet data networks and applications, December 2013. Available from: www.3gpp.org.
[10] 3GPP Work Group RAN2, R2-132628 discussion on MTC idle states for power saving, September 2013. Available from: www.3gpp.org.
[11] 3GPP Technical Report 37.869 v12.0.0, Study on enhancements to machine-type communications (MTC) and other mobile data applications; radio access network (RAN) aspects, September 2013. Available from: www.3gpp.org.
[12] 3GPP Work Group RAN2, R2-132893 summary of email discussion [82#13] [Joint/MTCe] evaluation of extended DRX cycles for UEPCOP. Available from: www.3gpp.org.
[13] 3GPP TS 25.304, v. 12.2.0, UE procedures in idle mode and procedures for cell reselection in connected mode, July 2015. Available from: www.3gpp.org.
[14] 3GPP TS 36.304, v. 12.1.0, E-UTRA, UE procedures in idle mode, July 2015. Available from: www.3gpp.org.
[15] 802.16-2009, IEEE standard for local and metropolitan area networks, part 16: air interface for broadband wireless access systems, 2009.
[16] H. Chao, Y. Chen, J. Wu, Power saving for machine to machine communications in cellular networks, IEEE GlobleCom Workshops, December 2011.
[17] 3GPP TS 36.133 v12.1.0, E-UTRA; requirements for support of radio resource management, September 2013. Available from: www.3gpp.org.
[18] M. Martsola, T. Kiravuo, J.K.O. Lindqvist, Machine to machine communication in cellular networks, in: 2nd International Conference on Mobile Technology, Applications and Systems, November 2005, 2005.
[19] Y. Chen, Y. Yang, Cellular based machine to machine communication with un-peer2peer protocol stack, in: Proceedings of the 70th IEEE Vehicular Technology Conference Fall (VTC 2009-Fall), September 2009, 2009.
[20] S. Sesia, I. Toufik, M. Baker, LTE—The UMTS Long Term Evolution: From Theory to Practice, John Wiley and Sons, Ltd., Publication, United Kingdom, 2009.
[21] 3GPP TS 36.331, v. 11.5.0, E-UTRA, RRC protocol, September 2013. Available from: www.3gpp.org.
[22] H. Chao (Alcatel-Lucent Shanghai Bell Co. Ltd.), Method to distribute MTC group messages. Chinese patent application 201210378852.5, September 2012.

Increasing power efficiency in long-term evolution (LTE) networks for machine-to-machine (M2M) communications

16

T. Tirronen
Ericsson Research, Jorvas, Finland

16.1 Introduction

Long-term evolution (LTE) is the most recent 3GPP radio access technology and provides high data rates for end users (peak data rate 3 Gbps in the downlink with the latest specification releases, also referred to as LTE-Advanced), low access latency, and good spectral efficiency. LTE itself characterizes the radio access part of the 3GPP system, which may include other radio interfaces as well. In addition to the actual radio interface and radio access network, the core network part takes care of providing IP connectivity and setting up the actual bearers for different types of data transmission. The whole system including the user equipment (UE), LTE access, and the core networks is called Evolved Packet System (EPS).

In this chapter, we will study the performance of the LTE access and radio interface in a machine-to-machine communication context. We will address the power consumption aspects and discuss what kind of measures could be taken to improve the energy efficiency of LTE for M2M communication. The issues will be discussed from the user equipment (UE) point of view and network equipment power consumption is not considered. We will treat all possible devices as UEs, meaning that any imaginable "machine" involved in M2M communication could be a UE in the LTE system if it includes the LTE radio modem part.

We will not directly discuss possible physical layer or hardware improvements, which is one way to achieve higher power efficiency. Instead, the focus will be on LTE procedures, which would be involved when relatively small amounts of data are transmitted infrequently. One key concept in this area is discontinuous reception (DRX), which means the mechanism of turning off the receiver part of the UE radio modem when nothing is expected to be received. In cellular systems, the UE needs to listen to the paging and control signaling sent by the network in order to be reached. The less time the UE needs to have its receiver circuitry on, the less energy it will eventually consume. This is also the key mechanism we will study in this chapter.

Most of the issues presented in this chapter have been discussed in 3GPP to some extent. Some of the solutions, such as the power-saving mode, will be specified, and

Machine-to-machine (M2M) Communications. http://dx.doi.org/10.1016/B978-1-78242-102-3.00016-2

the agreed changes to specifications are expected to be frozen during the latter part of 2014, when the 3GPP Release 12 is expected to be completed. Thus, starting with Release 12, it is already possible to attain high power efficiency for some Internet of Things or M2M scenarios.

The work in 3GPP is expected to continue into the future beyond Release 12 as well, as M2M becomes a more prevalent communication scenario. Note that in 3GPP, the acronym MTC, which stands for machine-type communication, is used to refer to M2M communication scenarios.

16.2 M2M scenarios

Typical connected devices in many forecasted M2M scenarios are sensors and actuators. Different use cases have different requirements, but the traffic characteristics in many cases can be seen to have similar properties. Especially, battery-operated devices designed for long lifetimes can be anticipated to have traffic patterns composed of the components listed below.

In uplink:

1. Sensor readings (e.g., temperature and utility meters)
2. Status indications (e.g., the status of a sensor or actuator)
3. Keep-alive responses
4. Bulk readouts of the device state

In downlink:

1. Control and command messages
2. Keep-alive queries
3. Firmware updates or other similar large data transfer events

We will focus on discussing and analyzing traffic patterns composed of the above components. With the exception of large data transfers in downlink, such as possible firmware updates, all of the messages are expected to fit into one IP packet or a few successive packets at most. See Table 16.1 on some general M2M scenarios with different requirements.

In EPS, data packets are conveyed over (logical) bearers from the UE up to the PDN[1] gateway, which allocates IP addresses and works as an anchor point for IP connectivity. The network architecture is fully IP-based, and as long as a UE is attached to the network, there is an always-on IP connection available. Thus, UDP and TCP are fully supported, and it is possible to use future protocols designed for M2M communications based on these transport protocols. One promising option is Constrained Application Protocol (CoAP) [1], which runs over UDP and can be seen as a lightweight version of HTTP. CoAP supports two message types, request and responses with low header overhead. Thus, it is suitable for the types of traffic we consider, for example, reading sensor data. However, ultimately, it does not matter what the

---

[1] Packet data network, PDN, in LTE context refers to any IP-based network. The PDN acronym is used mostly for historical reasons.

Table 16.1 Examples of M2M device categories with varying requirements on bit rate, power consumption, and latency tolerance (how fast the device should be able to be reached)

Application/ device type	Bit rate	Acceptable power consumption	Latency tolerance
Sensors	Low, medium for firmware updates	Low	High (i.e., delay-tolerant, minutes)
Utility meters (water, electricity, gas)	Low	Low	Medium (seconds)–high
Remote tracking	Low	Low	Medium
Actuator control	Low	Low	Low (i.e., should be reached fast, milliseconds scale)
Remotely controlled surveillance cameras	High	High	Low–medium

overlaying protocol and architecture is, but how to achieve efficient transactions using as few messages, and ultimately bits, as possible. Some typical transaction applications for CoAP or similar protocol are listed below:

1. Configuring sensors. The sensors are configured with reporting rules and operation rules, such as how often measurements are made and reported in uplink. The configuration message itself could be followed by an acknowledgment. Configuration messages can be delay-tolerant.
2. Actuator control. Short messages to control or configure actuators. It may be followed by acknowledgements in uplink. In critical applications, low latency is required, which proposes additional design challenge.
3. Unsolicited sensor readouts. Sensor sends readings in uplink based on an event, either triggered or periodic.
4. Solicited sensor readouts. Sensor readout in a CoAP response after a CoAP request message.

As a cellular technology, LTE has been designed to provide high data rates for today's smartphone users and mobile broadband. Hence, it may not be very efficient from the power consumption perspective when the above listed transactions are considered. For small user data, the signaling overhead can be considerable in size, and procedures, such as requirements that a UE should be reachable by the network and perform cell reselections, increase the power consumption. Further, mobile devices need to perform handovers as would stationary devices near to cell edges where the signal quality might be poor.

More data-heavy applications, such as video surveillance, are more suitable for devices with constant power supply. For the rest of the chapter, we will focus on the traffic type consisting of the components and transactions described for battery-operated sensors or actuators above. We will highlight the possible inefficiencies and improvements for LTE to handle small and relatively infrequent data transmissions better.

In addition to the data volume and frequency, the mobility of the device plays role in how efficient the LTE procedures will be. In a cellular system, the network needs to know the location of the device in order to deliver traffic through the base station (eNodeB or eNB in LTE) currently serving the device. If the device is stationary, handovers between network nodes need not be done and, the UE could perform less frequent measurement of neighboring cells without affecting the handover performance. An exception would be when the UE is near to the border between two adjacent cells: It may happen that the UE performs frequent handovers (and, correspondingly, consumes a lot of energy) because of its unfortunate location. Moreover, a moving device would need to go through frequent handovers, which in turn would result in high power consumption even if the actual handover procedure would be power-efficient.

Based on the above discussion, the traffic type we consider in this chapter will consist of traffic from a static machine-type device, with traffic consisting typically of couple of IP packets at most in the uplink direction. The traffic in uplink is also relatively infrequent, corresponding to interarrival time of minutes or more.

16.3 3GPP status and work

16.3.1 LTE power consumption-related work in 3GPP

LTE energy and power consumption aspects have been studied in 3GPP in different working groups. As a result, 3GPP Release 12 is expected to bring some improvements for machine-type communications. As M2M is a very popular topic in communications engineering and research at the moment, it is expected that further improvements will be approved for subsequent 3GPP releases. 3GPP has done work on M2M architecture and solutions before. For an overview of these earlier activities, see Ref. [2].

In 2013, a RAN WG2 study item called "Study on RAN aspects of Machine-Type and other mobile data applications Communications enhancements" on enhancements for machine-to-machine applications was conducted, and the corresponding work item started. RAN WG2 is responsible for radio interface architecture and layer 2 and 3 protocol aspects. The conclusions of the study are reported in 3GPP TR 37.869.[2] The study consisted of two parts: small data and device triggering enhancements (SDDTE) and UE power consumption optimizations (UEPCOP), aiming at signaling overhead reduction and UE power consumption optimizations, respectively.

---

[2]"Study on Enhancements to Machine-Type Communications (MTC) and other Mobile Data Applications; Radio Access Network (RAN) aspects", 3GPP TR 37.869, v.12.0.0.

TR 23.887[3] is another report where the SDDTE and UEPCOP solutions have been studied, but from core network perspective. For interested readers, the discussion in RAN WG2 took place in meetings RAN WG 2 #81bis, #82, and #83. The core network aspects have been mostly discussed in SA WG2, starting in meeting SA WG 2 #91 in 2012, and the discussions are still ongoing at the time of writing. The contributions presented in the meetings can be found at ftp.3gpp.org.[4]

The UEPCOP solutions mainly included solutions where discontinuous reception (DRX) cycle lengths are prolonged and a related power-saving mode solution, which will be discussed in Section 16.5 of this chapter. The extended DRX solutions are not going to be in 3GPP Release 12, but the power-saving mode has been agreed to and is going to be introduced in the specifications.

16.4 Introduction to basic LTE procedures

In this section, we will briefly present LTE procedures and mechanisms that have an effect on the power consumption of the devices. The different UE Radio Resource Control (RRC) states are especially important to understand, as they play a role in the power consumption considerations. Furthermore, part of these procedures will be discussed further in Section 16.5 where we develop a model for the power consumption for M2M UEs, see Ref. [9].

16.4.1 Initial access procedures

After the UE has been turned on, it needs to perform a cell search to obtain time and frequency synchronization with a cell. In LTE, two synchronization signals, primary synchronization signal (PSS) and secondary synchronization signal (SSS), are transmitted, which are used by the UE to first acquire symbol and frequency synchronization and then frame timing and the cell identity. PSS and SSS are transmitted every 5 ms, and it should be possible for the UE to do the cell search over one such period by buffering the signal and then postprocessing it. The next step is to obtain the system information (SI) broadcasted in the cell. SI contains all necessary configuration parameters of the cell so that the UE can properly operate within the cell and in the network in general. SI is divided into master information block (MIB) and system information blocks (SIBs), which are transmitted using different mechanisms. MIB contains limited amount of information and a new MIB is broadcasted in the cell every 40 ms. SIBs contain more detailed information and they are transmitted over the shared data channels. At the moment, there are 16 different SIBs defined (see TS 36.331[5]), and for M2M devices, it is expected that at least SIB1 and SIB2

---

[3]"Machine-Type and other Mobile Data Applications Communications Enhancements", 3GPP TR 23.887, v.12.0.0.

[4]For RAN WG2, the meeting directory is ftp.3gpp.org/tsg_ran/WG2_RL2/, for SA WG2, ftp.3gpp.org/tsg_sa/WG2_Arch/.

[5]"Radio Resource Control (RRC); Protocol Specification", 3GPP TS 36.331.

are needed for proper operation. SIB1 has fixed periodicity of 80 ms, while the periodicity of the other SIBs is configured by the network.

After the UE has decoded the necessary system information, it can perform the random-access procedure and start to communicate with the network. Random access in LTE basically consists of two messages in uplink and two in downlink. In the case of initial access, the UE does a network attach procedure to establish the necessary bearers and IP connectivity. After the attach procedure, the UE is ready to receive paging or to initiate communication with the network itself. For more information on the details of core network procedures, such as the attach procedure, see Ref. [3].

16.4.2   Idle and connected mode

In LTE, access stratum (AS) protocols run between the eNodeB and the UEs. These protocols determine what radio-specific procedures are running. Most notably, the Radio Resource Control (RRC) state of the UE determines the UE behavior and thus also the power consumption characteristics. There are two different RRC states in LTE, RRC_IDLE and RRC_CONNECTED, referred to as the idle and connected mode, respectively.

In the idle mode, the UE's exact location (cell) might not be known to the network. The UE listens to the paging channel, and a paging message is sent toward the UE, for example, when the network wants to initiate communication. In EPS, the cells, which are paged when trying to reach the UE, are grouped into tracking areas (TAs), and the UE updates its location if it moves beyond its current TA list or using a periodic tracking area update (TAU) procedure when a TAU timer expires. In the idle mode, UE also measures the signal strength of the neighboring cells according to measurement rules in order to perform cell selection and reselection procedures. The procedures the UE follows in idle mode are specified in more detail in TS 36.304.[6]

In the connected mode, the network knows the UE's location (the eNB) and unicast communication is possible. The UE continues to perform measurements on signals from neighboring cells and reports the results to the eNodeB. Also, channel quality of the cell where the UE camps is measured and reported, along with buffer status reports, which indicate the status of UEs transmission buffers, to the eNodeB.

Compared to the idle mode, there are more procedures the UE needs to follow in the connected mode, such as reporting its buffer status and channel state information (CSI). In general, the UE power consumption in both RRC states should be similar, and the frequency and the details of the procedures the UE follows can be configured by the RRC protocol to some extent. However, in reality, the consumption in connected mode has been measured to be several times the consumption in idle mode (see, e.g., [4]). This is a UE implementation-specific issue, and from a specification point of view, it should be possible to design M2M devices where the power consumption in both idle and connected modes is close to each other.

Uplink transmissions in LTE need to be time-aligned when reaching the eNB to ensure the orthogonality of transmissions from different UEs. From a power

---

[6]"User Equipment (UE) procedures in idle mode," 3GPP TS 36.304.

consumption point of view, maintaining the timing alignment between the UE and eNB results in increased consumption, and this is one reason why, in the connected mode, the UE is allowed to go out of synch, when no uplink transmissions are being performed. When the uplink is not synchronized, uplink transmissions are not possible, but the UE can still receive downlink control and data channels. In order to get back to synchronized state and be able to transmit in uplink, the UE needs to go through the random-access procedure. The transitions from idle to connected mode happen when either the UE or the network has need for unicast communication. The network drops the UE back to idle mode when there are no more data to be communicated (and in some special circumstances). Under normal conditions, the LTE specifications do not dictate when this will happen—it has been left for implementation for different network equipment vendors and operators to configure the networks as they like.

16.4.3   UE mobility

The UE mobility is handled differently in idle mode and connected mode. In the idle mode, the mobility is UE controlled, that is, the UE makes the decision to in which cell it is camping (or tries to camp). Connected mode mobility, on the other hand, is controlled by the network.

16.4.3.1   Mobility in idle mode

In idle mode, the UE performs cell reselection to access the cell with the best radio signal quality. The network provides the UE a white list of neighboring cells that the UE needs to consider in measurements for possible cell reselection. Further, the network also uses different parameters, such as priorities, signal quality thresholds, UE capabilities, and various other things, to control the UE cell reselection process.

16.4.3.2   Mobility in connected mode

In the connected mode, the network decides in which cell the UE is camping using the handover procedure. Typically, the UE reports measurements to the eNB, which decides if and to which cell or eNB the UE needs to hand over to. A blind handover, without measurement reports, is also possible.

For more information on measurements and mobility, refer, for example, to Ref. [5] or the 3GPP specifications.

16.4.4   Discontinuous reception (DRX)

To conserve energy as much as possible, it is desirable for a mobile device to keep its receiver circuitry turned off when it is not needed. Discontinuous reception (DRX) is a method that is employed in various wireless technologies to allow the device to turn its receiver off during periods of inactivity. In LTE, the data are transferred in subframes, whose duration is 1 ms, and typically, the user data channels in most of the subframes are not of interest for a particular UE. This is especially so in the case of M2M traffic

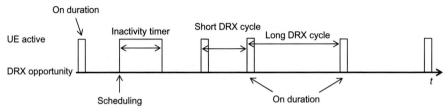

Figure 16.1 DRX cycles. In connected mode, the UE reads the control channel during "on duration" periods. If no UL or DL transmissions are scheduled, the UE has DRX opportunity afterward. In the case where there is a scheduling message, the UE starts the DRX inactivity timer and listens continuously until the timer expires. Afterward, the UE enters the short DRX cycle, if configured, and eventually the long DRX cycle.

we discussed in Section 16.2. Thus, LTE UEs can be configured to use DRX, where the UE needs to turn its receiver on sporadically. The UE needs to listen to control channel or paging once every DRX cycle. There are various parameters that can be tuned to optimize the performance of DRX. An example of such a parameter is the duration how long the UE needs to listen once it becomes active after DRX. In the current LTE specifications, the maximum DRX and paging cycle length is limited to 2.56 s.

DRX can be employed in both idle and connected modes in LTE, but the exact functionality differs to some extent. In idle mode, the DRX cycle is the same thing as the paging cycle, as the UE expects to only receive paging messages. The UE reads the paging channel at certain subframes within specific radio frames. The calculation of the possible paging occasions is UE-specific and the details can be found in TS 36.304.

In connected mode, the UE needs to monitor the physical downlink control channel (PDCCH) space for possible indication of incoming traffic. Without DRX configured, the UE would need to monitor the PDCCH search space on every subframe, that is, keep its receiver constantly open, resulting in high power consumption. Figure 16.1 shows an example of DRX in connected mode. The UE periodically checks if a transmission is scheduled and operates timers alternating between active time and DRX opportunities. A short DRX cycle may be optionally configured and it is used after the inactivity timer runs out. Finally, the long DRX cycle is entered after a short DRX timer runs out (if short DRX is configured).

To study the DRX in more detail, the DRX operation and the possible configuration values are specified in TS 36.321 (MAC protocol), TS 36.304 (idle mode), and TS 36.331 (RRC protocol). Bontu and Illidge [6] provide an overview of DRX in LTE with some configuration suggestions.

16.5 UE power consumption in LTE

In this section, we will discuss how power consumption ties into the procedures and traffic scenarios presented in the previous sections. We will explore different solutions on how the power consumption could be lowered and what kind of LTE system and specification impacts these changes would potentially have.

We present an example of a power consumption model to give tools to study the potential of different solutions and to study what is the technology potential to improve energy efficiency in LTE for machine-type devices. Similar models have been used and proposed in 3GPP UEPCOP studies,[7] but no single model has been agreed to be used in further evaluations so far. Some disagreements involve what procedures should be taken into account and what would be the realistic values for different time periods and power consumption parameters. However, regardless of the exact details of the models, different solutions can be evaluated to determine what kind of power consumption gains are potentially achievable. No model can exactly capture a realistic system, and in the end, a lot will depend on the details of UE hardware and protocol implementation.

One of the most significant debated issues is what would be a realistic ratio between the sleep, or deep sleep, power consumption and "active" state, where the UE is awake and listening to the radio interface. For today's smartphones, ratios of 1:100, meaning that if the UE consumes 1 unit of power per ms in active mode, then the sleep mode consumption would be 0.01 units per ms, are typical (see, e.g., [4]). For an optimized M2M device, however, the ratio could be 1:1000 or even 1:10 000. This ratio is one of the most important aspects affecting the possible gains that could be achieved with the power consumption optimizations. These aspects can be easily studied with the models proposed in 3GPP or in this chapter. Another important thing to note is that when considering infrequent uplink traffic consisting mostly of small data packets, some hardware optimizations, such as improvements in power amplifier efficiency, which would result in the modem consuming less power while transmitting, will actually not result in as significant savings. Naturally, for data-heavy applications, all such improvements are important, but not so much when considering the M2M traffic scenarios described in Section 16.2.

16.5.1   Example power consumption model

In the following, we present an example of power consumption model, which can be used to study the power and energy consumption of LTE modems for machine-type devices. We will assume that the modem hardware and procedures are optimized for M2M operation and for the traffic scenarios described in Section 16.2.

16.5.1.1   Hardware model

To understand the power consumption aspects without going deeply into the actual design of radio circuitry, our model for an M2M optimized device will contain four distinct parts, which can be turned on or off independently. The four parts are a branch for reception (RX), a branch for transmission (TX), an accurate clock, and a low-power clock. In addition, a real implementation of a device would additionally include parts dedicated for the actual operation of the device depending on its purpose, for example, measuring some sensor data and processing the measured data. We do not explicitly

---

[7]See, e.g., contribution R2-132893, "Summary of email discussion [82#13][Joint/MTCe] Evaluation of extended DRX cycles for UEPCOP," Huawei (Rapporteur), 3GPP TSG-RAN WG2 #83, 19–23 Aug 2013, Barcelona, Spain.

model these parts as we will focus only on what can be achieved from specification point of view, that is, the power-saving potential of the LTE radio modem part. The presented model is based on Ref. [7] and 3GPP RAN WG2 contribution R2-132805.[8]

The power consumption values and purpose of the four parts are listed below. We assume that the different parts can be turned on or off quickly, within tens of microseconds, that is, almost on LTE OFDM symbol basis, making these transition times negligible in the final model:

1. RX branch[9]: power consumption is 100 mW when turned on. RX branch is on always when the UE needs to listen for downlink transmissions. This time is also referred to as "active state" for the UE. Examples of procedures the UE will perform in the active state are measuring intra- or interfrequency cell signal strengths, receiving information on the control or paging channels, receiving (unicast) user data, and acquiring system information.
2. TX branch: power consumption is 300 mW when the UE is transmitting. The TX branch can be divided into smaller parts but model TX branch as one part consisting of, at least, a mixer, a digital-to-analog converter and power amplifier. We assume the used transmission power is in the range of 0–10 dBm, and the resulting power amplifier (PA) power consumption is included in the total. Higher TX power would result in higher PA power consumption and the relation is nonlinear. This part is turned on when the UE is actively sending data toward the eNB.
3. Accurate clock capable of keeping synchronization to the air interface: its power consumption is 10 mW. The accurate clock is activated when the UE is in DRX active mode or getting in synchronization with the network. When the UE is transmitting, it needs to be uplink-synchronized; similarly, when the UE is receiving data, it needs to be able to follow the frame and symbol timing; thus, the accurate clock is on always when either TX or RX branches are on.
4. Low-power 32 kHz crystal-powered clock: power consumption is 0.03 mW. The low-power clock is on all the time and covers the total consumed power when the UE is in sleeping mode, including the power consumption caused by leakage current.

The RX and TX branch consumption values include also the consumption of the necessary baseband processing. During the sleep time, we assume that most of the baseband processing can be turned off, resulting in very low power consumption. The digital baseband band could take up to 30% of the total power consumption, but it should be possible to turn much of the functionalities off, especially if the UE is expected to be sleeping for long periods of time (more details can be found, e.g., in [7]).

The above values are exemplary, and we will study the effect of some values, such as the power consumption of crystal-powered coarse clock later on.

16.5.1.2   State model

The device can be in three different states. Note that the low-power clock is always on.

Active state, where the RX branches and the accurate clock are on. This state is used for synchronization, SI acquisition, and receiving paging and in general when the UE is listening, receiving, and processing data.

---

[8]"Power consumption gain for DRX cycles longer than SFN range," Tdoc R2-132805, Ericsson, 3GPP TSG-RAN WG2 #83, 19–23 August 2013, Barcelona.

[9]Note that current LTE modems have (at least) two RX branches. However, 3GPP Release 12 will add support for a new low complexity UE category with only one RX branch.

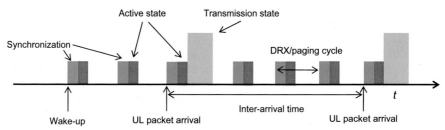

Figure 16.2 State model illustrated. Synchronization occurs before every reception or transmission event. UE can be in the active state before going into transmission state, for example if SIB1 needs to be read or idle to connected mode, state transition needs to be performed. In that case, the length of the active state may be longer than the actual transmission state, thus the illustration is not scaled properly for all scenarios.

Transmission state, where the RX branches, TX branch, and the accurate clock are on. This state is used for sending uplink data.

Sleep state. For all other times, only the low-power clock is on.

The state model is depicted in Figure 16.2. Synchronization could be additionally seen as a separate state, but we include the time it takes the UE to synchronize to the network timing in the active state. Especially for the future low complexity LTE modems, synchronization times can be longer compared to full-featured smartphone or mobile broadband LTE chips.

The UE follows two separate cycles: communication cycle and DRX (or paging) cycle. The communication cycle ends after the UE has sent data from its uplink buffer and received the possible response from a server. Thus, the interarrival time (IAT) for uplink data determines the length of the communication cycle. The length of the DRX cycle is determined by LTE system information and RRC configuration, where the lengths of the paging cycle and the possible DRX cycle are signaled.

We do not take explicitly into account possible retransmissions, nor do we model the related procedures such as HARQ or ARQ retransmissions on the MAC or RLC layer. Also, the DRX model is a simplified version of a possible real implementation.

Further, we assume a futuristic scenario where the device and traffic use suitable application protocols for MTC (such as CoAP over UDP) and the device is optimized for this type of communication. This means that the model is not directly comparable to what the current systems and smartphones could achieve, but rather shows the future technology potential.

The default model for UE actions is presented below:

1. UE wakes up from sleep state and synchronizes with the system. As default, this takes 34 ms in our model (aligned with the discussion in RAN2 WG, e.g., R2-132893). Note that for shorter DRX cycles, such as the ones currently specified for LTE, the synchronization should not take this long, and for UEs in bad coverage, this time could be considerably longer. Additionally, necessary measurements can be seen to be performed during this step (*active state*).
2. UE needs to be sure it has the correct system information. We will study two scenarios, the UE either assumes it has correct SI or alternatively uses 80 ms for reading SIB1 and checking

the value tag. We do not consider what will happen if the UE does not have correct system information. Note that in the model we use, the UE does not acquire full system information but merely checks it has not changed (*active state*).

3. If the UE has uplink data, then either
   a. the UE is in connected mode and out of synch. The random-access procedure needs to be done, and we assume 3 ms extra for both transmission and reception time (the RA messages + acknowledgements). The next step is 4b (*active state* and *transmission state*). or
   b. the UE is in idle mode and it needs to establish signaling and radio bearers, that is, it initiates the idle to connected mode transition (including RA procedure) (We assume 50 ms of reception and 5 ms of transmission for carrying out the procedure [7]. Note that this does not explicitly include ACK messages. The next step is 4b (*active state and transmission state*).).

4. UE either
   a. searches the PDCCH search space for paging indication, which would take 10 ms (includes DRX On duration time) (*active state*), and reads the PDCCH or
   b. if there is uplink data in the buffer (once every IAT), sends the data, which would take 50 ms with our assumptions (*transmission state*).
      i. NOTE: although not explicitly modeled, the uplink transmission time can be seen to include the time for receiving HARQ/ARQ feedback.

5. (*Optionally*) UE stays in connected mode before eNB tells it to release RRC connection. The UE may also use a shorter DRX cycle before following extended DRX in sleep state.

6. UE goes to sleep state (*sleep state*). Continue from 1 after one DRX/paging cycle.

Note that in our results in later sections, we use the traffic model from Section 16.2, that is, there is no downlink data; thus, the UE is not actually paged. As another remark, the total time of how long the RRC state transition procedure in step 3b takes, including the service request processing in the core network, would depend on the network configuration and implementation.

Additional procedures, not explicitly described above, can be incorporated into the model by increasing the time UE spends transmitting and receiving data over one DRX cycle. Step 4a includes the UE reading PDCCH, that is, receiving, and this time can be increased by the duration of the modeled additional procedure. Similarly, in step 4b, the time UE spends transmitting can be increased accordingly.

For example, if we want to model cell selection or reselection, we could add an extra 100 ms for UE to keep its receiver open (*active state*) to step 4a and some time for the UE to send uplink signaling and RRC messages (*transmission state*) to step 4b. The exact, explicit, procedure is not important as the result of any such procedure would be to extend the time UE spends communicating over the air interface. We will point out the effect of modeling some additional procedures later on.

It could be possible to optimize the procedures with regard to the time the UE spends in active state. For example, checking the SIB1 value tag for possible system information change does not necessary mean the UE needs to have its receiver on during 80 subframes. The interval of SIB1 transmission is 80 ms, and the UE only needs to have its receiver open for the actual reception of the SIB, not necessarily for the duration of whole 80 subframes. This could be seen as a form of "micro-DRX" during the SI reading procedure, leading to lower power consumption. Similar options exist for other procedures as well, such as allowing "micro-DRX" during the

RRC state change procedure, especially while waiting for the network to respond for the service request submitted during the procedure. We will use the values provided above for the RRC connection setup, thus implicitly assuming some "micro-DRX" functionality.

16.5.2   Way forward toward reduced power consumption

Now that we have a model in place, we discuss different options of actually making the UEs more power-efficient in the M2M context.

There are two clear paths that could be followed: either reduce the power consumption through hardware optimization, i.e., design platforms, which consume less power, or then modify the existing LTE procedures and protocols to make it possible for the UEs to consume less energy. Both of these paths could be divided into further categories and there are multiple points in both that could be addressed.

LTE UEs consist of many different functional parts and each of the part has different effect on the total power and energy consumption. If the LTE specifications were changed to allow simpler procedures for machine devices with smaller buffer sizes, the cost and power consumption of the digital baseband part, for example, could be lowered. Similarly, UE and chipset manufacturers can optimize their implementations of LTE protocol stacks to achieve even better power consumption figures.

In the following sections, we will study specific means to lower the power consumption for M2M communications.

16.5.3   Reducing tail energy consumption

For example, in Ref. [4], the authors identify "tail energy consumption" being one reason for relatively high measured power consumption in LTE smartphones. The tail refers to the time after the UE has finished its communication event but still stays in the connected mode before it is released to the idle mode. When this is combined with the fact that in smartphone implementations of the connected mode, the average power used can be several times the average power consumption in the idle mode, the result is "wasted" energy while staying in the connected mode waiting for a possible continuation of the communication. How long the UE actually stays on connected mode can be controlled by the eNB. The RRC connection can be released when it is not needed anymore, forcing the UE to go to idle mode.

This behavior of staying in connected mode is not necessary and will hurt the power efficiency especially for infrequent, small data traffic. Thus, a way to reduce the power consumption would be to introduce mechanisms in the network to deduce the traffic pattern and treat different services in a way to obtain good trade-offs between latency, signaling load, and power consumption. An additional way to reduce the tail energy consumption would be to introduce a method for the UE to indicate it would like to have fast transition to the idle mode after a transaction (e.g., similarly as fast dormancy is used in 3G W-CDMA / UTRAN networks). For example, for interactive applications, it is beneficial to stay in connected mode for a while to wait for possible continuation to avoid excessive signaling. On the other hand, periodic uploads of sensor data are finished after the initial upload; thus, it is not needed to stay in connected mode if the power consumption is higher in this mode compared to idle mode.

In our M2M model, we assume the device can turn most of circuitry off after communication and that the average power consumption in both idle and connected modes are similar. Thus, we do not actually see the tail energy problem using the presented power consumption model.

16.5.4   Extending the DRX cycle lengths

As it is already possible to use DRX in LTE for energy saving purposes, a promising way forward for further power savings would be to extend the length of the DRX cycles in connected mode and, similarly, the maximum paging cycle lengths in idle mode. The current LTE specifications limit the maximum paging and DRX cycle length to 2.56 s. This means that when the UE is attached to the network, it can only sleep for approximately two and a half seconds before it needs to wake up for reading the paging channel or PDCCH. Thus, currently, it is not possible for the UE to stay attached and sleep for extended periods of time. For mobile devices and current smartphone applications, the current behavior is desirable, and additionally, measurement procedures ensure that a UE will always be connected to a cell with reasonable signal strength and quality. However, for many M2M scenarios, this kind of behavior will only consume extensive amount of energy over the lifetime of the device.

The extension of the DRX cycle will have a cost in terms of delay. When the UE is sleeping, the network cannot reach the UE using paging but has to wait until the next paging occasion to reach the UE. The maximum cycle length thus depends on the reachability requirement of the UE. If the extended DRX cycles are to be used, the delay in reaching the UE has to be taken into account and balanced with the potential energy saving benefit. The suitable length of the DRX cycle depends on the traffic the UE is expected to receive.

Some advantages of the extended DRX solutions are the good configurability of the cycle length to traffic patterns and very good potential for energy savings. The UE reachability and power-saving gains can be balanced for each UE depending on the need.

16.5.4.1   Results: aggressive optimization assumptions

We study the best possible scenario with aggressive assumptions to see what is the ultimate technology potential of the LTE radio interface, given the hardware and state model described above.

Our aggressive assumptions are the following:

- Synchronization time is 10 ms.
- Paging or listening for downlink assignments is not considered at all—the UE only sends data in uplink periodically.
- The UE does not read any system information before sending data (i.e., the UE has received system information before and assumes it is correct).

• The UE stays in connected mode; thus, it does not perform the idle to connected transition. The UE needs to go through random-access procedure to obtain UL synchronization and scheduling grant for uplink.

In this scenario, the UE would have to perform the following reception events: 10 ms of synchronization, 1 ms of reading random-access response, and reception of the HARQ response for the sent uplink data packet. Reading the HARQ covers reception of the PHICH channel, which covers typically the first one or three OFDM symbols in a subframe. Thus, we use 200 μs as approximation for reading the HARQ response. In total, reception would thus take 11.2 ms.

We assume the UE sends a data packet that fits into one subframe worth of uplink resources. In addition, the UE sends the initial random-access preamble, which (for preamble format 0) takes 1 ms. Thus, the total uplink events take 2 ms in total.

In addition, the accurate clock needs to be on during the whole procedure. Let us assume the following: sync takes 10 ms and then a pause of 1 ms, after which the UE sends the random-access preamble (1 ms). After this, there is 4 ms in between transmission and reception, that is, 4 ms before random-access response (RAR), 1 ms of reading RAR, 4 ms preparing the UL packet, 1 ms for sending the packet, and finally 4 ms before reading the HARQ response after which the clock is turned off. In total, we would then have 27 ms for accurate clock on-time per DRX/paging cycle.

Using the values provided in the hardware model, we can then calculate the total energy consumption of one communication cycle as 11.2 ms × 100 mW + 2 ms × 300 mW + 27 ms × 10 mW = 1.99 mJ. If the UE sends a data packet every 10 min, then, over 1 year, the communication cycles would consume 365 × 24 × 6 × 1.99 mJ = 104.6 J. The base consumption would total to 365 × 24 × 60 × 60 s × 0.03 mW = 946.1 J. An off-the-shelf 1.5-V AAA battery with capacity of 6500 J would thus last for 6.2 years with our assumptions. If the base consumption would be halved to 0.015 mW, the lifetime would increase to over 11 years.

This is one example where the optimization of the base power consumption (from the low-power clock) shows its effect: When the actual traffic is rare and the UE procedures are optimized for it, it is very beneficial to try to reduce the sleep state power consumption as much as possible. Just reducing the power efficiency of transmissions (i.e., the TX branch) would have a more limited effect: Halving 300 mW to 150 mW would result in 1.69 mJ consumption per communication cycle, or 88.8 J per year; thus, the lifetime would be around 6.4 years. Similarly, the effect would be not very significant for reducing the accurate clock or active state power consumption, when we assume the highly optimized procedures.

In addition to highlighting the importance of optimizing the sleep mode power consumption, these results also indicate that the OFDM-based LTE radio interface technology has the potential to reach very long battery lifetimes.

16.5.4.2   Results: more realistic assumptions

As the starting point, we use the parameters introduced in the hardware and state model descriptions in Section 16.3.1 with a 6500 J AAA battery (this battery size is assumed from now on). Figure 16.3 shows how the average power consumption behaves when the DRX cycle length is increased and the UE stays in connected mode

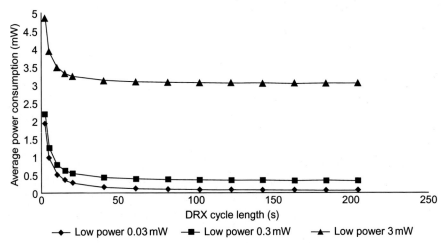

Figure 16.3 Example of average UE modem power consumption as the function of DRX cycle length.

after transmission. The interarrival time for uplink traffic is 60 min. In our model, the measurement requirements have been relaxed and the UE is assumed to perform the measurements during the active state. This could be a realistic way of operation for UEs that are static and under good radio coverage but requires that the UEs would be configured to perform infrequent mobility measurements. The results are shown for three different power consumption values for the low-power clock: 0.03 mW, 0.3 mW, and 3 mW. The corresponding increase in battery lifetime is shown in Table 16.2. These values show the effect of assuming different active/sleep power consumption ratios for M2M devices. The case corresponding to today's smartphones (i.e., 3 mW for base power consumption) would have only limited benefit from very long DRX cycles, while the other extreme, a highly optimized M2M hardware, would have a significant potential of extended battery lifetime.

With relatively rare uplink traffic, the effect of the actual data transmission in uplink is very low—in this case, optimizing the actual transmitter chain hardware would have a negligible effect on the power consumption as explained in the previous section. However, optimizing the downlink chain would this time have significant benefits, where the best gains would be achieved for shorter DRX cycle lengths. This can be seen, for example, by shortening the time required to achieve synchronization

Table 16.2 Battery lifetimes (AAA, 6500 J) corresponding to the average power consumption values in Figure 16.3

Low-power clock	DRX 2.56 s	DRX 61.44 s	DRX 204.2 s
0.03 mW	39 days	657 days	1268 days
0.3 mW	34 days	196 days	229 days
3 mW	16 days	24 days	25 days

or the time spent for reading the control channels for possible paging indication. For example, if the paging reading time would be 1 ms instead of 10 ms, the first row for 0.03 mW in Table 16.2 would be 49/765/1381 days, giving increase of 9–25% in lifetimes.

On the other hand, if we consider the UE to stay in idle mode between the transmissions, resulting in extra reception and transmission times through RRC signaling (step 3b), the first row in Table 16.2 would be 39/646/1228 days. Thus, in this case, there is insignificant change if we use idle more instead of connected mode. For shorter IAT, the difference between idle and connected modes would be larger.

If the DRX cycle length is increased, there will be some changes required to the RAN specifications in addition to just extending the cycle length value. In LTE, system frame number (SFN) is used to keep track of time, and today, the SFN wraps around after 10.24 s. In RAN WG2, some ideas to use, for example, GPS or UTC time, which are part of some system information messages, have been presented in addition to various ways of extending the actual SFN range.

Going beyond 2.56 s may result in the UE requiring to read the system information or to at least check SIB1 if the system information has changed during the time the UE spends in DRX. If the UE has to read SIB1 every time after waking up, this would actually result in increased power consumption with the longer cycle lengths. The actual system information may change only at specific radio frames determined by a modification period. The maximum modification period length is restricted by the SFN range; thus, currently, the UE needs to check for potential change in SI at least every 10.24 s, more frequently if the configured modification period is shorter.

An example of the effect of needing to read SIB1 after every wake-up beyond 10.24 second DRX cycle is shown in Figure 16.4. We see that in this example, the power consumption will actually be larger with 10.24 s than with the 2.56 s cycle. When going beyond 10.24, the power consumption will fall rapidly as a function of the DRX cycle length; thus, the problem does not exist for longer DRX cycles.

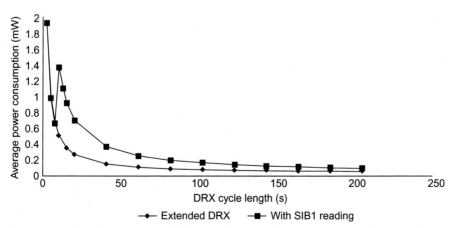

Figure 16.4 The effect of reading SIB1 every time with DRX cycles extended beyond 10.24 s. The average consumption increases at cycle length of 10.24 after which it declines again.

These results show that it is important to consider different system aspects when modifying the supported parameters ranges in order to not actually worsen the system performance. However, in this particular problem, a solution for M2M devices where SIB1 is not read every time would solve the issue. This could be a realistic option, as the system information is not expected to change very often.

For further study, some of the RAN WG2 contributions provide results and discussion on longer DRX. A good starting point is some of the papers submitted to RAN WG2 meeting #83, such as R2-132560 "Super power saving by very long DRX" by Qualcomm and R2-132517 "Power consumption analysis for extended DRX cycle" by MediaTek. The meeting contributions can be found at ftp://ftp.3gpp.org/tsg_ran/WG2_RL2/TSGR2_83/Docs/.

16.5.5 Power-saving mode

3GPP has not reached consensus on the necessity of the extended DRX for power saving, but something similar in the form of power-saving mode (PSM) will be specified in 3GPP Release 12. The power-saving mode solution gives the UE possibility for very low power consumption by keeping the UE attached to the network but staying in idle mode for long periods of time. The UE is required to perform periodic tracking area updates (TAUs) so the network can keep track of the UE's location (on a rough scale) and reach the UEs using paging after TAU. This TAU cycle duration is configured by the network and a periodic timer is run both in the network and in the UE.

The UE is not reachable while it is on the power-saving mode as it does not listen to paging or perform any other radio interface-related procedures (such as measurements). For TAU, the UE needs to perform random access and perform the necessary RRC signaling to enter connected mode. After the UE has performed the periodic TAU, and the UE has gone back to idle, there is a period during which the UE will listen to paging, configured by an activity timer (see Figure 16.5). During this time, the network can deliver possible downlink traffic intended toward the UE, received

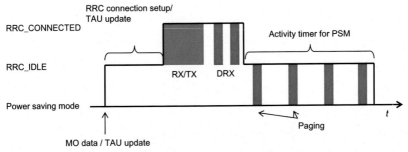

Figure 16.5 Power-saving mode depicted. When mobile-originating (MO) data are to be sent or the TAU timer expires, the UE first performs random access and RRC connection setup or the TAU procedure. Afterwards, the UE will be available for paging while the active timer is running. Note that if there is no additional UL or DL traffic, the UE does not need to stay in connected mode after TAU has been carried out.

when the UE was in the power-saving mode. Otherwise, the UE is not available for mobile terminating traffic. After the activity timer expires, the UE moves into the power-saving mode wherein the power consumption is as low as possible. Comparing to the power consumption model in Section 16.5.1, this would equal to having only the low-power clock on (note that in some power consumption models, there can be a separate deep sleep state for PSM but we have not modeled such explicitly).

Comparing the average power consumption and the battery lifetime with extended DRX, the power-saving mode solution has worse figures when the TAU cycle equals to the extended DRX cycle. This is because the added power consumption during the activity timer and the signaling required to perform the state changes between the connected and idle modes and to carry out the TAU procedure. For our modeling purposes, we assume that the TAU procedure includes 50 ms of extra reception and 5 ms of transmission time, thus corresponding to the UE entering connected mode in step 3b of the model on every TAU and going back to idle mode after the procedure has finished. In reality, the TAU procedure would take some additional time depending on the processing speed of the TAU request in the core network.

Figure 16.6 shows comparison between the power-saving mode and extended DRX. The power consumption of the low-power clock is 0.03 mW in this example. The interarrival time for uplink traffic is 60 min. For the power-saving mode, the activity timer is set to zero, that is, there is no paging after the TAU update and the UE goes to idle mode immediately. If we add some activity timer value, the figures will be worse for PSM in favor of extended DRX.

As an example comparison, using a TAU cycle of 2 min, the lifetime would be 555 days. Using extended DRX with 2-min DRX or paging cycle would result in

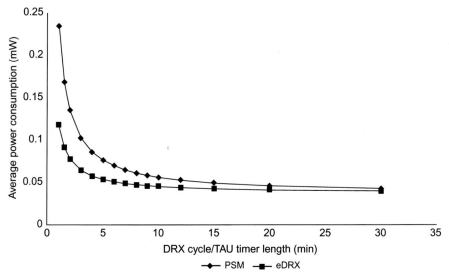

Figure 16.6 Power-saving mode and extended DRX compared. The average power consumption is similar with both solutions, when the TAU time and DRX cycle length are equal. The PSM consumption is a bit higher due to the extra signaling associated with the PSM.

1017 days for idle mode extended DRX and 1045 days for connected mode DRX. When the activity timer is set to 15 s with paging cycle of 640 ms, an evaluation would give lifetime of approximately 450 days when using the PSM. Thus, the cost for reachability using PSM would be extensive when compared to extended DRX cycle length in this case.

The PSM solution is preferred only for medium to long inactivity cycles as there is a signaling cost involved compared to the extended DRX solution. This signaling includes both RRC state changes and additionally the TAU updates, especially if the periodic TAU timer is short. Furthermore, the solution is mainly applicable for stationary or relatively slow moving UEs, as a fast-moving UE would update its location too rarely.

The advantages are that this solution has minimal impact on the RAN protocols and procedures and the main changes are on the core network side. The lifetime of battery-operated devices with suitable traffic pattern can be significantly extended, and having the options to use PSM in Release 12 specifications already enables power-efficient use of LTE for many IoT and M2M scenarios.

Further results and discussion can be found, for example, in RAN WG2 contribution R2-132794 "Further evaluation and way forward of selected UEPCOP solutions." The power-saving mode functionality for LTE is described in TS 23.682 and TS 23.401.[10]

16.5.6   Attach/detach

One method to save power in some M2M use cases is to simply detach the device from the network when there is nothing to communicate and then attach again when the need arises. When the device has detached from the network, it can turn itself off (i.e., turn off all 3GPP modem-related functionality) or use any other means it can to save energy.

This solution requires no changes to the current specifications. Existing procedures for both attach and detach can be used. When the UE is detached or turned off, it cannot listen to paging or perform cell selection or reselection related procedures, and the possible reattach happens solely based on the UE need to connect to the network (e.g., for uplink traffic).

One of the drawbacks of the attach/detach solution using existing mechanisms is the potentially high signaling load it causes, if many devices constantly perform these procedures in a cell or group of cells served by the same core network nodes. Also, the applications used by the devices should take into account this behavior, as the devices cannot be paged when they are detached. This leads to the UEs being unreachable, if they don't perform periodic attach/detach cycles.

---

[10]"Architecture enhancements to facilitate communications with packet data networks and applications," 3GPP TS 23.682 and "General Packet Radio Service (GPRS) enhancements for Evolved Universal Terrestrial Radio Access Network (E-UTRAN) access", 3GPP TS 23.401

16.5.7   Other methods improving power efficiency

For 3GPP Release 12, there have been discussions of a couple of other power consumption-related mechanisms as well. Core network-assisted eNodeB parameter tuning is one of them, where the eNodeB would be allowed to configure RRC connection and parameters based on information provided by the core network. The details, such as what kind of information the network provides, have not yet been completely agreed to.

Another proposed improvement is to introduce a MAC control element, which would directly send the UE to long DRX cycle. Currently, similar behavior is possible, but the UE first uses short DRX cycle (if configured) before changing to the long cycles. This functionality has been agreed to be introduced in 3GPP Release 12.

Additional power savings can be achieved by optimizing the radio protocols and processing, as has been hinted throughout this chapter. For example, optimizing the RRC procedures, random access, paging, and measurements so that the UE needs to spent less time having its receiver on would have noticeable benefits in the battery lifetime, as seen in some of the provided examples. Also, the network and RRC configuration options can be used to alleviate the power consumption to some extent.

16.6   Discussion and conclusion

16.6.1   Discussion

We have presented different options on how to attain lower average power consumption for various M2M scenarios. The basic idea is to extend the time the UE may spend in sleep mode—or equivalently reduce the time the UE spends in active state. Power-saving mode does this by disabling radio procedures for UEs spending long times in idle mode. Another tool to achieve this is the DRX mechanism, which can be used and whose parameters can be tuned to obtain good balance between the UE power consumption and reachability. The problem for M2M devices, especially battery-operated devices foreseen in many various Internet of Things scenarios, is that the desirable sleep time should be multiple of the current maximum DRX cycle. Moreover, the networks are often configured to keep the UEs in connected mode with short DRX cycle after transmission for possible continuation of the communications. This is desirable for interactive applications, but not for the typical M2M devices.

A preferable solution, as indicated by the use of the presented power model and the results in Section 16.5, would be to extend the DRX cycle to several times longer in comparison to the current maximum. Typical use cases would include, for example, sensors, remote tracking, and surveillance applications presented in Table 16.1. It would also be beneficial to keep the UE in connected mode, to avoid unnecessary signaling and also power consumption. The observed consumption in connected mode has been noted to be higher than in idle mode, but this is not a requirement dictated by the specifications, but more of a UE and network implementation and configuration issue. Thus, in practice, it would be beneficial to implement extended DRX in both idle and connected modes. The extended DRX cycle solutions have the benefit that the DRX cycles can be tuned to accommodate different scenarios with different requirements. Furthermore, with possible future additions and optimizations to the procedures, it could be possible to achieve extreme lifetimes using the LTE radio access technology as shown in Section 16.5.4.1.

The power-saving mode solution is a power-saving solution to be introduced in 3GPP Release 12 to be frozen in late 2014. In a way, PSM can be seen to be a form of infinite paging cycle, where the UE can initiate uplink transmissions at will, but the UE is not reachable by the network unless it performs a TAU. The benefit of this mode is that the UE can sleep beyond the maximum paging cycle, as it does not listen to paging while in PSM. However, there are applications where the network will want to page the UEs and control the reachability by setting a suitable DRX cycle value and not just rely on the TAU functionality of PSM. For example, remote tracking applications (Table 16.1) might require the UE to be paged once or twice per minute to determine the location. Similarly, some vital utility meters, such as gas meters, might require the utility company to be able to reach the device, for example, every 30 s. It is not viable to configure the UE to perform TAU every minute, especially as the TA is not expected to be changed so frequently. Excessive amount of messages related to TAU procedure causes unnecessary signaling load in the network. Further, the PSM means the UE goes to idle, before entering the PSM (while staying in idle from RRC point of view). There are added benefits of actually keeping the UE connected, but having longer DRX, as discussed above.

The attach/detach solution would work well for UEs that want extremely low power consumption and transmit only very rarely, such as sensors transmitting in uplink only few times a day. The drawback is that the UEs would be detached from the network while sleeping, the network cannot reach the UEs at all. Another downside is the signal load that is very high per transmitted bit, especially if the data are small in size.

For further reading on LTE power-saving functionality for mobile applications not necessarily restricted to MTC context, see, for example, Ref. [8].

16.6.2 Conclusion

As the envisioned future communication scenarios will have machines as the sole communicating entities, it is vital to research and evolve current wireless technologies in addition to other potential technologies enabling M2M communication. Power consumption aspects of any such technology should be carefully evaluated, especially as the devices, or machines, may not have constant power supply.

For 3GPP LTE, the energy efficiency of the radio modems in UEs can be increased by allowing more sleep time, that is, reducing the amount of time the modem needs to have its receiver (and transmitter) circuitry on. For the traffic types envisioned for many M2M scenarios, this is doable by supporting such behavior in the 3GPP specifications.

For 3GPP Release 12, some improvements in the energy efficiency have been agreed to be specified, such as introducing the power-saving mode. Some other options, such as extending the DRX and paging cycle lengths, could result in even greater gains but still need to be studied further and agreed to in the 3GPP working groups. The extended DRX solutions would allow most flexible configuration and be suitable for different M2M scenarios depending on the application requirements.

There is technology potential in using extended DRX cycles to reach extreme battery lifetimes (in the range of 10 years or so), but this would require additional work on allowing very optimized procedures and also modem hardware with lower power consumption figures.

Researching additional mechanisms for reducing the power consumption for M2M communications will remain an important and interesting problem. The 3GPP specifications will evolve over time, and it is expected that in the future releases, the power consumption aspects will be addressed further.

References

[1] C. Bormann, A.P. Castellani, Z. Shelby, CoAP: an application protocol for billions of tiny internet nodes, IEEE Internet Computing 16 (2012) 62–67.
[2] J. Puneet, P. Hedman, H. Zisimopoulos, Machine type communications in 3GPP systems, IEEE Commun. Mag. 50 (2012) 28–35.
[3] M. Olsson, S. Sultana, S. Rommer, L. Frid, C. Mulligan, SAE and the Evolved Packet Core: Driving the Mobile Broadband Revolution, UK, Academic Press, Oxford, 2009.
[4] J. Huang, F. Qian, A. Gerber, Z.M. Mao, S. Sen, O. Spatscheck, A close examination of performance and power characteristics of 4G LTE networks, in: MobiSys'12, June 25–29, 2012, Low Wood Bay, Lake District, UK, 2012.
[5] S. Sesia, I. Toufik, M. Baker, LTE—The UMTS Long Term Evolution: from Theory to Practice, Second ed., West Sussex, UK, John Wiley & Sons Ltd, 2011.
[6] C.S. Bontu, E. Illidge, DRX mechanism for power saving in LTE, IEEE Commun. Mag. 47 (2009) 48–55.
[7] T. Tirronen, A. Larmo, J. Sachs, B. Lindoff, N. Wiberg, Machine-to-machine communication with long-term evolution with reduced device energy consumption, Trans Emerging Tel Tech 24 (2013) 413–426.
[8] M. Gupta, S.C. Jha, A.T. Koc, R. Vannithamby, Energy impact of emerging mobile internet applications on LTE networks: issues and solutions, IEEE Commun. Mag. 51 (2013) 90–97.
[9] E. Dahlman, S. Parkvall, J. Sköld, 4G LTE/LTE-Advanced for Mobile Broadband, UK, Academic Press, Oxford, 2011.

Energy and delay performance of machine-type communications (MTC) in long-term evolution-advanced (LTE-A) 17

M. Gerasimenko, O. Galinina, S. Andreev, Y. Koucheryavy
Tampere University of Technology (TUT), Tampere, Finland

17.1 Introduction

17.1.1 Motivation and scope

Machine-type communications (MTC), also referred to as machine-to-machine (M2M) communications, has recently become a key communications technology. Industry reports indicate the huge potential of the MTC market, with millions of devices to be connected in the coming years and predicted revenues of up to $300 billion [1]. Moreover, since traditional voice service revenues continue to shrink, mobile network operators have become increasingly interested in MTC-based applications as the means to bridge the growing revenue gap [2]. Consequently, the European Telecommunication's Standards Institute (ETSI) has started new activities with the goal of defining an end-to-end MTC architecture [3], whereas emerging IEEE 802.16p proposals address enhancements for IEEE 802.16m technology to support MTC applications [4]. Our recent analysis [5] indicates that smart metering is likely to become one of the key use cases for MTC. Along with that, other important use cases in the areas of vehicular telematics, healthcare, and consumer electronics are being considered by the telecommunication industry.

Cellular technologies, such as 3GPP LTE and IEEE 802.16, are expected to play a pivotal role in enabling such applications. Reacting to this, 3GPP has defined several work items on MTC, primarily with respect to RAN overload control [6,7]. This is particularly relevant to smart-metering applications since (contention-based) random-access mechanisms in such cellular systems were not originally designed for the dense deployments that are typically encountered. The 3GPP services group is also interested in MTC-related improvements for LTE Release 12 within the context of mobile data applications [8].

17.1.2 Research background

In the last decade, the evaluation of RACH capacity for legacy 3G (CDMA) cellular networks, from both a simulation perspective and an analytic perspective, has been a popular research direction [9]. Originally, RACH served as an uplink

Machine-to-machine (M2M) Communications. http://dx.doi.org/10.1016/B978-1-78242-102-3.00017-4

contention-based channel to carry control information from client devices to the base station [10]. More specifically, the transmission of a random-access request from a network client can be decomposed into two stages. At the preamble transmission stage, a power ramping technique is used to adjust the transmit power to channel conditions (see the related analysis with respect to the blocking, throughput, and delay [11]). Next, a message is transmitted to the base station for the purposes of initial network access or bandwidth requesting.

The improved version of RACH within the 4G LTE-A system has also attracted significant research attention. For instance, the probability of successful transmission and the associated throughput of RACH were studied in [12]. An alternative approach to the throughput and access delay evaluation of RACH has been pursued as well [13], providing several options for enhanced RACH resource utilization. Many recent works also focus on performance analysis in overloaded RACH scenarios. For overloaded RACH scenarios, the number of contending devices per cell may reach the astonishing number of 30,000 (30K), as originally estimated by Vodafone [14], borrowed by some publications [15], and reused by 3GPP [16] to conclude on the expected device densities. Such high numbers of competitors may lead to prohibitive collision probabilities and quickly deteriorate system resources. Therefore, the 3GPP has recently been very active on evaluating the causes and consequences of such overloads. Reflecting initial discussions in 3GPP, identifying the key impact of RAN overload, some literature [17] has reviewed potential solutions and technology options to enhance the capability of LTE-A to handle numerous requests from MTC devices. Alternatively, other research works [18] have compared the two most likely (as per ongoing 3GPP discussions) candidate solutions for random-access preamble allocation and management.

However, previous work on RAN overload control has rather been a set of candidate proposals while 3GPP was actually evaluating those requiring minimal changes to LTE specification. More recent research [19] concluded on some of these efforts by detailing the officially approved 3GPP evaluation methodology produced within the work item on RAN overload control [16].

Due to the fact that the MTC devices are typically small-scale and battery-powered, accounting for their energy consumption is of paramount importance [20]. In what follows, we seek to augment a validated evaluation methodology fully compatible with the 3GPP test cases with an in-depth analysis of RACH performance in overloaded MTC scenarios. By including energy consumption in our framework together with the traditional performance metrics (such as access delay and success probability), we aim at providing a complete and harmonized insight into MTC device operation.

17.2 Technology background

17.2.1 Review of LTE-A signaling

The random-access (RA) procedure in 3GPP LTE-A is summarized in Figure 17.1. First, the user equipment (UE) sends a random-access preamble (Msg1) to the base station via the physical random-access channel (PRACH). Such message is randomly

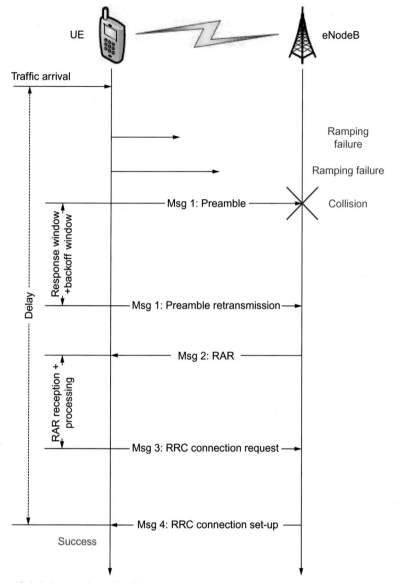

Figure 17.1 RA procedure signaling.

chosen out of the maximum of 64 available preamble sequences [21] (fewer preambles may actually be available, depending on the network configuration). A collision occurs at the base station when two or more UEs choose identical preamble sequences and send simultaneous requests. Preamble transmission may also fail due to insufficient transmission power.

If the preamble is correctly received, the base station (eNodeB) acknowledges it by sending a random-access response (RAR) message (Msg2) to the UE within the response window. The message sent over the physical downlink control channel (PDCCH) is actually a pointer to the actual resource in the physical downlink shared channel (PDSCH) where the full RAR can be received [22]. Furthermore, up to three pointers for different UEs can be aggregated into a single PDCCH message to further improve the efficiency of resource usage.

After processing the RAR message, the UE transmits an RRC connection request message (Msg3) via the physical uplink shared channel (PUSCH) using the resources granted by Msg2. The RA procedure ends with a successful reception of the RRC connection setup message (Msg4) from eNodeB.

The main bottleneck of the reviewed signaling procedure, especially when the number of UEs (and/or requests) is large, turns out to be the growing collision probability of the preamble (see Figure 17.2(a)). Besides, RAR delivery within the response window may also fail because of limited PDCCH resources (not considered in this chapter), and furthermore, there is also some probability of unsuccessful reception of messages Msg3 and Msg4. Clearly, all these events are leading to an increased network access delay.

17.2.2 3GPP evaluation methodology and system assumptions

As discussed in the previous sections, a comprehensive evaluation methodology for the LTE-RA procedure has been defined by the 3GPP [16]. The motivation is twofold: to identify the main system parameters to be used in simulations and to provide calibration data for a trustworthy comparison of the results obtained by various 3GPP member companies.

Table 17.1 summarizes the main system parameters and their settings [16,23]. Of particular importance is the so-called PRACH configuration index, which has a direct

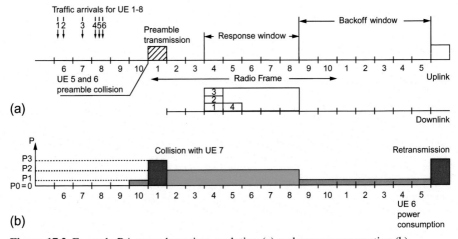

Figure 17.2 Example RA procedure: time evolution (a) and power consumption (b).

Table 17.1 Review parameters from several simulation methodology documents [16,23]

	Parameter	Value
–	Cell bandwidth	5 MHz
–	PRACH configuration index	6
s	Total number of preambles	54
L_1	Maximum number of preamble transmissions	10
–	Number of UL grants per RAR	3
–	Number of CCEs allocated for PDCCH	16
–	Number of CCEs per PDCCH	4
–	ra-ResponseWindowSize	5 ms
–	mac-ContentionResolutionTimer	48 ms
W_{max}	Backoff indicator	20 ms
π_3/π_4	Probability of successful delivery for Msg3/Msg4	0.9/0.9
L_3	Max. number of HARQ Tx for Msg3 and Msg4 (nonadaptive HARQ)	5
M	Number of MTC devices	5K, 10K, 30K
N	Number of available subframes for device activation	10K, 60K
b	Periodicity of PRACH opportunities	5 ms
K	RAR response window	5 ms
K_1	Preamble transmission time	1 ms
K_0	Preamble processing time at eNodeB	2 ms
t_{pr}	Processing time before Msg3 transmission	5 ms
t_{tx}	Time of transmission of Msg3, waiting, and reception of Msg4	6 ms
P_0	Power consumption in inactive state	0.0 mW
P_1	Power consumption in idle state	0.025 mW [24]
P_2	Power consumption of processing and Rx	50 mW [24]
P_3	Power consumption during Tx	50 mW [24]

impact in a number of system parameters of Msg1, namely, subframe numbers (where a UE can attempt preamble transmissions) or the preamble length. Likewise, the mac-ContentionResolutionTimer, which determines the maximum number of subframes the UE can wait after Msg3 transmission and before declaring a failure in the RA procedure, is an important parameter too.

The total number of (identical) MTC devices per cell is M (the definition of all the variables used in subsequent paragraphs can be found in Table 17.1). A device randomly chooses a subframe for its uplink activation following the uniform distribution (traffic type 1) or beta distribution (traffic type 2) over $[1, N]$. Whenever a device has data to send, it begins to contend with other such devices for the available PRACH resources. In particular, PRACH allocates periodic RA opportunities (at subframes $1, b+1, \ldots, b \cdot i+1, i \in Z^+$) for devices to send their randomly chosen preamble sequences (Msg1). A collision occurs whenever two or more MTC devices select the same preamble out of the s that are available. Otherwise, the transmission of the preamble sequence is successful with probability $1 - e^{-i}$ due to power ramping, where i is the transmission attempt index [16] and L_1 denotes the maximum number of attempts (the interested reader is referred to [25] for further details on the ramping procedure for power control). After K_0 subframes of pausing, which are needed to avoid intersection with the previous Msg2 reception period (response window), a new response window of size K starts (see Figure 17.2(a)). Within the response window, the eNodeB sends RAR messages (Msg2) in the subframe uniformly distributed over $[1, K]$. If the MTC device does not receive Msg2, that is, transmission fails due to the collision or insufficient power, then the associated devices restart the random-access procedure after a random time W, where W is the backoff timer, chosen uniformly within the backoff window size of W_{max}.

When the MTC device successfully receives the RAR message, it starts processing Msg3 for transmission. Next, it sends Msg3 and waits for $t_{tx} - 1$ to receive Msg4 (see Figure 17.1). Msg3 and Msg4 are successfully delivered with probabilities π_3 and π_4, respectively (as suggested in the 3GPP methodology document [16]). The maximum number of Msg3 transmission attempts allowed is L_3.

To conclude this section, we will briefly comment the considered traffic patterns. As the analysis in subsequent sections reveals, only traffic type 2 (beta distribution) actually causes overloads. This overloaded scenario, however, is difficult to evaluate analytically, and thus, we must resort to computer simulations (Section 17.4). On the contrary, performance in the presence of traffic type 1 (uniform distribution) can be accurately predicted with the analytic approach presented in Section 17.3. Still, traffic type 1 is primarily used for calibration purposes.

17.3 Analytic performance assessment

17.3.1 Delay analysis

In this section, we focus on the overloaded RACH scenario with traffic type 1 (uniform activation pattern) and conduct an analytic performance assessment in terms of, primarily, average network access delay.

We consider J different device classes with $M_1, \ldots, M_J$ devices in each. The total number of devices in the network is thus $M_1 + \cdots + M_J = M$. For the sake of analytic tractability, the traffic pattern at the particular device in a class j, $1 \leq j \leq J$, is replaced by stationary packet arrival flow characterized by a Bernoulli distribution with probability of arrival $\pi_j^{(0)}$ (a constant within the class). For uniform traffic, we have $\pi_j^{(0)} = 1/T_j$, where T_j is an activation time period.

We split the overall delay into two independent components associated with the processing of messages Msg1 and 2 and Msg3 and 4, respectively:

$$E[\tau] = E\left[\tau^{(1)}\right] + E\left[\tau^{(2)}\right] \tag{17.1}$$

In the expression above, $E[\tau^{(1)}]$ accounts for the expected time between the device activation and the reception of the RAR message; and $E[\tau^{(2)}]$ is the expected time elapsed between the end of the subframe where the RAR message was received and the completion of the processing of Msg4.

17.3.1.1   Expected delay of Msg3 and Msg4

The computation of the expected value of the random variable $\tau^{(2)}$ is straightforward and it reads

$$E\left[\tau^{(2)}\right] = t_{pr} + t_{tx} \cdot \bar{n}_3, \tag{17.2}$$

where $\bar{n}_3$ is the expected number of Msg3 and Msg4 transmissions addressed below, while parameters t_{tx} and t_{pr} denote the transmission and processing time, respectively. The expected number of Msg3 and Msg4 transmissions can be computed as follows:

$$\bar{n}_3 = \pi_{tx} \frac{\sum_{n=1}^{L_3} n(1 - \pi_{tx})^{n-1}}{1 - (1 - \pi_{tx})^{L_3}} = \frac{1}{\pi_{tx}} \frac{\left[1 - (1 - \pi_{tx})^{L_3}(1 + L_3 \pi_{tx})\right]}{1 - (1 - \pi_{tx})^{L_3}}, \tag{17.3}$$

where $\pi_{tx} = \pi_3 \pi_4$ denotes the probability that both Msg3 and Msg4 are successfully received and L_3 stands for the maximum number of Msg3 and Msg4 transmission attempts. Since the probability of unsuccessful reception $Pr\{\text{Msg3/Msg4 loss}\} = (1 - \pi_{tx})^{L_3}$ is negligible, at this stage, we do not consider retransmissions of preambles due to Msg3/Msg4 failures.

17.3.1.2   Expected delay of Msg1 and Msg2 without collisions

To analyze $\tau^{(1)}$, first we consider a random-access system without collisions. Hence, retransmissions are only related with the power ramping mechanism. In the case of successful preamble transmission at the first attempt, the service time consists of the preamble transmission time, the preamble processing, and the RAR response time. By also taking into account the average time elapsed between the device activation and the first preamble transmission attempt $b/2$, we have

$$E\left[\tau^{(1)} \mid \text{success at the 1st attempt}\right] = b/2 + K_1 + K_0 + (K+1)/2, \tag{17.4}$$

where K_1 is the preamble transmission time, K_0 is the pausing time, and K is the RAR response window size (in ms) as described above. Here, $(K+1)/2$ stands for the average RAR response time since we assume that the processing starts immediately after receiving the RAR response; it is obtained as the expectation of discrete uniform distribution over $[1, K]$.

As mentioned earlier, the probability of a successful preamble transmission at the ith attempt is given by $(1 - e^{-i})$. Further, we average the sum of the backoff time and additional waiting time until the next b-th subframe denoting the aggregate value as $\bar{w}$. The expected number of transmissions in the system without collisions can be calculated as follows (see "Appendix" for details):

$$\bar{n}_0 = \sum_{n=1}^{L_1} n \left(1 - \frac{1}{e^n}\right) \prod_{i=1}^{n-1} \frac{1}{e^i}. \tag{17.5}$$

Due to straightforward dependence of delay on the number of transmissions, we obtain the expected time before the beginning of Msg3 Tx in the system without collisions as

$$E\left[\tau_0^{(1)}\right] = (K_1 + K_0 + K + \bar{w})(\bar{n}_0 - 1) + \frac{b}{2} + K_1 + K_0 + \frac{K+1}{2}, \tag{17.6}$$

where $\bar{w}$ is the average waiting time between preamble retransmissions (see "Appendix") and $\bar{n}_0 = \sum_{n=1}^{L_1} n\left(1 - \frac{1}{e^n}\right)\prod_{i=1}^{n-1} \frac{1}{e^i} \cong 1.42$, for $L_1 > 4$. Clearly, $E[\tau_0^{(1)}]$ is a lower bound of the actual average delay for processing messages Msg1 and Msg2 (denoted by $E[\tau^{(1)}]$) that will be analyzed in the next section.

17.3.1.3   Expected delay of Msg1 and Msg2 with collisions

The analysis of a system with collisions constitutes a more challenging task, and an accurate solution is difficult to obtain due to the property of memory and system features such as random backoff time, constant timings, and, especially, large number of preambles. For example, in the classical multiuser system with one preamble and without power ramping, the approximate delay values can easily be obtained as has been done for ALOHA [26]. For our system, however, the use of that popular technique does not give a good approximation and we resort again to analytic approaches [27]. In order to abstract away the memory property and estimate $E[\tau^{(1)}]$ *with* collisions, we make the following assumptions:

1. We assume a Bernoulli activation flow with devices generating a new connection request per subframe with equivalent probability $\pi_j^{(0)} = 1/T_j$, where T_j is an activation time period in class j. The probability of at least one arrival in class j between two RA opportunities (b subframes) is $\pi_j = 1 - (1 - \pi_j^{(0)})^b$. We treat multiple requests as one, if the RA procedure has not been initiated yet.
2. If the first transmission fails due to a collision or insufficient power, the transmitting device is backlogged until it successfully transmits its preamble or exceeds the maximum allowed number of transmission attempts. We omit explicit consideration of the waiting interval and the backoff window and, instead, we assume that, at every preamble transmission (every b-th subframe), a backlogged device activates with probability $\pi_0 = b/(K_0 + K_1 + K + \bar{w})$. Basically, this means that the backlogged device activates once over $(K_1 + K_0 + K + \bar{w})/b$ preamble transmission opportunities.
3. The probability of successful departure is μ, that is, the request is served with a certain probability μ in the subframe of preamble transmission; otherwise, the device attempts to access the channel in the next available preamble transmission subframe.
4. Finally, we abstract away the maximum number of preamble transmission attempts in our analysis.

Within the simplified equivalent system model, an approximation of the mean network access delay can be obtained (see derivations in "Appendix"). Further, we consider the system being in the state $(i_1, \ldots, i_J)$, where i_k is the number of backlogged devices of type k, and θ_j is the steady-state probability of being in the state j (see formula (A.6) in the Appendix). According to that, the estimated expected number of preamble transmissions can be expressed as

$$\bar{n} = \sum_{(i_1, \ldots, i_J)} \theta_{(i_1, \ldots, i_J)} n_j$$
$$= \sum_{(i_1, \ldots, i_J)} n_j \binom{M_j - 1}{i_j} \rho_j^{i_j} \left(1 - \rho_j\right)^{M_j - i_j} \cdot \prod_{k=1, k \neq j}^{J} \binom{M_k}{i_k} \rho_k^{i_k} (1 - \rho_k)^{M_k - i_k},$$

(17.7)

where $\left\{ \theta_{(i_1, \ldots, i_J)} \right\}$ is the steady-state distribution, $\rho_k = \pi_k / \mu$ is the system load, and parameters μ, n_j are given by formulas (A.4) and (A.5) in the Appendix. Finally, expected service time can be readily computed as

$$E\left[\tau^{(1)} \right] = \bar{n} \cdot (K_1 + K_0 + K + \bar{w}) + \frac{b - K + 1}{2} - \bar{w}$$

(17.8)

17.3.2 Discussion

The analytic approach presented in the previous section is valid for stationary systems only. This, of course, holds for traffic type 1 (uniform distribution of the activation time). For traffic type 2 (beta distribution), however, one should take into account the dynamic changes of the parameters $\pi_j(t)$, $\pi_0(t)$, and $\mu(t)$, which is rather tedious. Instead, one can obtain an upper bound of the access delay by replacing beta distribution traffic by a uniform one with the same intensity.

However, our approach allows for a broad range of important practical extensions, such as overload control mechanisms, for example, initial backoff (see Section 17.5 for details). It will produce changes to the derivations above concerning the probability to avoid collisions, that is, the probability to collide should be set to $\pi_j \cdot \pi_0 \cdot 1/s$ (the probability to activate times the probability corresponding to backoff times the probability to select the same preamble; see details in the Appendix, Section A.4), for all inactive devices due to the device activation before its first transmission attempt. Here, all the activated devices are treated as backlogged starting from the activation moment.

17.3.3 Energy consumption analysis

Here, we enhance the analytic approach presented in the previous sections to also encompass energy-related metrics. To that aim, we consider four different power states of the MTC device (see Figure 17.2(b)):

1. P_0 —Inactive state. Here, power consumption is the lowest possible since there are no data to transmit.
2. P_1 —Idle state. The device is activated, but it does not transmit in the current subframe.
3. P_2 —Rx state. The device is either waiting for Msg2/Msg4 or processing those messages.

4. P_3 —Tx state. The device is transmitting Msg1/Msg3. Here, power consumption is the highest one.

Next, we compute the percentage of time spent in each of the aforementioned states. To recall, N stands for the total number of subframes for device activations. The percentage of time spent in the Tx state can be easily computed as

$$q_3 = \frac{1}{T}\left\{ K_1\bar{n} + \frac{1}{\pi_{tx}}\left[1 - (1-\pi_{tx})^{L_3}(1+L_3\pi_{tx}) \right] \right\}, \tag{17.9}$$

where $\bar{n}$ is an estimate of the average number of preamble transmission attempts, $K_1\bar{n}$ accounts for the time devoted to preamble transmission, and the second term in the summation accounts for the average number of Msg3 transmissions; T is an activation period for the devices of a certain class. Next, the percentage of time spent in the Rx state reads

$$q_2 = \frac{1}{T}\left\{ K(\bar{n}-1) + \frac{K+1}{2} + t_{pr} \right.$$

$$\left. + \frac{t_{tx}-1}{\pi_{tx}}\left[1 - (1-\pi_{tx})^{L_3}(1+L_3\pi_{tx}) \right] \right\}, \tag{17.10}$$

where $K(\bar{n}-1)$ is the time spent waiting for RAR messages, $(K+1)/2$ is the mean index of the response from eNodeB at the successful attempt, and the remainder corresponds to the processing and receiving of Msg3 and Msg4. Finally, the percentage of time spent in the idle state can be computed as

$$q_1 = \frac{1}{T}\left\{ \frac{b}{2} + K_0\bar{n} + (\bar{n}-1)\bar{w} \right\}, \tag{17.11}$$

where $K_0\bar{n}$ is the time for the eNodeB to process the preamble after its reception and $\frac{b}{2}$ is the average idle time between the activation and the beginning of the preamble transmission. The approximate average number of preamble transmission attempts is given by (17.7).

Bearing all the above in mind, the total energy expenditure of one MTC device can be estimated as

$$\epsilon = P_0(1-q_3-q_2-q_1) + P_1q_1 + P_2q_2 + P_3q_3 \tag{17.12}$$

In Figure 17.5, we present some analytic and numerical results (obtained with the simulator to be presented in the next section) on the power consumed by the MTC device with traffic type 1. Clearly, the analytic approximation is very accurate even for scenarios with a large number of devices.

17.4   Performance assessment via simulation

First of all, we provide an overview of our advanced protocol-level simulator of RACH operation. This simulator was developed to overcome the limitations of other existing simulation platforms such as NS2 (limited functionality) or OPNET (cost and complexity). And also due to the fact that, in general, those platforms were found to be inadequately slow for extensive simulations or simply lacking the necessary signaling support. Besides, one more advantage of our simulator lies in its flexibility in the choice of the parameters of interest, including the number of devices, signaling timings, and processing mechanisms, or system settings such as the number of preambles and backoff window size. More importantly, our simulation tool allows for simple integration of the extended components, such as overload control mechanisms and power consumption measurements. Finally, the software is supplied with flexible statistics collecting and processing functions that largely facilitate the evaluation of various parameters of interest ranging from access latency/probability to fine-grained energy-related metrics. All the messages transmitted over the same channel are multiplexed and processed jointly with explicit modeling of collision behavior.

In Figure 17.3, we depict a simplified block diagram of the simulator. There are three core classes implemented in C++: traffic generator, UE, and eNodeB. The traffic generator has support for three basic distributions, uniform, Poisson, and beta, which are configured for all the UEs at the beginning of a simulation run. Full-buffer (saturated) model is also available as a separate option. Each device has a dedicated traffic generator implementing the chosen traffic pattern.

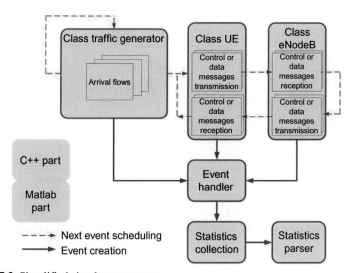

Figure 17.3 Simplified simulator structure.

The UE class supports operations related to the transmission of Msg1 and Msg3 and reception of Msg2 and Msg4. Other functionalities such as power ramping are implemented as well. The eNodeB class is responsible for the detection of Msg1 failures due to collisions or insufficient transmission power. After the detection procedure, a decision on whether to send Msg2 is made.

In our event-driven simulator, each event is processed by the event handler and can trigger another event of the same or different type. For example, a traffic arrival event triggers Msg1 transmission mechanism at the appropriate UE, which in turn schedules Msg2 transmission at the eNodeB if Msg1 has been successful. At the same time, a traffic arrival event causes the formation of another traffic arrival event at the same device based on the traffic arrival patterns discussed above. After Msg2 reception, Msg3 transmission is scheduled. This process is repeated until the successful reception of Msg4, which is enabling the statistics collector. Finally, simulation results are saved into a file that is delivered to a MATLAB parser for the purposes of visualization.

17.4.1   Calibration of the simulator

In order to validate our simulation tool against the 3GPP test cases, we conducted an in-depth calibration. In particular, we used the available reference data approved by the 3GPP in the technical report TR 37.868 [16], for traffic type 1. In Table 17.2, we compare the original results in TR 37.868 (rows labeled with "Methodology"), our simulation results ("Simulation"), and, for convenience, some analytic results ("Analysis") as per Section 17.3. A close match between all of them can be observed. Moreover, it appears that traffic type 1 does not cause significant network congestion (see specific values in Table 17.2) due to its low-intensity and uncorrelated arrivals. This is not the case for traffic type 2, as the next section illustrates.

In addition, the CDFs of initial network entry delays for 30K MTC devices (both uniform and beta traffic patterns) are presented in Figure 17.4. Finally, we calibrated our simulator with analytic predictions in terms of power consumption (in Figure 17.5) in the uniform traffic-based scenario.

Table 17.2 Calibrating with methodology

Number of devices	5K	10K	30K	Results origin
Collision	0.01	0.03	0.22	Methodology [16]
Probability (%)	0.01	0.03	0.23	Simulation
Number of preamble	1.43	1.45	1.50	Methodology [16]
Tx attempts	1.43	1.44	1.50	Simulation
	1.44	1.47	1.57	Analysis (17.7)
Access delay (ms)	25.60	26.05	27.35	Methodology [16]
	25.70	26.00	27.10	Simulation
	25.90	26.40	28.45	Analysis (17.8)

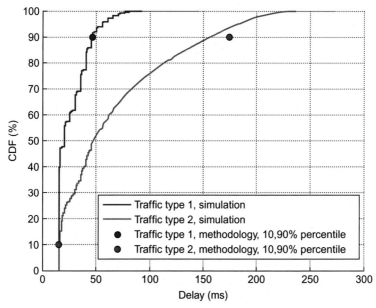

Figure 17.4 Access delay CDF for 30K MTC devices (simulation results).

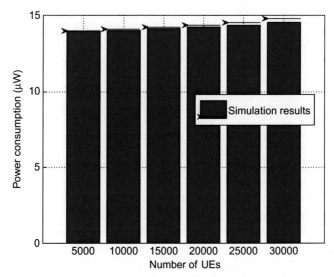

Figure 17.5 Simulation results for power consumption in a uniform traffic scenario; parameters are summarized in Table 17.1.

17.5   Numerical results

In this section, we assess the performance of a mechanism for RAN overload control in the presence of traffic type 2 (beta distribution) that, as discussed in Section 17.2.3, yields more correlated network access attempts. To that aim, we resort to the simulator presented in Section 17.4. This assessment nicely complements the *analytic* study conducted for traffic type 1 in the previous sections (to recall, the applicability of the analytic approach for traffic type 2, which is nonstationary, is very limited).

Our main interest here is to assess the feasibility of the *candidate* RAN overload control solutions (see relevant document [19]). Specifically, we consider the *simultaneous* use of an initial backoff [28] mechanism (e.g., prebackoff, when a device defers its channel access already before the first transmission attempt) and an MTC-specific backoff [29] (when MTC devices and UEs employ different backoff window values, with MTC-specific values typically being higher). The main idea is that the backoff time is invoked not only after any unsuccessful preamble transmission attempt but also at the very beginning of every RA procedure (prebackoff). By doing so, it is possible to decorrelate the surge in channel access attempts from many MTC devices at least for large enough values of the backoff indicator (BI). For non-MTC devices, the maximum value of the BI according to 3GPP specs [21] is 960 ms. For MTC devices, three additional BI values are allowed: 1920, 3840, and 7680 ms. By doing so, the probability of successful network access can be increased at the expense of an additional delay that can be acceptable for delay-tolerant MTC applications.

In Figure 17.6, we depict the probability of collision and the probability of successful access (after all retries) for a range of BI values. Interestingly, for BI = 960 ms (maximum value for non-MTC traffic), the probability of successful access is roughly 80%, which is barely acceptable for many MTC applications. For large values (those specifically designed for MTC), on the contrary, the probability of success virtually reaches 100%. Complementarily, Figure 17.7 illustrates how using higher BI values has a direct impact in the reduction of power consumption.

In summary, the simplistic overload control mechanism encompassing prebackoff and MTC-specific backoff alleviates congestion for traffic type 2. However, when designing overload control mechanism to handle the correlated network entry attempts, we should not negatively impact the regular MTC operation.

17.6   Conclusion and further research directions

In this chapter, we have discussed the lack of comprehensive evaluation frameworks for the performance assessment of the RACH mechanism within the 3GPP LTE-A technology. Moreover, previous evaluation results are often difficult to compare due to the fact that a unified 3GPP calibration methodology has only been finalized very recently.

To circumvent that, we have developed our own advanced protocol-level RACH simulator and validated it with the latest reference (calibration) data approved by 3GPP. On that basis, we have conducted an in-depth analysis of the case when the

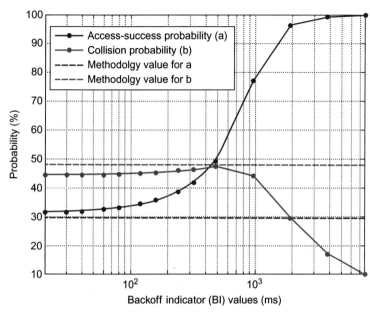

Figure 17.6 Overload control performance: collision probability (a) and access–success probability (b); simulation results, traffic type 2; parameters are summarized in Table 17.1 (methodology values are constant and do not depend on BI).

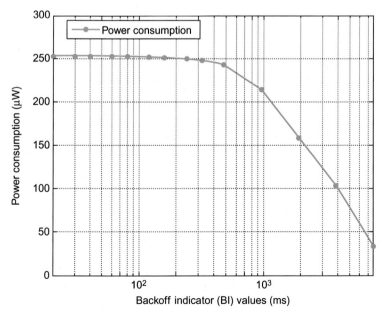

Figure 17.7 Overload control performance: power consumption (simulation results).

RAN is facing a surge in near-simultaneous network entry attempts from a large number of MTC devices. We have proposed a RAN overload control mechanism that does not require major protocol change and combines a prebackoff technique with the usage of larger MTC-specific backoff values.

In addition, we also found that the analytic approach presented in this chapter is a powerful tool that can be used to extend the 3GPP RACH calibration methodology [16]. One such improvement accounts for the power-related metrics of an MTC device to conclude on all the aspects of the random-access procedure, including its energy efficiency.

As far as future research directions are concerned, our analytic approach is also applicable for studying other MTC-related enhancements within LTE-A, such as sending scheduling requests via PUSCH, extended access barring (EAB) scheme [30], and extended wait timer (eWaitTime) mechanism [31]. Another challenging research direction is the consideration of scheduling-based approaches [32] for MTC, which can also be incorporated into our framework.

References

[1] Harbor Research Report, Machine-To-Machine (M2M) & Smart Systems Forecast 2010–2014, 2009.
[2] G. Wu, S. Talwar, K. Johnsson, N. Himayat, K. Johnson, M2M: from mobile to embedded Internet, IEEE Commun. Mag. 49 (4) 2011 36–43.
[3] ETSI, Machine-to-Machine communications (M2M); M2M service requirements, TS 102 689, 2010.
[4] H. Cho, J. Puthenkulam, Machine to Machine (M2M) Communication Study Report, IEEE 802.16p-10/0005, 2010.
[5] N. Himayat, S. Talwar, K. Johnsson, S. Mohanty, X. Wang, G. Wei, E. Schooler, G. Goodman, S. Andreev, O. Galinina, A. Turlikov, Informative text on Smart Grid applications for inclusion in IEEE 802.16p Systems Requirements Document (SRD), IEEE C802.16p-10/0007r1, 2010.
[6] 3GPP, System Improvements for Machine-Type Communications, TR 23.888, 2011.
[7] 3GPP, Study on enhancements for MTC, TR 22.888, 2012.
[8] 3GPP, Machine-Type and other Mobile Data Applications Communications Enhancements. TR 23.887, 2012.
[9] I. Vukovic, Throughput comparison of random access schemes in 3GPP, in: Vehicular Technology Conference (VTC), 2003.
[10] Y. Yang, T. Yum, Analysis of random access channel in UTRA-TDD on AWGN channel, Int. J. Commun. Syst. 17 (3) 2004 179–192.
[11] Y. Yang, T. Yum, Analysis of power ramping schemes for UTRA-FDD random access channel, IEEE Trans. Wireless Commun. 4 (6) 2005 2688–2693.
[12] I. Koo, S. Shin, K. Kim, Performance Analysis of Random Access Channel in OFDMA Systems, Systems Communications, 2005.
[13] P. Zhou, H. Hu, H. Wang, H. Chen, An efficient random access scheme for OFDMA systems with implicit message transmission, IEEE Trans. Wireless Commun. 7 (7) 2008 2790–2797.

[14] 3GPP, TSG RAN WG2.RACH intensity of Time Controlled Devices. R2-102296, 2010.

[15] A. Maeder, D. Staehle, P. Rost, The challenge of M2M communications for the cellular radio access network, in: 11th Wurzburg Workshop on IP: Joint ITG and Euro-NF Workshop, 2011, 2011.

[16] 3GPP, Study on RAN Improvements for Machine-type Communications. TR 37.868, 2011.

[17] S. Lien, K. Chen, Toward ubiquitous massive accesses in 3GPP machine-to-machine communications, IEEE Commun. Mag. 49 (4) 2011 66–74.

[18] K. Lee, S. Kim, B. Yi, Throughput comparison of random access methods for M2M service over LTE networks, in: International Workshop on Machine-to-Machine Communications, 2011.

[19] M. Cheng, G. Lin, H. Wei, Overload control for machine-type-communications in LTE-advanced system, IEEE Commun. Mag. 50 (6) 2012 38–45.

[20] S. Andreev, O. Galinina, Y. Koucheryavy, Energy-efficient client relay scheme for machine-to-machine communication, in: Global Telecommunications Conference (GLOBECOM), 2010.

[21] 3GPP, Evolved Universal Terrestrial Radio Access (EUTRA); Medium Access Control (MAC) protocol specification. TS 36.321, 2007.

[22] C. Johnson, Long Term Evolution in Bullets, CreateSpace, Northampton, 2010.

[23] 3GPP, Feasibility study for Further Advancements for E-UTRA (LTE-Advanced). TR 36.912, 2011.

[24] M. Dohler, J. Alonso-Zrate, T. Watteyne, Machine-to-Machine: An Emerging Communication Paradigm, Wireless World Research Forum, 2010.

[25] 3GPP, Evolved Universal Terrestrial Radio Access (EUTRA); Physical layer procedures. TS 36.213, 2011.

[26] L. Kleinrock, S. Lam, Packet-switching in a multi-access broadcast channel: performance evaluation, IEEE Trans. Commun. COM-23 (4) 1975 410–423.

[27] M. Sidi, A. Segall, Two interfering queues in packet-radio networks, IEEE Trans. Commun. COM-31 (1) 1983 123–129.

[28] 3GPP, Access barring for delay tolerant access in LTE. TSG RAN WG2 Meeting 74. R2-113013, 2011.

[29] 3GPP, Backoff enhancements for RAN overload control. TSG RAN WG2 Meeting 73bis. R2-112863, 2011.

[30] J. Cheng, C. Lee, T. Lin, Prioritized Random Access with dynamic access barring for RAN overload in 3GPP LTE-A networks, in: GLOBECOM Workshops, 2011, 2011.

[31] 3GPP, Discussion on the UE behaviour when receiving the eWaitTime in LTE. TSG RAN WG2 Meeting 73bis. R2-112202, 2011.

[32] Y. Elias, D. Zaher, Uplink scheduling in LTE systems using distributed base stations, Eur. Trans. Telecommun. 21 (6) 2010 532–543.

Appendix

Below, we provide some important calculations, which have been omitted from the main text for the sake of brevity.

A.1   Calculation of the average number of Msg3 and Msg4 transmissions

The distribution of the number of Msg3 and Msg4 transmissions is given as follows:

$$Pr\{n_3 = 1\} = \pi_{tx},$$

$$Pr\{n_3 = 2\} = (1 - \pi_{tx})\pi_{tx},$$

$$Pr\{n_3 = L_3\} = (1 - \pi_{tx})^{L_3 - 1}\pi_{tx},$$

where $\pi_{tx} = \pi_3\pi_4$ is the probability that both Msg3 and Msg4 are transmitted successfully (complementary, $(1 - \pi_3\pi_4)$ is the probability that either Msg3 or Msg4 is lost) and L_3 is the maximum number of allowed Msg3 and Msg4 transmission attempts. Here, we take into account only the successful transmissions. Since the loss probability $Pr\{\text{Msg3}/\text{Msg4 loss}\} = (1 - \pi_{tx})^{L_3}$ is negligibly small, at this step, we do not consider retransmissions of preambles lost due to Msg3/Msg4 failure.

Therefore, the expected number of Msg3 and Msg4 transmissions can be established as follows:

$$\bar{n}_3 = \pi_{tx}\frac{\sum_{n=1}^{L_3} n(1 - \pi_{tx})^{n-1}}{1 - (1 - \pi_{tx})^{L_3}} = \frac{1}{\pi_{tx}}\frac{\left[1 - (1 - \pi_{tx})^{L_3}(1 + L_3\pi_{tx})\right]}{1 - (1 - \pi_{tx})^{L_3}} \tag{A.1}$$

A.2   Calculation of the average number of Msg1/Msg2 transmissions (system without collisions)

The distribution of the service time $\tau_0^{(1)}$ in the system without collisions for Msg1/Msg2 can be given as follows:

$$Pr\left\{E\left[\tau_0^{(1)}\right] = \frac{b}{2} + K_1 + K_0 + \frac{K+1}{2}\right\} = \left(1 - \frac{1}{e^1}\right),$$

$$Pr\left\{E\left[\tau_0^{(1)}\right] = \frac{b}{2} + (K_1 + K_0 + K + \overline{w}) + K_1 + K_0 + \frac{K+1}{2}\right\} = \frac{1}{e^1}\left(1 - \frac{1}{e^2}\right), \dots$$

$$Pr\left\{E\left[\tau_0^{(1)}\right] = \frac{b}{2} + (n-1)(K_1 + K_0 + K + \overline{w}) + K_1 + K_0 + \frac{K+1}{2}\right\}$$

$$= \left(1 - \frac{1}{e^n}\right)\prod_{i=1}^{n-1}\frac{1}{e^i},$$

where $b/2$ stands for the time between the arrival and the beginning of the first preamble transmission attempt and $(K_1 + K_0 + K + \overline{w})$ is the component, which is added every time when transmission fails. The expected number of transmissions in the system without collisions may be calculated as follows:

$$\bar{n}_0 = \sum_{n=1}^{L_1} n\left(1 - \frac{1}{e^n}\right) \prod_{i=1}^{n-1} \frac{1}{e^i}$$

A.3 Calculation of the average time that the preamble is waiting for retransmission

We continue with calculation of the average time $\bar{w}$ that the preamble is waiting for retransmission. Let us find the time interval, during which the preamble is waiting for retransmission if the backoff window is zero. One may obtain the length of this interval as follows:

$$c_1 = \left\lceil \frac{K_0 + K_1 + K}{b} \right\rceil b - (K_0 + K_1 + K)$$

Furthermore, we note that if the backoff window is not greater than c_1, then the preamble is waiting for c_1. If the backoff window equals $c_1 + 1 \ldots c_1 + b$, the waiting time for the preamble is $c_1 + b$, and so on up to the maximum value $W - 1 = c_1 + k_0 b + l_0$, which corresponds to $c_1 + (k_0 + 1)b$ of waiting. Therefore, we may define

$$k_0 = \left\lceil \frac{W - 2 - c_1}{b} \right\rceil, \quad l_0 = W - 1 - c_1 - b\left\lceil \frac{W - 2 - c_1}{b} \right\rceil$$

Taking into account uniform distribution of backoff window size, we may find the average waiting time before transmission as

$$\bar{w} = (c_1)^2 + b(c_1 + b) + b(c_1 + 2b) + \ldots + l_0(c_1 + k_0 b + b)$$

$$= (c_1)^2 + bc_1(k_0 - 1) + \frac{b^2 k_0(k_0 + 1)}{2} + l_0(c_1 + k_0 b + b) \tag{A.2}$$

A.4 Calculation of the average number of Msg1/Msg2 transmissions (system with collisions)

Here, we describe the necessary details for the calculation of the average number of Msg1/Msg2 transmissions for the simplified system model introduced in Section "Expected delay of Msg1 and Msg2 with collisions", which takes collisions into account.

We continue by actually accounting for collisions. Let us consider one preamble transmission opportunity and assume that a particular (tagged) backlogged device of class j has generated a request and also selected a preamble.

In the state $(i_1, \ldots, i_J)$, for all $\sum_{k=1, k \neq j}^{J} i_k$ backlogged devices, the probability of accessing the channel and selecting the same preamble as the tagged has is $\pi_0 \cdot 1/s$ (the probability to activate times the probability to select the same preamble). For the inactive $M_j - i_j$ devices of class j, the corresponding probability is $\pi_j \cdot 1/s$ (the probability of arrival in a subframe times the probability to select the same preamble), and for other inactive devices, $\sum_{k=1, k \neq j}^{J}(M_k - i_k)$ of other classes is $\pi_k \cdot 1/s$, respectively.

Thus, the probability π_j^* to avoid collision for the tagged device of class j in the state $\theta_{(i_1, \ldots, i_J)}$ can be calculated as follows:

$$\pi_j^* = \left(1 - \sigma \frac{1}{s}\right)^i \left(1 - \pi_j \frac{1}{s}\right)^{M_j - i_j} \prod_{k=1, k \neq j}^{J} \left(1 - \pi_k \frac{1}{s}\right)^{M_k - i_k},$$

where i stands for the total number of backlogged devices $i = \sum_{k=1}^{J} i_k$.

Further, we account for the power ramping effect. The probability to avoid collision at the attempt n is given as follows:

$$Pr\{\text{1st successful}\} = \left(1 - \frac{1}{e}\right)\pi_j^*,$$

$$Pr\{\text{2nd successful}\} = \left(1 - \left(1 - \frac{1}{e}\right)\pi_j^*\right)\left(1 - \frac{1}{e^2}\right)\pi_j^*,$$

$$Pr\{\text{nth successful}\} = \left(1 - \frac{1}{e^n}\right)\pi_j^* \prod_{i=1}^{n-1}\left(1 - \pi_j^*\left(1 - \frac{1}{e^i}\right)\right),$$

Here, we also neglect all the lost preambles as we did before, averaging by successful transmissions and replacing the sought expectation with the conditional one. The average number of attempts can be obtained as

$$\overline{n}_{(i_1, \ldots, i_J)} = \pi_j^* \sum_{n=1}^{L_1} n\left(1 - \frac{1}{e^n}\right)\prod_{i=1}^{n-1}\left(1 - \pi_j^*\left(1 - \frac{1}{e^i}\right)\right) \tag{A.3}$$

Taking into account the effect of power ramping, we establish the probability μ_j of successful request i transmission:

$$\mu_j = \left(\overline{n}_{(i_1, \ldots, i_J)} \cdot (K_1 + K_0 + K + \overline{w}) + \frac{b - K + 1}{2} - \overline{w}\right)^{-1}, \tag{A.4}$$

where $\overline{n}_{(i_1, \ldots, i_J)}$ is given by (A.3). The estimated expected number of preamble transmissions may be obtained by averaging over all possible states:

$$\overline{n} = \sum_{(i_1, \ldots, i_J)} \theta_{(i_1, \ldots, i_J)} n_{(i_1, \ldots, i_J)} = \sum_{(i_1, \ldots, i_J)} n_{(i_1, \ldots, i_J)} \prod_{k=1}^{J} \binom{M_k}{i_k} \rho_k^{i_k} (1 - \rho_k t)^{M_k - i_k}, \tag{A.5}$$

where $\{\theta_{(i_1, \ldots, i_J)}\}$ is the steady-state distribution and $\theta_{(i_1, \ldots, i_J)}$ is the steady-state probability of being in the state $(i_1, \ldots, i_J)$.

In order to obtain the stationary distribution $\{\theta_{(i_1, \ldots, i_J)}\}$ defined above, we would need to consider all the state transitions and solve the corresponding matrix equation of dimension $M_1 \cdot \ldots \cdot M_J$. To reduce the complexity of such calculations, we omit more complicated transitions between the states and average $\theta_{(i_1, \ldots, i_J)}$, using binomial distribution, by

$$\theta_{(i_1, \ldots, i_J)} = \prod_{k=1}^{J} \binom{M_k}{i_k} \rho_k^{i_k} (1 - \rho_k)^{M_k - i_k}, \tag{A.6}$$

where $\binom{M_k}{i_k} = \frac{M_k!}{i_k!(M_k - i_k)!}$, and $\rho_k = \pi_k E[\tau^{(2)}]$ is probability to be in active, Rx or Tx state for the device of class k (in other words, not to be in inactive state), assuming that the service time is equal for all types of devices.

Part Four

Business models and applications

Business models for machine-to-machine (M2M) communications

18

J. Morrish
Machina Research, Reading, UK

18.1 Introduction

This chapter focuses on commercial aspects of machine-to-machine (M2M) solutions. Section 18.2 focuses on an overview of M2M from a commercial perspective, including the discussion of M2M as a "horizontal" capability that cuts across a range of different markets and applications. We also position the market for M2M as an incidental consequence of wider endeavors to improve products and services.

Section 18.3 outlines a brief history of M2M, including discussion of the roots of M2M back in the SCADA systems of the 1970s and 1980s and the current-day situation including new elements of the environment such as M2M/IoT Application Platforms. We also discuss the near-term future of M2M, including consideration of how our "connected" future will emerge.

In Section 18.4, we draw from Machina Research's published forecasts for M2M connectivity, including discussion of the total opportunity worldwide, the market sectors that drive growth, and key applications in the M2M space. We also discuss the role of alternative connectivity technologies and regional differences in the adoption of M2M.

Next, we outline at a high level the benefits of M2M, including cost efficiency, new business models, improved quality, and the potential to support pre-emptive action and also eco-friendly aspects of M2M technologies. We then move on to discuss the business models associated with M2M, including identification of funding parties and rationale for funding M2M solutions.

Finally, we discuss the concept of return on investment as applied to M2M, including the drivers of widespread adoption of M2M and the positioning of M2M as initially a competitive differentiator but ultimately a qualifier for participation in competitive markets. We conclude that the adoption of M2M connectivity across all aspects of our modern lives and in the infrastructure and products and services that support our lives is simply a matter of time.

18.2 An overview of M2M from a commercial perspective

Much of the content of this book is dedicated to the technical, platform, and standard considerations of M2M. These aspects are critical to the successful development and deployment of M2M solutions. However, from a commercial perspective, M2M is not

Machine-to-machine (M2M) Communications. http://dx.doi.org/10.1016/B978-1-78242-102-3.00018-6

about a technical solution. It is about applications that leverage connectivity, either to perform new functions or to perform old functions better. These applications could be stand-alone solutions such as a connected navigation device within a car, or they could be stitched into wider processes. An example of the latter category might be fleet management solutions that support job dispatch to a field force, ensure optimal routing for field force vehicles, and can also monitor driver behavior, potentially to stimulate a more fuel-efficient driving style or reduce insurance premium costs.

Ideally, from a commercial perspective, all of the complex (and necessary) technical considerations attendant on the specification and deployment of M2M solutions would be completely transparent to anyone looking to develop and deploy M2M solutions. In fact, from a commercial perspective, there is no such thing as a "market for M2M," rather M2M is a horizontal capability that may, or may not, play a role supporting any given product or service across multiple vertical markets. From a commercial perspective, the only consideration of any relevance is that some products and services are "better" than others. Coincidentally, some of these "better" products and services may leverage M2M technologies as part of an overall solution.

As the costs and other barriers to the deployment of M2M solutions fall, we can expect to find M2M solutions deployed in ever more diverse situations, supporting an ever-wider array of commercial (and other) solutions. But there is no aspect of our modern lives that will be immune in any way to the effects of M2M, and, conversely, M2M technologies may potentially find application in almost any aspect of our future society. The reality is that the "market" for M2M is extremely fragmented: as fragmented as all of the systems solutions in all industries worldwide today and more besides. But, from a commercial perspective, that "market" is a purely incidental consequence of seeking to make products and services "better" in some way.

18.3   A brief history of M2M

18.3.1   The roots of M2M

M2M connectivity is not a new concept. In fact, the roots of M2M extend back over several decades and include basic fleet management solutions and SCADA (supervisory control and data acquisition) solutions. Traditionally, M2M solutions have been conceived and deployed as "stovepipe" (or stand-alone) solutions with the aim of improving (or enabling) a specific process and without consideration of how these solutions might one day be integrated into a wider business context. In the case of some early fleet management solutions, the need was to improve the monitoring of high-value shipments (particularly deliveries of cash), often for the purposes of driver security. In the case of many SCADA deployments, the need has been to improve the oversight and control of a range of industrial processes, ranging from manufacturing production lines to nuclear power plants and from electricity distribution to wastewater collection and processing.

In these early days of M2M, any aspiring application developers would have had to essentially create an entire solution stack to support their intended application.

Relatively few of the components needed to develop specific solutions were available "off the shelf," and relatively few solution components could be shared between different solutions. In all, an early M2M developer would have had to contend with three key elements of an overall M2M solution:

- Devices: Consideration of the capabilities of an individual device, in terms of operating environment, use cases, sensing, actuating, processing, and human interface capabilities.
- M2M application environment: This includes a diverse range of considerations and restrictions relating to any applicable operating systems, carrier communication requirements, and the applicable capabilities of any potential M2M middleware providers.
- M2M application logic: As the element of the overall solution that actually solves the developer's problem, this is the piece the application developer would have actually wanted to build, although the functionality and flexibility of any specific application would have been hindered by the need to reflect any desired application capabilities through the application environment described above.[1]

This historic environment was clearly inefficient as it required applications to be built from scratch. This somewhat limited the potential of M2M applications. The harder it is to develop an M2M application, then the more expensive it is, and because most M2M solutions were justified on the basis of a "business case," so demand and addressable markets for M2M solutions were limited.

Such an environment provided fertile ground for M2M platforms, which are designed to enable the rapid development and deployment of M2M applications. M2M platforms have transformed the M2M market by making device data far more "accessible" to application developers and also by offering well-defined software interfaces and making APIs available so that application developers can readily integrate information sources and control parameters into their applications. Over the past decade, the M2M platform space has developed rapidly and now includes the following broad platform functions[2]:

- *Connectivity support:* This encompasses all of the most fundamental tasks that must be undertaken to configure and support a machine-to-machine connection. In a mobile environment, such tasks include connection provisioning, usage monitoring, and some level of support for fault resolution.
- *Service enablement:* Defined from a telecoms perspective,[3] service enablement platforms are flexible and have extensive horizontal capabilities in terms of solution support, reporting, and the provision of a software environment and APIs to facilitate solution development. Together, connectivity support and service enablement functions represent the "horizontal" elements of the M2M platform industry.

---

[1] For instance, it simply is not possible to build capabilities into an M2M application without building corresponding capabilities into M2M middleware so that appropriate information is made available to the M2M application.

[2] Often, these platform types will correspond to the market positioning of participants in the M2M platforms sector, but the relationship is often not clear-cut: the sector is far more complex than these five bullet points might suggest.

[3] Since service providers, systems integrators, and many platform vendors would apply the term "service enablement" to industry specific solutions. Today's service enablement platforms (as defined here) typically aspire to support full M2M applications, but generally only play a supporting role.

- *Device management:* Device management platform functionality has typically been aligned to single-device manufacturers and will potentially support devices connected through multiple networks. Device management platforms essentially exist to facilitate sales of devices (and device-centric solutions) and usually for a single vendor, where those devices typically require some form of nonstandard systems support (reporting, management, etc.).
- *Application support:* Application support platforms will typically be targeted at niche customer sectors. They are characterized by the provision of tailored and often specialized solutions, encompassing connected devices potentially of multiple types, potentially connected with multiple technologies, and potentially connected to the networks of multiple CSPs. Axeda's work in the healthcare vertical is a classic example.
- *Solution provider:* Platforms used by systems integrators to support turnkey and client-specific solutions. In most cases, a solution provider M2M platform should be regarded as an enabler for a wider systems development initiative, rather than as a stand-alone offering.

It should be noted at this point that none of these broad platform types specifically addresses the historic stovepipe nature of the M2M market: they simply aim to speed the development of the same kinds of stovepipe applications as have historically been developed.

Of course, while all elements of the functionality described above are required for any specific M2M application, it is not always necessary for such functionality to be provided by a third-party platform provider. Larger users of M2M may often decide to bring higher-level functionality in-house, on the basis that this gives more control over a solution and at relatively little cost in terms of lost scale. Smaller-scale users of M2M solutions would tend to be more likely to make use of outsourced platform solutions of different types.

18.3.2 The present

While the platform market structure as described above is broadly fit for purpose for current-day M2M applications, it is not fit for purpose for emerging Internet of Things (IoT)-type applications. The reason is that IoT-type applications must draw on (and control) a range of data inputs from diverse sources.

Historically, M2M applications have been generally stovepipe in nature: M2M connections have been deployed in support of a specific M2M application, with little consideration of how any specific M2M application might potentially integrate with other applications (unless such functionality is specifically designed-in). Broadly speaking, leading-edge M2M applications have evolved over time from simple device-centric solutions (of varying degrees of sophistication) to process-centric solutions (again, of varying degrees of sophistication) to the point where the most advanced developers are seeking to combine data streams drawn from a range of diverse data sources (not necessarily all M2M-connected) and ally this with flexible control capabilities. We term this our hierarchy of M2M applications and it is illustrated in Figure 18.1.

This emerging need to analyze information (both core application information and "data exhaust") from a diverse range of applications, and also to control those

	Stage	Description	Comments
Device centric M2M	1	Reactive information	• Devices can be polled for information, or provide information according to a set timetable
	2	Proactive information	• Devices communicate information as necessary
	3	Remotely controllable	• Devices can respond to instructions received from remote systems
	4	Remotely serviceable	• Software upgrades and patches can be remotely applied
Process centric M2M	5	Intelligent processes	• Devices built into intelligent processes
	6	Optimised propositions	• Use of information to design new products
	7	New business models	• New revenue streams and changed concept of 'ownership'
	8	The Internet of Things	• Publishing information for third parties to incorporate in applications, control commands from diverse sources

Figure 18.1 Machina Research's hierarchy of M2M.
Source: Machina Research, 2013.

applications, has given rise to the need for a new kind of M2M platform: the M2M/IoT Application Platform. The key function of an M2M/IoT Application Platform is to abstract across a diverse range of data sources and also more traditional M2M platforms, so enabling application developers to efficiently create solutions that incorporate, and draw from, multiple M2M devices (and other information sources). In essence, the emergence of M2M/IoT Application Platforms enables the open application development environment that presages the emergence of the Internet of Things, including multidomain applications and mash-ups.

18.3.3 The future

The future of M2M must also embrace near-field communications (NFC). To be clear, at this point, I am referring to near-field communications as a concept, rather than NFC as a technology. Right now, the overwhelming focus in the area of NFC is around payment solutions and NFC as a technology. But that is missing the point. There are many, many, more things that can be done with NFC technologies, not least support for identity verification and bootstrapping for higher-level technologies such as Wi-Fi and Bluetooth. And there is a wide range of technologies that can support NFC applications, even including, and for some applications, passive RFID and printed codes.

Ultimately, connectivity will become engrained in our daily lives and it is an individual's interaction with their local environment and the way in which that local environment reacts to the presence of that individual that will characterize our "connected" future. The concepts that will support and define our interaction with this future connected world will be the Internet of Things and big data, both of which, unfortunately, lie outside the scope of this book.

18.4   The potential for M2M

This section focuses on the potential for M2M, which Machina Research defines as "Connections to remote sensing, monitoring and actuating devices, together with associated aggregation devices." Our forecasting approach is set out in Box 18.1 for background information.

Overall, we anticipate that there will be around 18.3bn M2M connections worldwide by 2022, growing from a total of around 3.2bn at the end of 2013.[4] It is worth

---

Box 18.1   Machina Research forecasting approach

The Machina Research Forecast Database is a comprehensive guide to the global M2M market opportunity. Our segmentation of the market is based on 12 M2M sectors, plus a thirteenth non-M2M sector 'PCs, tablets, and handset data.' Within each of these 13 sectors there is a diverse set of devices, applications and services. As a result we base our analysis on individual applications, of which there are over 180 covering all of M2M. We then roll these applications up into application groups, of which there are 61.

It is almost impossible to give a definitive methodology for the Forecast Database. Each of the application groups has different dynamics requiring particular attention. As a result, the methodology for each application group will be completely different. However, there are some general things that we tend to take account of when compiling each of the application group forecasts, including:

- Current adoption
- Regulation
- Demographics
- Sector-specific statistics
- Service deployment plans
- Value chain positioning
- Technology availability
- Evolving bill of material costs

Once we have built this bottom-up forecast we then undertake a process of tallying this with available figures from operators, regulators and other industry bodies around the world. This is specifically relevant to wireless wide-area network connections. This way we can ensure that our bottom-up picture of the current market tallies with our understanding of the total size of the market globally and on a country-by-country basis.

Source: Machina Research 2013.

---

[4]Mostly local area network infrastructure, connected security alarms, audiovisual sources and displays, smart metering, and personal multimedia devices.

noting that growth is forecast to be strong throughout the period to 2022, and, in particular, the forecast growth in 2022 is around 14%. Also worth noting is that the total number of connected devices in 2020, or 2022, is often put forward as a figure to illustrate the growth potential for M2M, this approach is in fact misleading since growth rates in the early 2020s are generally forecast to be strong. Ultimately, there will be far more M2M connected devices than the 18.3bn that we forecast for 2022, and the impact that M2M will have on society and our daily lives will be far greater than that figure might suggest. Figure 18.2 illustrates the growth in the total number of M2M connections worldwide through to 2022.

In terms of the verticals and sectors that comprise this figure of 18.3bn connections by 2022, then the main contributors are consumer electronics and intelligent building applications. This illustrates the importance of the absolute size of addressable markets in M2M. Consumer electronics and intelligent building devices are purchased by individuals and households, and there may be several of any specific kind of device per household. It should not be surprising therefore that such applications are expected to comprise the bulk of M2M connected devices in the future. Conversely, consider the case of connected ambulances. A connected ambulance is a very compelling proposition, allowing hospital accident and emergency facilities to be prepared for the arrival of a patient and to begin analysis of that patient's vital readings. However, it is estimated that there are only around 48,000 ambulances in the entire United States. Clearly, connected ambulances are never likely to feature prominently in any analysis of the total number of M2M connected devices in any market. Much the same logic applies to many "industrial" M2M applications, and this further underlines the degree to which M2M markets are fragmented. Figure 18.3 illustrates the share of the total connections that is represented by some of the most significant sectors.

Unsurprisingly, the M2M applications accounting for the greatest number of M2M connections in 2022 can be expected to be driven mainly by adoption by consumers in

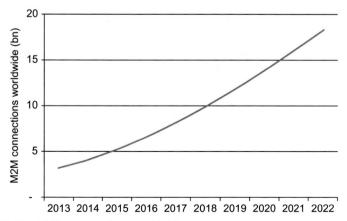

Figure 18.2 Total M2M connections worldwide.
Source: Machina Research, 2013.

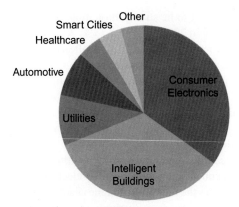

Figure 18.3 M2M connections by sector, 2022.
Source: Machina Research, 2013.

their homes: building automation, audiovisual sources, security, audiovisual displays, smart metering, and network infrastructure. Many of these M2M applications share a similar topology, including a single hub (or control) device with multiple associated subsidiary devices. For instance, a home alarm is composed of a single control unit with multiple associated sensors. The contribution of these identified M2M applications to the total estate of connected devices in 2022 is illustrated in Figure 18.4.

The profile of M2M connections highlights the degree to which many M2M connected devices can be expected to be found in homes and other buildings where short-range connectivity is often the preferred choice. Between them, short-range connectivity technologies such as Wi-Fi, ZigBee, and Powerline can be expected to support around 73% of M2M connections in 2022, mostly due to the cost advantages of short-range technologies when compared to wide-area technologies. Cellular connectivity can be expected to support a significant number (14%) of connections, due to its ubiquity, potential to support out-of-the-box connected experiences, and, not least,

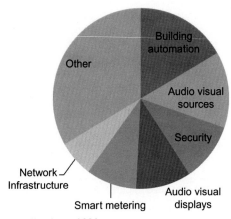

Figure 18.4 Key M2M applications, 2022.
Source: Machina Research, 2013.

ability to support mobile M2M applications (such as vehicle-based solutions). Metropolitan area network (MAN)-style connections can also be expected to be significant, mostly driven by the use of powerline communications and also a range of dedicated wireless solutions for smart metering. The profile of M2M connections by technology in 2022 is illustrated in Figure 18.5.

In terms of the regional distribution of connections, then by 2022, the Emerging Asia Pacific can be expected to be the biggest region, followed by Europe and North America. While North American and European regions will still clearly be relatively more "developed" than Emerging Asia Pacific in 2022, per capita wealth can be expected to be higher and the propensity to adopt M2M connected solutions can be expected to be greater, as highlighted in Figure 18.6, these effects are outweighed by the sheer weight of population in emerging Asian markets.

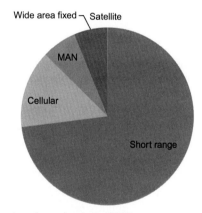

Figure 18.5 M2M connections by technology, 2022.
Source: Machina Research, 2013.

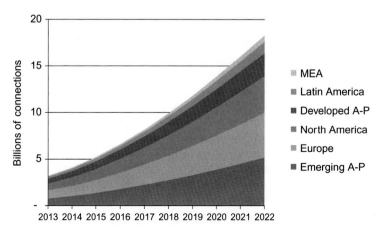

Figure 18.6 Regional differences in M2M adoption. A-P, Asia Pacific.
Source: Machina Research, 2013.

18.5   The benefits of M2M

The benefits of M2M are manifold. As I stressed in the opening section of this chapter and from a commercial perspective, the "market" for M2M is a purely incidental consequence of seeking to make products and services "better" in some way. I also described how fragmented the "market" for M2M will be. The diversity of M2M solutions that will be stitched into a range of commercial (and other) processes is limited only by human imagination. It is therefore somewhat of a challenge to do justice to a discussion of the benefits of M2M within just one chapter of a book. However, broad categories of benefit flowing from M2M connectivity include the potential for M2M to:

- reduce the cost of specific processes and enable more efficient operations,
- enable new business models,
- improve quality,
- enable pre-emptive action,
- save the planet.

One of the main benefits flowing from M2M, and one of the main drivers for the development and deployment of M2M solutions, is the fact that M2M solutions can reduce costs and enable more efficient operations. In cases where M2M reduces cost, then this is often due to the M2M solution displacing human effort: Processes are automated, and associated employment costs are reduced. It is no surprise, therefore, that these kinds of M2M solutions will tend to be adopted more rapidly in developed markets where labor is more expensive (given that technology and hardware costs tend to vary less between developed and developing markets). Another example is fleet management, where M2M solutions can allow for real-time optimization of delivery truck routes, reducing not only associated employment costs but also overall fleet costs including vehicle-related costs.

M2M solutions can also assist suppliers of products and services to refine their product offerings so that they are appropriately engineered for the relevant target market. For instance, currently, many manufacturers of washing machines would overengineer most aspects of their machines given expected usage profiles as a kind of insurance policy: no matter how the machine is treated by an end user, it is unlikely to fail within a certain time frame. However, with the advent of M2M, manufacturers may be able to better monitor the use of their machines in the field and so focus their investments on the components that are most likely to wear and potentially save money by not overengineering other aspects of a machine. The costs of servicing such a machine could also be reduced if a manufacturer can remotely interrogate the machine to understand why any given fault had occurred.

The same washing machine manufacturer may also be able to turn a monitored device into a new business proposition: washing as a service, charged for "per wash" and with remote monitoring and regular servicing to guarantee service levels. It may take some time before such business models take off in the world of consumer white goods, but they are already very much a reality in industry. Wet leases have been prevalent in the aerospace industry for many years, and manufacturers of jet engines will often lease those engines on the basis of usage. As the costs and other barriers to the adoption of M2M solutions come down, so these same business models can be applied to more and more assets. Right now, M2M connectivity is being adopted to support shared and pooled-car schemes. In the future, similar concepts will be applied to ever-cheaper assets.

In many situations, M2M solutions can also improve the quality of monitoring and so the quality of decisions taken on the basis of the corresponding data. The most compelling and stark example of this is in the case of medical patient monitoring. Patients that need to monitor their own vital signs have a habit of doing so in a way that is inconsistent over time, and often, patients will "fake good" in order to proffer "better" results to their doctors. Collecting equivalent data automatically, via an M2M solution, increases both the quality and consistency of data and allows medical professionals to take better decisions on the basis of those data. The medical case is an extreme example, but the potential of M2M monitoring to improve quality manifests itself in many different situations, ranging from the monitoring of wastebins to ensure that they are only emptied when full to the closer monitoring of livestock increasing yield, reducing the prevalence of disease, and benefiting animal welfare.

The potential for M2M to enable pre-emptive action is closely related to many of the use cases described in the last paragraph. When complemented by data mining and sophisticated data analytics, the data provided by an M2M connected solution can yield results that enable service providers to intervene before the devices that are being monitored break down. For instance, a particular kind of vibration sensed in a wind turbine gearbox may presage a gearbox failure. Equally, information provided by a vehicle platform may enable the manufacturers of that vehicle to recommend that it be taken for a remedial service. The most compelling example of M2M enabling pre-emptive action is probably in the sphere of medical monitoring, where remotely monitored patients with, for example, heart disease can be instructed to attend hospital before a coronary event occurs. Admitting a heart patient to hospital before a heart attack is far cheaper and less stressful for the patient than would be an emergency admission after a coronary event. Indeed, the vast majority of the benefit for clinical remote monitoring tends to derive from the avoided need to wrap an expensive accident and emergency facility around a patient after something has gone wrong.[5]

And M2M can help save the planet. Reduced resource usage is a corollary of all of the benefits that have been discussed in the preceding paragraphs. The less resource we use in the course of our daily lives, generally, the better that is for the planet.

However, even given all the potential benefits listed above, it is important to recognize that M2M is just one component of an evolving technological landscape. Solutions that leverage M2M must evolve in-step with the contemporaneous environment, including consideration of related technologies, systems, and markets. Again, the key to understanding M2M is to recognize that it is just one element of an armory of techniques that appropriately motivated people might leverage to make products and services "better."

---

[5]The reduced costs of ongoing monitoring supported by M2M tend to be offset by the fact that remotely monitored patients live longer. But avoided emergency hospital admissions are a clear win.

18.6   Business models for M2M

The most fundamental point about M2M business models is that connectivity is irrelevant: it is what service providers can do with connectivity that matters. End users do not buy or deploy "connected" solutions simply because they are connected; they opt for "connected" solutions because they are better propositions in some way than equivalent nonconnected solutions.

Secondly, it is worth recalling my assertion from earlier in this chapter that M2M connected solutions will ultimately pervade all aspects of our lives, including the environments with which we interact and also every aspect of the economic and other activities that underpins our daily lives. Really, it is better to think of M2M as being a next level of capillarity of systems integration: a technique that assists with extending the reach of current IT systems and connected environments into the field. M2M will become all pervasive and can be expected to become integrated into every aspect of our society and daily lives.

Accordingly, any discussion of business models for M2M must be somewhat unbounded and open-ended. M2M will pervade all aspects of our future lives, and, in every instance, the business model associated with any specific connection will reflect how that connection can support an application that is somehow beneficial to the extent that somebody might be motivated to pay for it. It is hard to escape the conclusion that complexity and fragmentation of business models for M2M must at least reflect the complexity and fragmentation of every aspect of our modern society. In fact, business models for M2M will be more complex and fragmented, since M2M connectivity ushers in the potential for new business models.

However, it is possible to identify a relatively few underlying business models that will be particularly relevant for M2M. These include the following:

*   M2M solutions purchased by an end user as a fundamental part of a product or service
*   M2M capabilities funded by an end user as an optional add-on to a product and/or service
*   M2M solution funded by a content provider or other suppliers
*   M2M solution funded through increased efficiencies

The first of these business models is the simplest. End users (either consumer or nonconsumer) purchase a connected solution outright (either directly or with the assistance of some form of financing). Use cases include devices where connectivity is a fundamental aspect of a connected device (e.g., a media server) or where connectivity adds significant value to a device (e.g., in the case of a connected games station). This business model also includes use cases where connectivity is included as a secondary consideration (e.g., Nest's thermostat that was primarily designed to be a user-friendly thermostat, with connectivity essentially as a secondary consideration). In all these cases, connectivity costs, including hardware, any costs of capacity, and also costs of cloud and server infrastructure, are simply funded by the end user. A relatively minor tweak on this business model sees an end user funding the purchase of an M2M connected device through regular payments or a subscription to a related service.

A closely related business model sees M2M capabilities funded by an end user as an optional add-on to a product and/or service. Examples include laptops that are distributed with embedded mobile connectivity that can be activated by an end user simply entering payment details. Clearly, in this case, the underlying product would work perfectly acceptably in the absence of the potential M2M capability (mobile connectivity, in this case), but end users are easily able to avail of that capability should they so wish. Revenues from end users that do activate such capabilities effectively cross subsidize the increased device costs for those that do not. In fact, in the example above and potentially independently of whether the connection has actually been activated or not, laptops with embedded mobile capabilities will communicate a stream of useful information to mobile network operators and laptop suppliers, including where a laptop is typically used. This is potentially useful since, for example, the owners of laptops that are often used in the same location may represent opportunities to sell larger external screens. This brings us neatly to the third business model listed above.

M2M solutions can also be funded by content providers or other suppliers, for instance, in the case of the sale of a book to a Kindle user, where part of the revenue for the book is used to cover data carriage fees and also maintenance of the Kindle storefront environment. This business model can also be applied to physical goods and also services. For instance, pharmaceutical companies may choose to monitor remotely the patients who have been prescribed with drugs using connected medication dispensers. Ultimately, the funding for such an initiative may come either through increased sales of drugs to patients who are more likely to adhere to their prescriptions or through the increased efficacy of more closely monitored treatment regimens resulting in an increased propensity for physicians to prescribe drugs from a particular pharmaceutical company. This brings us to the fourth business model listed above.

In a business environment, M2M solutions are often funded on the basis of increased efficiency, particularly operational cost savings and particularly in developed markets (where labor is more expensive). In fact, it would be more accurate to say that in a business environment, M2M solutions are funded if the business case "costs in." That is, if the expected benefits outweigh the expected costs. Clearly, any potential for increased revenues or decreased risk would also figure in such an analysis. For example, the deployment of a usage-based insurance (UBI) device by an automotive insurance company could be expected to reduce costs through the enforcement of better driving practices. However, it could also be expected to result in the more accurate pricing of insurance policies for drivers and also the definition of new insurance products (for instance, policies that do not allow for the use of a car at night) and so have a positive impact on revenues. In essence, a company developing and deploying a UBI capability can expect this to rapidly become a competitive differentiator when compared with insurance companies with no such capability: both reducing costs and enabling access to new revenue streams through better pricing. And this scenario highlights another interesting dynamic. In a competitive industry and where one company deploys a technology that becomes a competitive differentiator, then competitors will be compelled to somehow match that capability or else risk being outcompeted. Essentially, the adoption of technology (any technology) in a commercial environment is an arms race leading to the ever more widespread adoption of technology in an attempt to gain an edge over the competition. This dynamic has the potential to very significantly drive the adoption of M2M solutions within industry.

Of course, none of the models listed above is mutually exclusive. For example, early Kindles were funded through a mix of up-front purchase, commissions from sales of eBooks. Your author also suspects that Amazon has generated advertising revenue from targeting adverts based on Kindle end user reading history and that some organizations will have invested in Kindle solutions as a way to increase efficiency (e.g., in situations where service engineers may require access to a very wide range of documentation).

18.7   The return on investment

In the earlier sections of this chapter, I have addressed the history of M2M and the motivations and business models that drive M2M today. I have made comments about connected solutions being adopted if they are "better" and if business cases cost in. The natural question to ask, therefore, is how to measure the return on investment.

There is, of course, a textbook answer to these questions: you assess the cost of deploying an M2M solution and the impact that the deployment of such a solution will have on free cash flows over time. For certain M2M connected solutions (e.g., Wi-Fi bathroom scales), this approach is perfectly suitable: a business case for a company developing and selling connected bathroom scales can be modeled essentially in isolation of other considerations.

However, in complex scenarios, and where M2M can impact the competitiveness of a company within a wider industry, then this textbook approach must be applied more carefully. Take, for example, the case of usage-based insurance. UBI has the potential to shift the basis of competition within the insurance industry. It is a competitive differentiator, in that it allows for insurance companies to better price the risks that they are underwriting and so win "good" business away from competitors and (better) leave "bad" business with those competitors that are unable to price the risk so accurately. So the adoption of UBI propositions by an insurance company has the very real prospect of increasing both revenues and margins. These effects would clearly need to be reflected in any calculation of ROI. However, the benefits of UBI would not go unnoticed by any of our insurance company's competitors, whose margins are likely to be reflecting the effects of a relative inability to accurately price risk. These companies may therefore feel compelled to adopt similar UBI solutions in order to remain competitive. Over a period of a few years, after it is first adopted in the market, UBI propositions can therefore be expected to be a commonplace element of any mainstream insurance company's offering.

So, in a competitive environment, the adoption of M2M solutions can be regarded as one element of an ongoing arms race where all market participants constantly strive to offer better products and services, and M2M is just one of the techniques that can be deployed in pursuit of this goal. And the question of ROI on M2M solutions should therefore be reframed in terms of *when* to invest in M2M in order to maximize ROI, rather than *whether* to invest in M2M solutions. In fact, this logic also applies to the bathroom scales example: One day, nearly all bathroom scales will be connected and manufacturers that continue to manufacture only nonconnected bathroom scales will most likely have gone bust.

Similar effects also drive public service and noncommercial deployments of M2M-based solutions, the main difference being that the feedback of public opinion is felt via the ballot box, rather than the revenues line of a P&L account.

But it is not all so depressingly processional. The adoption of M2M solutions does have the potential to generate real economic returns, for instance, by supporting "new" products and services that could not work without M2M connectivity. In the case of vehicle insurance, then UBI will allow for insurance to be extended to groups that would otherwise have been excluded from the market. In the case of shared car ownership schemes, then M2M is enabling many people that would not otherwise be car drivers to drive cars. M2M also affords individual companies the opportunity to extend up, or down, their traditional value chains allowing for increased efficiency. For example, M2M can help providers of capital equipment move into leasing or the provision of field service capabilities and remote monitoring and diagnostic capabilities. Overall, where M2M is deployed to reduce the cost of a product or service, then the volumes of sales of that product of service can typically be expected in increase as a consequence of price–volume elasticity.

Ultimately, however, in a competitive environment, the "benefits" of M2M should be expected to be reflected in reduced costs of services to end users, rather than in increased margins from providing those services. In a commercial environment, competition ensures that any advantages of reduced costs or increased margins are competed away. And a company that does not invest in any M2M solutions today should most appropriately be compared with a company that in the 1970s did not invest in any computing systems or IT capabilities. The reality is that the adoption of M2M connectivity across all aspects of our modern lives, and in the infrastructure and products and services that support our lives, is just a matter of time.

Machine-to-machine (M2M) communications for smart cities

19

I. Vilajosana[1], M. Dohler[1,2]
[1]Worldsensing, Barcelona, Spain; [2]King's College, London (KCL), London, UK

19.1   Introduction

More than 50% of the world's population lives in cities today (2014) [1]. To support such an urban population requires some enormous efforts from a city management point of view [2]. To add to these problems, the effect of urbanization on global climate has also come into the limelight [3]. Solutions are thus needed to address these challenges.

Smart city technologies, in the broadest sense, will be an integral part of such solutions. This has triggered major global ICT players to launch their respective smart city initiatives. For a good reason, the smart city market is estimated to be of hundreds of billions of dollars by 2020, with an annual spending reaching nearly $16bn [4].

Pike Research [4] defines a smart city as "the integration of technology into a strategic approach to sustainability, citizen well-being, and economic development." Viable smart city models thus ought to be "multi-dimensional, encompassing different aspects of smartness and stressing the importance of integration and interaction across multiple domains." A city, in the end, is a system of systems, and "any models that attempt to define its dynamic nature must also be able to represent the diversity of those elements."

There are clearly strong societal, industrial, and political drivers for smart cities and underlying technologies to be deployed. The aspect of urban life where improvements will be felt most is in transportation, be it public or private transport. A specific element of transportation is related to parking which is the root of several problems. First, citizens experience considerable problems when looking for parking, spending on average some 15 min per search. This is time taken off the citizens' daily routine, as well as great source of stress given the uncertainty in finding a spot. Second, the extra driving time in search of a parking spot, coupled with the fact that this extra traffic often causes 30% more traffic and with it traffic jams, causes significant levels of pollution, which is of major concern to cities. Third, some 30% of drivers do not pay for their parking spot, which is naturally a problem to parking operators and cities.

Given these problems for three stakeholders in the city, it is to no surprise that the concept of smart parking has been gaining ground recently. It refers to the process of installing reliable sensors in each parking spot that are able to detect the presence and absence of cars and relay this information to citizens, city authorities, and parking operators. As per Figure 19.1, it is clear that smart parking has become an economic drive in modern cities.

Machine-to-machine (M2M) Communications. http://dx.doi.org/10.1016/B978-1-78242-102-3.00019-8

Figure 19.1 Global press releases highlighting the potential and need for smart parking. The major argument throughout has been to offer citizens a more informed guess on vacant parking spaces, thus saving fuel, nerves, and time [5].

We will therefore review the underlying—mainly machine-to-machine—technologies, the economics behind the technology and smart city data flows in general. We will discuss some governance challenges related to it. To this end, we review the barriers of entry for this smart city market, which limit the smart cities ecosystem development. We propose a viable approach to scale business within the said ecosystem and describe a high-level model that exemplifies a self-sustainable way to develop city economies based on smart city infrastructures [6].

19.2 Smart city technologies

There are several new constituents emerging in the technology landscape of cities today [7–17]. Notably, cities start to roll out smart city "operating systems" or data platforms that rely on the input of data from sensors deployed in the field; data gathered from people via, for example, social media; or data taken from the Internet at large. The platforms then process the data, often in real time, and take appropriate decisions. The decisions could be taken in real time, such as changing a traffic light controller to ease congestion, or over a longer time horizon, such as adapting the parking policies in a city based on the past statistics of usage. The trend is thus clearly to close the big data cycle in that data are being gathered, processed, and acted upon.

In this chapter, we will focus specifically on the technology facilitating smart parking that is exemplified in Figure 19.2. It involves the following elements: sensors,

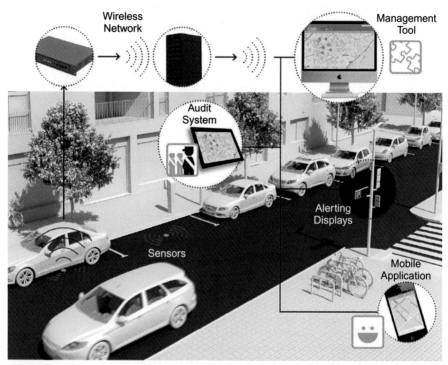

Figure 19.2 Example deployment of smart parking technology, including the sensors in the street, gateways, and informative panels, as well as smartphone applications for citizens (bottom right) and a management platform for parking operators and cities (top right) [5].

gateways, back-end platform, and front-end capabilities, as well as support of smartphones. These technologies are typically categorized by (a) M2M technology in the field, (b) a big data back-end platform, and (c) a client or user interface.

19.2.1 M2M technology in the field

As shown in Figure 19.2, smart city M2M technologies are generally composed of sensors, data loggers, wireless modems, and gateways. Some outdated technologies also use repeaters. With application to smart parking, all these constituents are now discussed in more detail:

- Sensors: Their role is to measure different physical quantities, such as pollution, noise, and magnetic fields. Generally speaking, analog and digital sensors are available today where the former are cheaper but the latter more performing. They are feeding into the datalogger, discussed below. In the case of smart parking, sensors are designed to detect the presence/absence of a car via optical, magnetic, or other sensors.

- Datalogger: This unit samples the output of analog sensors or takes the digital output directly. It stores the data, processes it, and passes it on to the wireless modem. High-precision dataloggers can be very expensive, but fortunately, most smart city-sensing applications do not require such a high sampling precision.
- Wireless modem: The modem receives the data from the datalogger and transmits it via the given air interface. While outdated IEEE 802.15.4-like technologies are still in use today, most successful smart city applications make use of low-power wide area (LPWA) or even cellular modems. In the case of smart parking, the difficulty is to ensure connectivity despite a car having parked on top, which typically yields a shadow of 40–60 dB. In smart city deployments, one rarely distinguishes the notion of sensor, datalogger, and modem—they are jointly often referred to as the sensor.
- Repeater: IEEE 802.15.4-like technology requires repeaters to be used to reach the Internet-enabled gateway. This is because the transmission powers are very low and the bandwidths used fairly high—thus yielding a poor link budget. Cities do not like repeaters because they mean more street furniture; in addition, the issues of location, electricity, and maintenance give smart city projects unnecessary problems. It was reported that one smart parking deployment in California, the United States, had repeaters in the trees!
- Gateway: The data from the sensor modem are received directly or via repeaters by the gateway that acts as a relay of information into and from the Internet. Gateway designs differ depending on the M2M radio technology used. IEEE802.15.4-enabled gateways would require a dense rollout since range is limited, whereas LPWA-enabled gateways require a significantly reduced rollout footprint. For instance, Worldsensing's Moscow smart parking deployment only required 3 gateways to cover the center of the Russian capital.

All this on-street equipment requires long lifetimes, maintenance, support, etc., and this needs to be factored into project budgets. It is thus clear that the amount of street furniture ought to be minimized generally.

19.2.2 Big data back-end platform

The data back-end platform is an important constituent of any M2M city in the smart city space. It is largely driven by storage and processing capabilities and support of (standardized) APIs. Delay, scalability, and reliability are key performance design indicators. Since M2M business is largely business-to-business (B2B) still (as of 2014), many clients wish to keep the platform in dedicated servers on the premises; however, cloud approaches are already emerging where we expect a complete coverage by 2020. These high-level elements can typically be found in a modern back-end platform:

- Data storage: The storage capabilities are of utmost importance since large volumes of data are being accumulated by M2M applications. Different standardized or pseudostandardized storage approaches are available.
- Process capabilities: Powerful, often virtualized and cloudified, processing capabilities need to be supported that allow the real-time crunching of data, in turn paving the way for big data paradigms.
- APIs: The interfaces between elements in the data back-end platform are important as they often allow gluing together different internal constituents of very different designs, as well as interacting with external platforms in a quick and scalable manner.

The development times and efforts of M2M back-end platforms should not be underestimated. As of 2014, it is recommended to use developed platforms on a platform-as-a-service (PaaS) basis. Given the importance of the platform and the API, we will discuss these in subsequent sections.

19.2.3 Client interfaces

Exemplified by Figure 19.3, client interfaces are important to interact with the final client. These elements can typically be found in smart city M2M systems:

- Front-end: The real-time and statistical data insights are typically displayed in an easy-to-understand form via graphs, videos, timelines, etc. This is the role of the front-end for which a wide variety of programs exist today (2014).
- Smartphones: The mobile smartphone and M2M trend has a similar growth curve, and it is thus to no surprise that both technologies appear on different ends of M2M industry solutions. In the case of smart parking, smartphones make parking information readily available to citizens, which allows them to find a parking spot quicker or reserve it.
- Special end devices: Finally, many industries rely on specialized devices that—in the case of the smart parking vertical—are for traffic wardens to issue parking tickets. Typically, these devices are also provided with information from M2M devices in the field.

Figure 19.3 Examples of advanced M2M smart parking front ends where on the left there is the dashboard for the city or parking operator and on the right the smartphone application that is used to guide citizens to the vacant parking space [5].

19.3   M2M smart city platform

19.3.1   Platform architecture

From a generalist point of view, all the existing platforms offer similar characteristics. They offer seamless interconnection with monitoring systems at infrastructure level. On top, there emerge big data structures allowing for storing and analyzing the generated information, which is eventually offered to 3rd parties through standardized interfaces in an open data fashion. In Figure 19.4, the general architecture with most of the available platforms is depicted.

Smart city platforms [18,19] are designed to provide heterogeneous services to support a variety of application domains. A capillary network layer is composed of sensor nodes, actuators, software components, and other devices that gather data from (or actuate on) the city infrastructures and citizens. To deal with the data generated by the individual utility provider capillary networks, a generic middleware providing both control and data management planes is usually proposed. The data management plane provides massive raw data storage for historical datasets and real-time data storage to support stream-computing applications [20]; this is decoupled from data filtering rules and postprocessing and interfaced by a generic service layer providing access, management, orchestration, and security services over the data. The control plane provides well-defined application programming interface (API) to manage the underlying capillary network infrastructures as well as the middleware components and configuration. More general services are built on top of the basic core services provided by the middleware. This second layer of services offers big data management services related to the smart city domain. The latter include analytic services and prediction engines, real-time and streaming interfaces, and standardized open data APIs exposed as web services.

19.3.2   Open data APIs and app stores

As of today (2014), we still see a certain immaturity in the open data delivery, especially when articulated by the public sector. They have neither the necessary operational structure nor the necessary expertise to deliver data services. Therefore, we see that corporations or public–private partnerships (PPPs), acting as IoT service operators, will handle the delivery of new open data interfaces in a corporate fashion. An example of a PPP in that context is London's Transport for London (TfL), which currently also acts as open data provider on the transport data. In that process, not only utilities and operators but also SMEs play a crucial role. We believe that the API delivery process will be articulated similarly as the markets and the app stores, which dominate the smartphone app scene. We thus define the new concept of API stores as platforms delivering standardized APIs to interact with the smart city infrastructure.

Thus, the delivery of open data interfaces will be regulated as an app market. The delivery of the information generated at infrastructure level to 3rd parties will be handled by corporations/PPPs who grant access to data following different approaches, which will be linked to different business models. Possible models are discussed below:

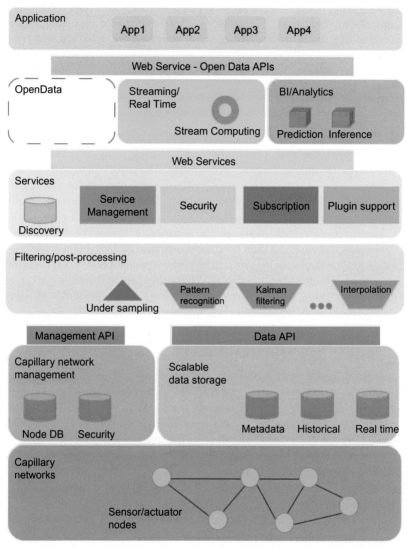

Figure 19.4 Generic platform showing the main components to support heterogeneous smart city applications, including storage, processing, and strong middleware capabilities [5].

1. App store-like model: Upon developer subscription (might involve some subscription fee), a set of verified API that grants access to useful data is delivered to the developer. Developers build their apps, analytic tools, and services taking advantage of the delivered information. The developed apps most likely will end up in Apple or Android app markets. IoT operators may want to capture a small percentage of the app sales in these markets. This will bring high-quality smart city services and apps into the mobile app revenue stream. Developers will have the advantage that their app will be automatically valid for all the cities running under the same IoT operator and processes to ensure quality and security of the apps will be in place.

2. Google Maps-like model: Some of the developed services may only make sense if the granularity, reliability, and authenticity of the provided information through the API store are high enough (e.g., traffic information). Thus, the percentage fee on the app sales price will be scaled according to the granularity of the queries made to the API. A low number of queries in apps will end up being free, which will help ramping up the introduction of new services.
3. Open data model: Some cities may want to grant access to some of the data as a classical open data vision, without charging any fee to developers.

A side effect of this model is that IoT operators/utilities/PPPs will have to update and maintain their APIs constantly offering new access to new capillary networks and fostering the interaction among the generated services. The latter will trigger closer interactions with IoT device developers, solution developers, and service provider.

The regulated delivery of information will ensure on the one hand the quality of service provisioning and on the other avoid fraudulent use or abuse of open data information. Moreover, the regulation by specific departments and the promising business model behind it will ensure market persistence, support, political independence, and standardization. Utilities running services in different municipalities will benefit from the new services developed by third parties, capturing part of the sales market. Freelance developers, SMEs, and 3rd parties will also have their opportunities to develop new analytic tools, new services, and new devices.

19.3.3 Interaction with stakeholders

If we look carefully at the established relations between the different stakeholders in the model presented above, we observe that corporate open data services, likely regulated through public–private partnerships, constitute a basic element to benefit all the stakeholders in a smart city deployment. For illustrative purposes, in Figure 19.5, the different interactions between the different stakeholders involved in a city are presented.

The interaction between smart device developers and industry is crucial for feeding information to corporate open data platforms. Industry will pay for sensing and actuating infrastructure but benefit from the data generated. Industry will push smart sensor device developers to create new and better solutions.

Developers from regulated and standardized interfaces (API stores) will be able to develop new services and analytic tools that will be used by citizens through the classical Android and Apple markets. Revenue sharing models have proven to be successful in the smartphone's scene; we thus believe that this will also be applicable to the smart city scenario.

Citizens and cities will also benefit from this structure. First of all, a new market place will be created. Freelance developers, app developers, and SMEs will benefit from a standardized and regulated access to the data, with the advantage of not having to deal directly with public bodies or utilities. Cities will benefit from a new market place, at the top of the value chain, which will attract new talent that in turn will generate new services that will have impact on citizen's quality of life. Utilities/service providers, apart from the operational benefits of better operating their infrastructures, will get into a new market by capturing small amounts of the crowd-sourced services developed.

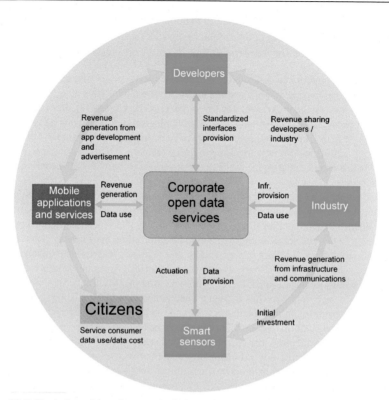

Figure 19.5 Depiction of data flows and relations among the stakeholders in the corporate open data service framework [5].

19.4 Financing M2M deployments in smart cities

19.4.1 Smart city rollout phases

Concerning the financial dimension, it is possibly the biggest factor preventing meaningful M2M rollouts and smart city deployments. Importantly, no clear business models exist, and only recently, Ref. [21] proposed a suitable platform to successfully evaluate the business models of new services offered by cities. However, the situation has further sharpened with the latest global economic downturn. Funding needs to be drawn from the scarce range of sources in order for smart cities to bloom; policy makers, technology and service companies, investors, and utilities alike wonder how to get this market going. The smart city market, in the sense of instrumenting, interconnecting, and making it more intelligent, clearly has enormous potential but, as of today, is a zero billion-dollar market.

Digging a little deeper, however, we were able to understand that some domains of the smart city market were very hesitant, while others are currently flourishing [22]. If organized well, the scarce public funding can be used to unleash private and philanthropic investments. However, access to funding requires a creative articulation of value cases using metrics that demonstrate social, economic, and political value [21]. Without going into too many economical, historical, or strategic details, we advocate clearly defined 3 stages for the deployment of smart city technologies and services:

1. The first phase ought to be dedicated to technologies and services that not only offer utility (in the sense that they are of great use and make the urban living truly smarter) but also offer very clear return on the investment (ideally after a very short breakeven time). It is an utmost important phase, since it essentially sets the technological basis (introducing the developed platforms described above) and guarantees a viable bootstrapping of the smart city market by generating cash flows for new investments. An example for this first phase has been discussed in this chapter, that is, the usage of smart parking to provide both benefit and significant revenues.

2. The second phase has scope of ramping up technologies and services requiring large upfront investments, showing longer return of the investment periods or not necessarily producing direct financial gains, but may just be of great use (or may yield second-order revenues). The said technologies and services are expected to be attracted by the finances generated in the first phase, which will attract private capital and take advantage of previously deployed infrastructures to lower its barriers of entry (i.e., platforms). An example here is the deployments of pollution, traffic, weather, transport, and other M2M data sources in the city, the potential of which only becomes apparent when pulled together in a single platform.

3. The third phase relies on data availability through standardized APIs, which are offered by the implemented platforms. Multiple services might be then offered by 3rd-party developers. This phase has the scope of making the system self-sustainable, by developing services on top of the existing smart cities' infrastructures and involving the whole value chain (though standardized APIs). This might produce a new tertiary sector exploiting data generated in the existing infrastructures that will be used to offer new services to cities, utilities, and citizens. An example here is the opening of all transport data in a given city, such as London, which in turn is provided through open APIs and thus able to power an emerging ecosystem of applications on top of that.

We thus observe that a second important issue towards a sustainable smart city development is a coherent 3-phase smart city introduction, where (1), in the bootstrap phase, services are offered, which yield utility and revenues and set the basis for future developments (i.e., generic platforms and capillary networks); (2), in the growth phase, previously deployed services generate cash to trigger new investments that will turn into new services showing higher barriers of entry; and (3), in the business-to-consumer (B2C) wide-adoption phase, services are offered by all stakeholders taking advantage of the existing infrastructures independently of offering revenues, utility, or fun dimension.

19.4.2 Barriers of entry

Although there are obvious reasons that justify the introduction of smart cities, they are not really taking off and not truly realizing the projected potentials. The need for affordable housing, increasing traffic congestion, the raising of energy costs, water scarcity, environmental targets, and regulations are strong enough reasons to justify the concept's introduction.

However, the need for policy changes, limited capital availability, and piecemeal funding structures are preventing investment in smart cities. In addition, there are political uncertainties, which are not setting a favorable environment for public and private investment. For example, the absence of long-term stability of carbon price or the lack of public incentives for low-carbon initiatives makes investment in low-carbon technologies unattractive. Moreover, the inconsistency in international, national, and regional rules and regulations related to environmental policies does not help to scale initiatives. Finally, there is a lack of appropriate and systematic methodologies and metrics for reporting and verifying the investment returns due to smart city technologies [21]. The actual economical context is not also helping to the introduction of the smart cities context. The depleted public finances from recession are slowing down public investments. The financial situation, the unavailability of credit, and the new pressures and regulations on financial institutions to reduce risk exposition by building stronger deposit bases are limiting the available cash flows, slowing down private investments.

Moreover, there are few alternative secondary markets to finance large smart city projects. Up to now, only grants coming from EU funds or small local initiatives and local philanthropic capital are allowed to run first trials [22]. Barriers of entry are not only economical but also political [23]. The geographic dispersion of the ongoing smart city project and the multiple and complex technologies involved and their small size are not helping to create the required critical mass that may help to show the viability of the smart city deployments. The latter is seen as an increased risk to investors, who find it risky and difficult to aggregate individual and small-scale projects into large-scale investment vehicles. All the above factors are translated into a certain immaturity of the market as viewed from the private sector, which in turn is enhanced by the complexity of the relationships with the public sector.

19.4.3 Bootstrapping the smart city market

As stated above, there are political, financial, and technological barriers, which need to be overcome to facilitate the introduction of the smart city business. Achieving grow-beyond-organic growth rates has three dimensions today: (1) a political dimension, calling for an establishment of smart city departments; (2) establishment of transversal and interoperable technological platforms to manage the huge amounts of data generated; and (3) a financial dimension, calling for a coherent self-sustainable business model [9].

Under the political dimension, historically, interactions between the private sector and the smart cities' ecosystem have been fairly complicated, in particular in European cities, which have grown over centuries. One often observes that, when it comes to decision taking, ownership, decision making, and responsibilities are heavily intertwined. This, in the best of cases, hinders smart city deployments, but often even prevents key players talking to each other at all.

It has hence become clear that the entire decision and execution process within smart cities needs to be institutionalized, and many cities have indeed commenced forming their "smart city departments" with their own decision-making infrastructure and procurement processes. This development bears similarities with the process that occurred a few decades back when, with the occurrence of the first computers, IT departments slowly emerged. Indeed, while it may have been very political back then, the choice of a specific computer within a city hall today to facilitate better management is under the sole auspices of the IT department and is not questioned at all by the managing office. The smart city (SC) departments are, compared to the IT departments, still far from this obvious modus operandi.

We thus observe that the first important issue towards a sustainable smart city development is an independent smart city department (or equivalent), which clearly decouples the political element of the improved city servicing from the underlying technologies and old-fashioned procurement processes. Notably, formal relationships can be impeded by the complexity of public sector procurement regimes, which are simply not adapted to new technologies and as a result are too expensive and time-consuming for the new ecosystem of high-technology companies (sometimes very small) to engage with. By simplifying procurement practices and adapting those to, for example, servicing models based on quantifiable metrics, the level of engagement between cities and private ecosystem could be increased.

Although local entrepreneurs, SMEs, and innovators have brilliant ideas and marketable technologies, they often have limited resources. On their own, they will not be able to implement smart city infrastructures as water saving systems, lighting control systems, or real-time transport information systems. But collectively, if the smart city department takes into account their ideas and involves them (stakeholder groups) to define the business case, it may happen that the local authority, a utility, or both may invest in such infrastructures.

Thus, utilities will definitely play an important role in smart cities uptake. Generally, they will offer smart city services using some ICT infrastructure, which ought to be optimized for resource distribution, for prevention of resource outages, for easy and rapid maintenance actions, etc. Sustainable smart city approaches ought to simplify and accelerate service delivery and therefore reduce the operational cost and also enhance the return of investment in a shorter period of time. Being continuously in interaction with the customers, they will provide customized services for customer-specific needs and preferences. By engaging utilities, entrepreneurs and the public sector attract a portfolio of potential innovations, and outcomes can be created to demonstrate the value of such investments.

19.4.4 Smart city value chain

In Figure 19.6, the value chain and the different stakeholders under the smart city deployment are presented. At the bottom of the chain, there are cities and citizens as not only prime enablers but also prime consumers of the generated services. Citizens as consumers enjoy various smart services provided by utility companies and city authorities, which aim to enhance their life in terms of security (e.g., better street lighting, smart parking guidance services, and prevention of accumulations of rubbish), health and well-being (e.g., reduced CO_2 emission), and economics (e.g., resource optimization).

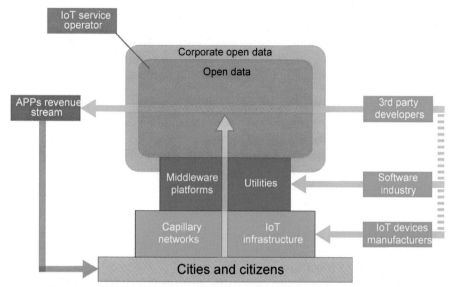

Figure 19.6 Representation of the value chain and relevant stakeholders in a smart city deployment process, including data sources and the various industries involved.

Today, around 80% of the European citizens live in urban areas at a very high density. Cities have to sustain the growth of the population and the consequent strain on the utility providers and services that are necessary to facilitate daily life. Cities own the prime infrastructures that utilities operate. IoT devices and capillary networks are becoming essential technologies to monitor and control infrastructures operated by utility providers for improving daily operations and service delivery. This is attracting IoT device manufacturers and 3rd-party developers to take part in new sensing and communicating device developments. Often, IoT device manufacturers are in the form of start-ups offering vertical solutions targeting very specific problems (e.g., smart parking). They are usually evangelizing not only the corresponding departments from different city halls but also utilities who are, in most of the cases, their potential customers. Utilities build their transversal platforms on top of multiple services. They offer middleware solutions that integrate multiple solutions with the scope of better service delivery and cost reduction.

19.5 The ten smart city challenges

What follows is a review of some of the challenges we face as of the design year 2014; the focus here is mainly on the use and deployment of smart city infrastructure—such as M2M technologies—rather than the social, citizen, and other aspects.

Designing smartness into cities requires major infrastructure upgrades; in a sense, we are constructing the brain of the city. We have done something similar before: we have built highways, the Internet, and the smart grid. These prior connectivity exercises have two things in common: (i) they have meshed and connected entities (computers, cities, etc.) to the point that connectivity is secondary, while the ability to provide services (Facebook, supply chains, etc.) has become the prime point of interest, and (ii) they were built without a specific or proven return-of-investment (ROI) model in mind, with all of them having returned their investment by orders of magnitude.

These connectivity initiatives however differ in one critical point compared to smart cities, in that they had significant governmental support that allowed them to become operational fairly quickly. Cities worldwide are generally short of money and governments historically do not want to interfere too much with cities, the result being that there is very little financial support to make smart cities happen.

The biggest challenge of this early part of the twenty-first century is thus to take advantage of the undoubtedly high-potential of the smart city market in cities that are essentially broke.

With the experience of building and running the smart city pioneering company Worldsensing, we figured there are roughly three types of cities: (1) ROI-driven, (2) carbon-driven, and (3) vanity-driven. As for (1), the aim of rolling out smart city technologies is to generate income that pays for its deployment and more. There are many cities in the Western hemisphere that fall into this category, such as Los Angeles and London. As for (2), the aim here is to reduce the carbon footprint and ideally become carbon-neutral in the long term. These are mainly cities in middle and northern Europe, such as Luxembourg and Helsinki. Finally, "vanity"-driven cities are mainly driven by events where the entire world is watching and they want to be perceived as "modern."

The deployment strategies in all three city types are of course very different. Take the example of smart parking, which is one of the product lines of Worldsensing: in city type (1), one would sell the business process behind the system that is able to spot infringers who have not paid their parking ticket; the ability to guide people to vacant parking spaces was rarely discussed during the sales meetings. In city type (2), one would sell the ability of the system to guide drivers quickly to an empty parking spot or advise drivers that certain areas are completely full, thus reducing unnecessary traffic, traffic jams, and thus pollution. The ability to fine citizens had to be kept very quiet. In city type (3), while budgets are often not a problem, the deployment rarely went through public procurement that required a technical and vision alignment with the prime partner for the smart city project.

What follows are ten specific design challenges that smart cities face today (2014), grouped into political and procurement issues, stakeholder dynamics, and data-driven decision taking.

19.5.1 Political cycles and decision taking

The first challenge pertains to political cycles in cities. No matter how beautiful and useful your M2M or other products' design, most of the things one sells into the city will go through some form of public approval or procurement. The problem is that these procurements are, in many countries around the world, heavily coupled into political activities.

In Spain, for instance, there are the city elections, the regional and the government elections: plus/minus 4–6 months of each of them; during this time the political, and thus executive and financial firepower is completely paralyzed. If timing is unfortunate, this can leave you with only a 20% sales window opportunity. Can smart city business thrive under these conditions?

A recommendation to overcome this is to introduce governance mechanisms at city, regional, and national levels that decouple the political cycles from the technological ones and thus facilitate a proper uptake of smartness.

The second challenge pertains to political decision taking. Since there is virtually no precedence of large-scale smart city rollouts, decisions to use or at least to trial new technologies are very political in the sense of personal relationships with town halls, political agendas, etc. That is not a good turf for smartness to grow in.

A recommendation here is to ensure deployment mechanisms, even for early adaptors, which are based on fair market drivers and associated competition. Similar to what happened in the IT industry some decades back, we must decouple the political element from the technological one, where choosing a smart city technology ought to be as straightforward as choosing a new computer.

19.5.2 Procurement and finances

The third challenge pertains to intelligent procurement. Most of the smart city technologies will need to rely on a procurement process. Procurement today, by its very essence, chooses the set of technologies that fulfills the minimum requirements and then chooses the cheapest one. This is of course a recipe for failure and certainly does not allow for sustainable growth and implementation of smart technologies.

A recommendation here is to enforce mechanisms that meaningfully evaluate the future capability of the technology under consideration and not base the decision on pure pricing alone. We thus need to shift from a cost-driven approach to a long-term purpose-driven procurement approach.

The fourth challenge pertains to the lack of finances in a city. The lack of financial power of cities is very visible and very problematic. In the ROI-driven cities and economies, there is a strong feeling that the only way of getting smart cities going is to properly bootstrap the market.

A recommendation is to concentrate strictly on ROI-healthy solutions, and there are some of them in the smart city context. Examples can be found in transportation (e.g., smart parking), smart street lightening, smart bins, etc. Another important shift that needs to be invoked relates to the change from CAPEX-driven city deployments to OPEX-driven approaches; the paradigm here would be a smart city as a service (SCaaS).

19.5.3 Established and complex stakeholder system

The fifth challenge is the well-established stakeholder system. In contrast to the common beliefs among the newcomers in the smart city community, the city space is serviced by a very established stakeholder system. It is run by companies most of us in the IT business have never heard of, but these are companies that mainly provide the infrastructure, that is, the visible part of the city. They are not the stakeholders that provide the intelligence, that is, the ability to make the city really smart.

A recommendation here is to ensure there is a real dialogue between the established stakeholders and the emerging stakeholders without the former feeling threatened about their space and without the latter believing they are able to do it all alone.

The sixth challenge pertains to the city legacy system. Cities are thousands of years old. Arguably, the biggest technical challenge in any smart city endeavor is to retrofit smartness into these cities with a strong political, cultural, and technical legacy.

A recommendation here is to make sure that smart city solutions not only are perfect standalone ideas and products but also are actually able to be deployed and retrofitted; that is, not only pursue design for a perfect end purpose but also make sure you know of the exact steps of getting it out, deployed, and used.

The seventh challenge pertains to the complex ecosystem. Cities are extraordinary complex in terms of stakeholder composition and interaction. To get anything meaningful done, also at a global scale, under these circumstances is an arduous task.

A recommendation here is to take lessons from prior and related design efforts. Indeed, it is not the first time that we faced the design of such complex systems with a global footprint. The design of successful systems has typically undergone these processes:

1. *Standardization*: This plays an important role in ensuring scalable uptake of technology, interoperability, fair competition, and long-term availability. Therefore, smart city technologies ought to inherently be standards-compliant. First, global initiatives on smart city standardization are well under way, such as the ISO smart cities, which the author, M. Dohler, is involved with (as of 2014) on a global scale.
2. *Virtualization*: This is often overlooked but plays an important role in properly decoupling different stakeholders that in turn facilitate independent growth in each ecosystem. The computing industry has shown the way where the hardware, operating system, and application software ecosystems have evolved independently while always ensuring operability. Similarly, it is important to ensure that the smart city hardware, software, and service applications evolve as independently as possible.

19.5.4 Urban fab labs, data, and citizens

The eighth challenge pertains to urban fab labs. Manufacturing is returning to the Western hemisphere. Urban fab labs, that is the production and manufacturing through, for example, 3-D printing done locally, are a trending development. While improving product supply chain problems, other serious problems emerge that are related to an increase of pollution in fab lab areas, poorer waste recycling ecosystems, and the inferior supply chains of raw materials.

As with supply chains and many other verticals, a virtualization to manufacturing might be worth studying and considering, where—instead of dismantling them—macromanufacturing sites are shared in a cost-efficient manner by companies worldwide, to achieve, for example, cheap access to supply chain-optimized and waste/pollution management-certified 3-D printing. Another avenue worth exploiting is to use the well-honed supply chain of major chains, such as supermarket chains, to bring raw material into urban environments, and equally use their optimized waste-recollection system.

The ninth challenge, and arguably the most important challenge, pertains to our relationship to data. Reference to big, open, or private data appears in each presentation today. There are three important issues to be considered in the upcoming years:

1. *Big data*: There is no doubt that big data can give unique and unprecedented insights, mainly when cross correlated with data from different domains. However, we observed that most big data insights are very well known to those really working 24/7 in the concerned vertical. Big data should not be used to graphically improve presentations but to solve problems. Arguably, the trend that ought to drive this century is not big data but big action.
2. *Open data*: There is also no doubt that open data will be a major factor in making big data happen. Open data, however, cannot be enforced as those companies generating the data typically go through just-at-margin procurements—why would I release data on which somebody else will be capitalizing on?
3. *Privacy issues*: The recent (2014) privacy issues that have emerged across the globe have not helped, but building higher and stronger privacy walls is not the right way to go either. Nature does not work like this; imagine the brain, based on privacy argumentation, refuses to instruct the arm muscles to retract after the temperature sensors in the hand signal burning heat. The real big data opportunities come with private data used carefully.

Finally, the tenth challenge pertains to citizens. We have become aware of the broken value chain between citizens and the decision makers within cities. There is a strong trend of re-engaging the citizen in the design process of making a city a smarter place. However, there is a danger of becoming too obsessed about this re-engagement as (i) not everybody wants to be involved in designing urban space (in fact, most of us just want to get on with life; the heated discussions among like-minded individuals give us the impression that all want to be part of this process) and (ii) even if you involve an eager crowd, your engagement will be in the order of percentages and certainly not democratic.

Since people are by nature great at complaining, the process of delivering and acting upon these complaints ought to be made as efficient and transparent as possible. Simple feedback has always been the most effective way of improvement and is an efficient way of involving the crowds.

19.6 Conclusions

The total and addressable smart city market is estimated to be of three-digit billion dollars by 2020. Most of this market value is related to smart technologies, acting as facilitators for advanced services. Because of the cost and the unproven business models, the market is reported to have difficulties of taking off. In this book chapter, we have looked into why this is the case and proposed suggestions on how to help the ecosystem to grow.

An important constituent for growth is the right use and deployment of technologies, most of which will be M2M and data delivery (fiber, Wi-Fi, etc.) technologies in the field. We had thus dwelled in great details on the M2M technologies and exemplified this with the strongest growing market example of smart parking.

We then dealt in great details with the flow, value, and monetization of data. We underlined the importance of a middleware capability, which is able to gather different sensor data streams, and put this into a database where the data are stored and processed. We also showed that the app-store approach, similar to what we observe in the smartphone market, could be advantageous in the data market.

We also discussed the economic value of smart city M2M deployments by examining the stakeholder ecosystem and how it is interlinked. Along these lines, we outlined important design challenges related to M2M and smart cities.

All these insights are based on firsthand experiences we obtained by building and growing Worldsensing, an M2M smart city pioneer with the largest smart city M2M footprint to date (Q1 2014).

References

[1] United Nations Population Fund (UNFPA), "State of world population 2007. Unleashing the Potential of Urban Growth," Online report, 2007. Available: http://www.unfpa.org/public/publications/pid/408.

[2] M. Kinver, "The challenges facing an urban world," in *BBC News Website*, Online article, 13th June 2006. Available: http://news.bbc.co.uk/2/hi/science/nature/5054052.stm.

[3] M. Kinver, "UN report: Cities ignore climate change at their peril," in *BBC News Website*, Online article, 29 March 2011. Available: http://www.bbc.co.uk/news/science-environment-12881779

[4] Pike Research, "Smart Cities. Intelligent Information and Communications Technology Infrastructure in the Government, Buildings, Transport, and Utility Domains", Online report, 2011. Available: http://www.navigantresearch.com/research/smart-cities.

[5] Worldsensing website www.worldsensing.com; last accessed Q1 2014.

[6] I. Vilajosana, X. Vilajosana, J. Llosa, B. Martinez, M. Domingo-Prieto, A. Angles, Bootstrapping smart cities through a self-sustainable model based on big data flows, IEEE Commun. Mag. 51 (2013) 128–134, Special Issue on Smart Cities.

[7] B. Yang, G. Zhu, W. Wu, Y. Gao, M2M Access Performance in LTE-A System, Transactions on Emerging Telecommunications Technologies (ETT) 25 (2013) 3–10, Feature Issue on "Smart Cities—Trends & Technologies".

[8] A. Cimmino, T. Pecorella, R. Fantacci, F. Granelli, T.F. Rahman, C. Sacchi, C. Carlini, P. Harsh, The role of small cell technology in future smart city applications, Trans. Emerg. Telecommun. Technol. 25 (2013) 11–20, Feature Issue on "Smart Cities—Trends & Technologies".

[9] Gianluca Aloi, Luca Bedogni, Marco Di Felice, Valeria Loscri, Antonella Molinaro, Enrico Natalizio, Pasquale Pace, Giuseppe Ruggeri, Angelo Trotta, Nicola Roberto Zema, STEM-NET: An evolutionary network architecture for smart and sustainable cities, Transactions on Emerging Telecommunications Technologies (ETT) 25 (2013) 21–40, Feature Issue on "Smart Cities—Trends & Technologies".

[10] M.V. Moreno, M.A. Zamora, A.F. Skarmeta, User-Centric Smart Buildings for Energy Sustainable Smart Cities, Trans. Emerg. Telecommun. Technol. 25 (2013) 41–55, Feature Issue on "Smart Cities—Trends & Technologies".

[11] L. Guenda, E. Santana, A. Collado, K. Niotaki, N. Borges Carvalho, A. Georgiadis, Electromagnetic energy harvesting - global information database, Trans. Emerg. Telecommun. Technol. 25 (2013) 56–63, Feature Issue on "Smart Cities—Trends & Technologies".

[12] A. Manzoor, C. Patsakis, A. Morris, J. McCarthy, G. Mullarkey, H. Pham, S. Clarke, V. Cahill, M. Bouroche, CityWatch: exploiting sensor data to manage cities better, Trans. Emerg. Telecommun. Technol. 25 (2013) 64–80, Feature Issue on "Smart Cities—Trends & Technologies".

[13] C. Perera, A. Zaslavsky, P. Christen, D. Georgakopoulos, Sensing as a service model for smart cities supported by internet of things, Trans. Emerg. Telecommun. Technol. 25 (2013) 81–93, Feature Issue on "Smart Cities—Trends & Technologies".

[14] D. Rebollo-Monedero, A. Bartoli, J. Hernandez-Serrano, J. Forne, M. Soriano, Reconciling privacy and efficient utility management in smart cities, Trans. Emerg. Telecommun. Technol. 25 (2013) 94–108, Feature Issue on "Smart Cities—Trends & Technologies".

[15] T. Heo, K. Kim, H. Kim, C. Lee, J. Hong Ryu, Y. Taik Leem, J. Arm Jun, C. Pyo, S.M. Yoo, J.G. Ko, Escaping from ancient Rome! applications and challenges for designing smart cities, Trans. Emerg. Telecommun. Technol. 25 (2013) 109–119, Feature Issue on "Smart Cities—Trends & Technologies".

[16] S. Evenepoel, J. Van Ooteghem, S. Verbrugge, D. Colle, M. Pickavet, On-street smart parking networks at a fraction of their cost: performance analysis of a sampling approach, Trans. Emerg. Telecommun. Technol. 25 (1) (2013) 136–149, Feature Issue on Smart Cities—Trends & Technologies".

[17] M. Dohler, I. Vilajosana, X. Vilajosana, J. LLosa, Smart cities: an action plan, in: Barcelona Smart Cities Congress 2011, Barcelona, Spain, Nov–Dec 2011, 2011.

[18] P. Fritz, M. Kehoe, J. Kwan, "IBM Smarter city solutions on Cloud," White Paper, 2012. Available: http://www-01.ibm.com/software/industry/smartercities-on-cloud

[19] J. Hogan, J. Meegan, R. Parmar, V. Narayan, R.J. Schloss, Using standards to enable the transformation to smarter cities, IBM J. Res. Dev. 55 (1.2) (2011) 4:1–4:10.

[20] A. Biem, E. Bouillet, H. Feng, A. Ranganathan, A. Riabov, O. Verscheure, H. N. Koutsopoulos, C. Moran, IBM infosphere streams for scalable, real-time, intelligent transportation services, Proceedings of SIGMOD 10 (2010) 1093–1104.

[21] N. Walravens, P. Ballon, Platform business models for smart cities: from control and value to governance and public value, IEEE Commun. Mag. 51 (2013) 72–79, Special Issue on Smart Cities.

[22] R. Robinson, W. Webb, R. Robinson, W. Webb, What's so clever about Smart Cities? in: Ingenia Online Magazine, 50, The Royal Academia of Engineering, 2012.

[23] C. Mulligan, M. Olsson, Architectural implications of smart city business models: an evolutionary perspective, IEEE Commun. Mag. 51 (2013) 80–85, Special Issue on Smart Cities.

Machine-to-machine (M2M) communications for e-health applications

E. Kartsakli[1], A.S. Lalos[1], A. Antonopoulos[2], S. Tennina[3], M. Di Renzo[4], L. Alonso[1], C. Verikoukis[2]
[1]Universitat Politècnica de Catalunya (UPC), Barcelona, Spain; [2]Telecommunications Technological Centre of Catalonia (CTTC), Barcelona, Spain; [3]WEST Aquila srl, L'Aquila, Italy; [4]Laboratory of Signals and Systems (L2S), CNRS–SUPELEC–Univ. Paris-Sud XI, Paris, France

20.1 Introduction

Machine-to-machine (M2M) communications is an emerging technology that envisions the interconnection of machines without the need of human intervention. The main concept lies in seamlessly connecting an autonomous and self-organizing network of M2M-capable devices to a remote client, through heterogeneous wired or wireless communication networks. An intelligent software application is usually employed at the remote client to process the collected data and provide the end user with a set of smart services and a practical interface. Although the idea of telematics and telemetry applications is not new, the widespread use of the Internet, along with the trend for ubiquitous connectivity, especially via wireless communication systems, has placed M2M systems on the spotlight of attention for both academia and industry.

The increasing interest in M2M communications poses significant challenges that need to be met. A key issue to be handled is the large number of devices that must be supported in an M2M network, since market predictions estimate that the number of M2M-enabled devices with Internet connectivity will reach up to 50 billion by the end of 2020 [1]. Regardless of the exact figures, the growth rate is impressive, and major efforts are required to provide scalable solutions that support the increasing number of devices with diverse characteristics and requirements. Another challenge stems from the multitude of technical solutions that can be employed in M2M systems. Depending on the application deployment, different approaches may be adopted for the interconnection of M2M devices, such as wired or wireless technologies, short-range or long-range communications, and solutions based on existing open communication standards or proprietary technologies.

The above challenges stress the imperative need for standardization of M2M communications [2]. To this end, the European Telecommunications Standards Institute (ETSI) has established the M2M technical committee that aims to provide an end-to-end view of M2M standardization, focusing on the interoperability of M2M devices with existing standards. In July 2012, ETSI and six other major standards

Machine-to-machine (M2M) Communications. http://dx.doi.org/10.1016/B978-1-78242-102-3.00020-4

development organizations (ARIB and TTC of Japan, ATIS and the TIA of the United States, CCSA of China, and TTA of Korea) joined their efforts in the oneM2M initiative, with the goal of creating a single universal standard for M2M communications [3]. This global standardization effort is crucial to enable the integration of heterogeneous technologies in order to achieve seamless end-to-end connectivity, removing potential barriers to market growth.

The penetration of M2M solutions for monitoring and remote control in a wide range of markets, including building and industrial automation, security and surveillance, smart metering, energy management, and transportation, generates great business opportunities. The application of M2M-enabling technologies to the healthcare sector, in particular, is expected to be one of the major M2M market drivers: market projections forecast that more than 774 million health-related devices with M2M connectivity will be available by 2020, yielding a total revenue of 69 billion euros in that year [4].

The use of information and communication technologies to facilitate and improve healthcare and medical services, often referred to by the term e-health, is bringing a shift to healthcare delivery. The M2M paradigm in the context of e-health involves the use of appropriate sensor devices on patients to enable the remote monitoring of vital signals, the early detection of critical conditions, and the remote control of certain medical treatments [5]. The medical sensors, placed in the vicinity of, or inside, the human body, are usually interconnected through a short-range wireless technology, thus forming a wireless body area network (WBAN). An M2M-enabled gateway node collects all the sensory data from the WBAN and forwards them to a remote online server, where processing and integration with medical-related software applications take place. The connection of the gateway to the Internet is generally based on long-range communication access technologies for wireless local area networks/ wireless metropolitan area networks (WLANs/WMANs).

The emerging application scenarios are numerous, including the active management of diseases such as diabetes (e.g., by measuring blood sugar levels and controlling the insulin dosage accordingly), the support for independent aging to the elderly (e.g., by tracking their medication intake and their activity level), and the monitoring of personal fitness activities to improve health and well-being (e.g., by logging health and fitness indicators during workouts) [5]. Overall, e-health can offer significant benefits for both patients and healthcare providers, reducing the cost of healthcare services while ensuring enhanced quality, efficiency, and flexibility in healthcare delivery.

In the recent years, the research community has been motivated by the diversity of applications, the promising benefits, and the potential market opportunities of e-health M2M solutions. The main technological challenges for M2M communications, the most representative usage models, and the status of global standardization efforts are discussed in Ref. [6]. Focusing on the emerging M2M technologies for e-health applications, a technical discussion on the communication network design is given in Ref. [7], but generally, most of the related work adopts a high-level approach. In Ref. [8], some interesting challenges from the network perspective are identified, while interoperability issues and recent standardization efforts are presented in Ref. [9]. On a different level, a lot of research activity has been focused on the body area domain, on the design of medical sensor devices [10], and on the main advances and challenges in the field of WBANs [11–14]. In this chapter, we aim to provide a comprehensive overview of M2M systems for e-health applications from a wireless communication perspective. We discuss different aspects of the M2M ecosystem in the healthcare domain, focusing mainly on the end-to-end connectivity: First, the high-level ETSI architecture for M2M systems and the key elements for a healthcare application scenario are described in Section 20.2. Section 20.3 provides an overview of the enabling wireless communication technologies that can be employed in M2M systems, focusing on both open communication standards and proprietary solutions. Then, in Section 20.4, we offer an end-to-end perspective of M2M systems for e-health, focusing on the integration and convergence of different communication technologies, through both theoretical approaches and test bed implementation, and presenting the key security challenges that arise. The survey of existing works in the field of M2M communications for e-health is completed by a summary of current research projects in Section 20.5, whereas some concluding remarks are given in Section 20.6.

20.2 M2M network architecture

In recent years, ETSI has been actively engaged in the development of a standard for M2M systems, with the objective of ensuring interoperability between the diverse M2M components and the already existing technologies. To this end, ETSI proposes a high-level horizontal architecture, dividing the system into three domains: (i) the device and gateway domain, where the M2M devices communicate with a gateway through short-range area networks, (ii) the network domain that connects the gateway to the applications through long-range access and core communication networks, and (iii) the application domain, where various application services are defined depending on the use case [15].

Figure 20.1 illustrates an example of the M2M system architecture for wireless healthcare applications. The ETSI architecture consists of five key elements that are described below [15,16]:

- The *M2M devices*, which are devices capable of transmitting data autonomously or after receiving a data request. In the context of healthcare applications, the M2M devices are principally low-power medical sensor or actuator devices with embedded wireless communication modules.
- The *M2M area network*, also known as the capillary network, which is a short-range network that provides connectivity between the M2M devices and the gateway. In the considered scenario, the area network will also be referred to as WBAN, given that the M2M devices are deployed near or within the human body.
- The *M2M gateway*, which acts as a proxy between the M2M devices (interconnected through the WBAN) and the network domain. Practically, the gateway must be a portable device with advanced processing capabilities and multiple radio interfaces, able to operate in technologies employed by both the WBAN and the communication network. Typical examples of M2M gateways include smartphones, personal digital assistants (PDAs), and smart watches.

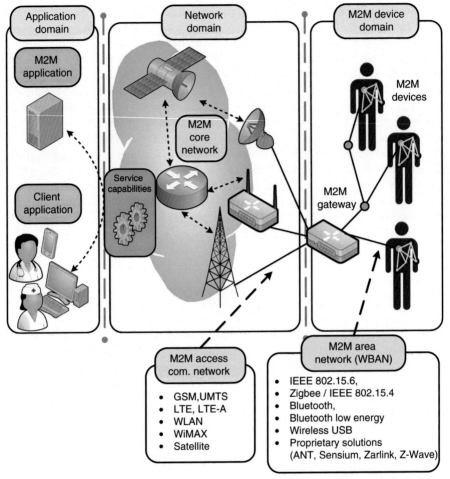

Figure 20.1 Simplified M2M architecture for wireless connectivity in healthcare application scenarios.

- The *M2M access communication network*, which connects the M2M gateway to the M2M application server via the Internet.
- The *M2M application server*, which is the middleware layer that provides data to the specific business applications.

20.3 Enabling wireless technologies: standards and proprietary solutions

The ETSI M2M architecture framework, described in the previous section, defines two levels of wireless connectivity: (i) short-range connectivity within the M2M area network (WBAN) and (ii) long-range connectivity within the M2M access communication network. In this section, we give an overview of the key enabling technologies for each type of network.

20.3.1 M2M area network

The main requirements for the candidate WBAN technologies are short-range connectivity, low-power operation, and quality of service (QoS) provisioning to support real-time monitoring and critical events in e-health applications. The presented technologies are classified into two categories: (i) open communication standards that ensure interoperability and compatibility among different vendors and (ii) proprietary solutions available in the market that are tied to specific vendors. Here, we briefly describe the most prevalent available technologies in both categories, giving particular emphasis to the IEEE 802.15.6 standard defined specifically for WBAN communications. The main technical features of the presented technologies for M2M area networks are summarized in Table 20.1.

20.3.1.1 Open standards

Bluetooth (http://www.bluetooth.com) is an industrial standard for short-range wireless communications. The lower layers of the Bluetooth protocol stack, namely, the physical layer (PHY) and the medium access control (MAC) layer, are specified in the *IEEE 802.15.1* standard [18], whereas a number of application profiles are defined at the upper layers to support application-specific tasks. Bluetooth operates at the unlicensed 2.4 GHz industrial, scientific and medical (ISM) band, using spread spectrum and frequency-hopping transmission techniques. It is mainly intended for peer-to-peer connections or small ad hoc networks formed by a master and up to seven slave devices. In the context of e-health, Bluetooth defines a specific profile for healthcare and fitness applications, called Health Device Profile. The latest version of the Bluetooth standard includes *Bluetooth low energy*, a technology that enhances energy efficiency by providing ultralow-power operation and reliable point-to-multipoint data transfer between M2M devices.

ZigBee (http://www.zigbee.org) is a protocol stack widely applied in remote control and sensor applications. ZigBee defines the application, the security, and the network layer specifications and is built on top of the PHY and MAC layers defined by the *IEEE 802.15.4* standard [19]. It mainly operates at the unlicensed 2.4 GHz ISM band, employing direct sequence spread spectrum techniques for interference tolerance. ZigBee defines the Personal, Home, and Hospital Care (PHHC) profile, ensuring device interoperability for secure and reliable monitoring of noncritical healthcare services, such as chronic disease management, obesity, and aging.

Wireless Universal Serial Bus (WUSB) is a short-range, high-bandwidth wireless radio communication standard that operates at the frequency range of 3.1–10.6 GHz [20]. WUSB is placed among the first candidates to be commercially available for short-range high-speed wireless interfaces, offering reliable and fast point-to-point links with a security level similar to wired communications. At its latest version,

Table 20.1 Enabling WBAN technologies for e-health applications [9,17]

	Technology	Frequency band	Data rate	Range (m)	Modulation	Topology
Open standards	Bluetooth low energy	2.4 GHz ISM	1 Mbps	10	GFSK	Star
	WUSB	3.1–10.6 GHz UWB	48–53 Mbps	3–10	MB-OFDM	Star
	ZigBee	868 MHz, 915 MHz, 2.4 GHz ISM	250 Kbps	30–100	O-QPSK	P2P, tree, star, mesh
	IEEE 802.15.6	Multiple, including 402–405 MHz MICS, 900 MHz ISM, 2.4 GHz ISM, and 3.1–10.6 GHz UWB	100 Kbps–12 Mbps	~3	Multiple, including $\pi/2$-DBPSK, $\pi/4$-DQPSK, and IR-UWB on–off	Extended star
Proprietary solutions	ANT	2.4 GHz ISM	1 Mbps	10–30	GFSK	P2P, mesh, tree, star
	Sensium	868–915 MHz	50 Kbps	5–25	BFSK	Star
	Zarlink	402–405 MHz MICS 433–434 MHz	200–800 Kbps	2	2FSK/4FSK	P2P
	Z-Wave	900 MHz ISM	9.6–40 Kbps	30	GFSK	Mesh

WUSB is expected to offer the same functionality as the standard wired USB devices, without the cabling restrictions.

The above technologies share some weaknesses when applied to e-health scenarios, mainly due to the fact that they have been designed with different target applications in mind. Bluetooth, for example, has limited scalability and QoS support and is not very energy-efficient. Bluetooth low energy reduces power consumption but the scalability and QoS are still an issue. ZigBee, on the other hand, is energy-efficient and scalable but provides lower data rates, whereas WUSB can only support to peer-to-peer connection topologies.

In an effort to overcome these limitations and tackle the specific requirements of WBANs, the *IEEE 802.15.6* standard has recently been issued for short-range wireless communications in the vicinity of, or inside, the human body. A key characteristic of IEEE 802.15.6 is that it supports operation at very low transmission powers and macroscopic and microscopic power management through hibernation and sleep modes, respectively, in an effort to increase battery lifetime and comply with the safety regulations limits on the specific absorption rate (SAR) level for in-body communications. It also provides QoS guarantees for the prompt delivery of alarms in emergency situations and employs robust security mechanisms to provide privacy and confidentiality protection of the medical data.

The standard provides specifications for the PHY and the MAC layers. With respect to the PHY, three different technologies are supported: (i) the narrowband (NB) PHY, which introduces low control overhead, very low peak power consumption, and robustness against interference; (ii) the ultra-wideband (UWB) PHY, based on a technology for transmitting information over a large bandwidth, offering high performance, robustness, low complexity, and ultralow-power operation; and (iii) the human body communication (HBC) PHY, which uses the human body as a means of propagation for the data transmission. Operation at multiple frequency bands is supported, starting from the medical implant communications service (MICS) band of 402–405 MHz, reserved for medical implant communication, up to the unlicensed 2.4 GHz ISM band.

With regard to the MAC layer, IEEE 802.15.6 defines eight levels of user priorities, with level 0 corresponding to the lowest priority class and level 7 being assigned to the highest priority traffic for emergency situations or medical implant event reports. The WBAN operates in an extended star topology, with all nodes connected directly, or through a single relay node, to the coordinating node, denoted as the hub. The standard considers both contention-based access and contention-free channel access.

Two random access schemes for contention-based access are supported, namely carrier sense multiple access with collision avoidance (CSMA/CA) and slotted ALOHA. The standard also supports three contention-free access modes that require centralized control from the WBAN hub: (i) improvised access, which is a polling scheme for immediate of future allocation for both uplink and downlink; (ii) scheduled access, which are periodic allocations for uplink, downlink, or bidirectional transmissions that are negotiated between the nodes and the hub during the association phase; and (iii) unscheduled access, which is a best-effort version of the scheduled access.

As far as security is concerned, the standard defines three different connectivity levels: (i) unsecured communication, (ii) authentication only, and (iii) authentication and encryption.

20.3.1.2   Proprietary solutions

Apart from the open communication standards, some proprietary solutions for WBANs are also available in the market. *ANT* (http://www.thisisant.com) is a wireless networking protocol and embedded system solution, operating at the 2.4 GHz ISM band, originally designed for general purpose wireless sensor networks. It provides ultralow-power consumption, low latency, and simplicity in implementation, while limited QoS support and low-data rates are its main drawbacks. The next generation of ANT, named *ANT+* protocol, focuses on the seamless interoperability between sensors and monitoring devices of different manufacturers (e.g., heart rate sensors with smart watches) and defines various health and fitness device profiles.

Sensium (http://www.toumaz.com) is an ultralow-power platform designed for healthcare and lifestyle management applications. The Sensium platform integrates various sensor devices with a processing unit and a low-power transceiver operating at 900 MHz. A master–slave topology is considered, where slave sensor nodes periodically send data to the Sensium platform that processes them and forwards them to a monitoring device through a gateway.

Zarlink (http://www.zarlink.com) has designed a series of ultralow-power transceivers for medical implants, operating at the 402–405 MICS band. The Zarlink devices achieve extremely low-power consumption, by spending most of their time in a deep sleep mode and being woken up by special wake-up signals transmitted at the 2.4-GHz band. Their drawbacks are low-data rates and limited QoS provisioning.

Z-Wave (http://www.z-wave.com) is a proprietary protocol designed for home control and monitoring. It provides low-power interconnection of home devices such as lights, thermostats, and door locks to provide a smart living environment. It operates at the 900-MHz band, thus avoiding interference from most short- and medium-range wireless technologies (operating at the 2.4-GHz band), but offers very low-data rates.

20.3.2   M2M access communication network

The M2M access communication network requires medium- or long-range wireless technologies with high-data-rate capabilities. Hence, wireless technologies for WLANs and WMANs are the main candidates: the first provide connectivity within a limited area, either indoors (e.g., a house or a hospital ward) or outdoors (e.g., a public hotspot), whereas the latter offer ubiquitous connectivity to an extended coverage area. Since the key employed technologies are widely used and well known, they are briefly stated next:

- The *IEEE 802.11* specification [21] for the PHY and the MAC layers, commonly referred to as *Wi-Fi*, is the predominant technology adopted in WLANs for the 2.4- and 5-GHz ISM bands.
- The *IEEE 802.16* standard [22], commonly known as *WiMAX*, defines a broadband access technology for WMANs, offering high rates and QoS provisioning.
- *Long Term Evolution (LTE)* and its evolved version *LTE-Advanced (LTE-A)*, are 3rd Generation Partnership Project (3GPP) mobile communications standards for high-data rates and spectral efficiency.

20.4   End-to-end solutions for M2M communication: connectivity and security

The goal of an e-health application is to provide a bridge between the patient and the medical personnel. Hence, the M2M system must provide end-to-end connectivity, connecting the medical sensor devices via the M2M gateway to the Internet and ultimately to the application server. Since standardization efforts on M2M communications are still underway, there are many different approaches to achieve end-to-end connectivity. However, despite their differences, most approaches partly follow the system architecture shown in Figure 20.2.

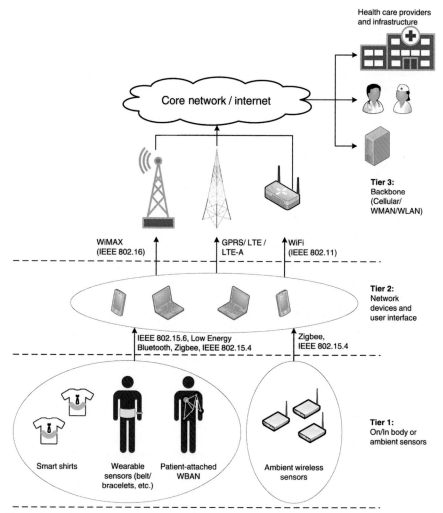

Figure 20.2 Example of an M2M system for e-health applications.

In this section, we examine end-to-end solutions for M2M communications, as well as security and privacy issues pertinent to e-health applications. In particular, we first discuss theoretical works that focus on the high-level integration of WLAN/WMAN access communication technologies with WBANs. Then, we give some practical examples of end-to-end solutions by presenting test bed implementations for health-care monitoring. Finally, we identify security and privacy issues that arise in M2M communication scenarios and are particularly related to the sensitive medical data handled by e-health services.

20.4.1 Technology integration for M2M communications

The integration of different wireless technologies into a unified end-to-end solution for e-health applications is an interesting topic that has attracted a lot of attention in the research community. Since there is no unique solution to this problem, in this section, we present different approaches proposed in the literature. Some works identify the key challenges and application scenarios for the seamless integration of different technologies, while others provide analytic frameworks for end-to-end performance evaluation.

The work published in Ref. [23] studies the challenges concerning the deployment of a Wi-Fi-based network within a healthcare facility such as a hospital unit. Practical guidelines for the design, dimensioning, and the installation of the network, as well as appropriate validation methods, are provided, in order to satisfy the specific e-health application requirements.

In Ref. [24], a two-tier network architecture is considered. The lower tier consists of the WBAN, where multiple sensor devices worn by a single patient are connected to a coordinating node by employing the CSMA/CA access mode of IEEE 802.15.4. At the upper tier, multiple WBAN coordinators (corresponding to multiple patients) located within a specific area, for example, a hospital ward, communicate with an access point through Wi-Fi. End-to-end packet delay and access time have been modeled as a function of the number of coexisting WBANs. An extension of this work in Ref. [25] has introduced service differentiation to prioritize high-rate data streams (e.g., electroencephalography (EEG) data) over low-rate streams (e.g., electrocardiography (ECG) data). The idea is to provide contention-free access to the high-priority data flows while maintaining CSMA/CA access for the lower-priority nodes.

In Ref. [26], the authors discuss the feasibility of employing a hybrid network based on Wi-Fi and WiMAX technologies as the access communication network in an M2M system for e-health services. The integration of the two technologies poses several challenges, mainly related to QoS provisioning, connection admission control, scheduling, and mobility management through seamless vertical handovers. The envisioned heterogeneous deployment scenario considers nodes with either single- (Wi-Fi or WiMAX) or dual-radio interface and aims to provide wireless connectivity between different subnetworks, including WBANs, home-care networks, mobile patients, and networks of healthcare providers (such as intranets of hospitals, clinics, and drugstores).

Another remote monitoring scheme that provides ubiquitous connectivity for mobile patients has been presented in Ref. [27]. In the proposed scheme, a patient-attached monitoring device collects the WBAN data, classifies them as high priority (e.g., critical data such as blood pressure, pulse rate, and heart rate) or normal priority (e.g., ECG signal), and forwards them towards the e-health provider through an heterogeneous Wi-Fi/WiMAX access communication network. The access technology is selected depending on the patient's location, considering that Wi-Fi hotspots cover only specific (mainly indoor) locations and WiMAX has a wider (outdoor) coverage. In addition, two types of connections are provided by the network operator: (i) low-cost reserved connections, allocated to patients for given amounts of time (e.g., weeks) and (ii) high-cost on-demand connections, employed when the available bandwidth for reserved connections is not enough to cover the traffic load. The authors approach this e-health scenario from the service provider's side, who has to buy in advance a certain number of reserved connections from the network operator to serve a given number of patients. Stochastic programming techniques are used to determine the optimal number of reserved connections for each wireless technology in order to minimize the provider's cost.

As far as LTE-based solutions for healthcare applications are concerned, the relevant works in the literature are limited. The impact of 4G communication technologies on e-health and the emerging challenges are discussed in Ref. [28]. In Ref. [29], a mapping between the QoS requirements of e-health services and the existing service classes defined by 3GPP standards is proposed, aiming to provide guidelines for network operators. A cross-layer design for QoS of medical video streaming over mobile WiMAX (IEEE 802.16e) and high-speed packet access (HSPA)[1] and a comparison between the performance of the two technologies are proposed in Ref. [30], opening the road to further investigation on LTE-based e-health applications.

20.4.2 Test bed implementation of M2M solutions

Research on end-to-end connectivity is not limited to a theoretical-only level. During the last years, research efforts have been devoted to the actual implementation of M2M networks. In this section, we present the most representative examples of implemented end-to-end solutions for e-health applications.

In Ref. [31], the authors introduce a new file format for the transfer of sensory data and implement a pilot test bed for an end-to-end patient monitoring application. Their contribution is twofold. First, they present an enhanced version of a standard protocol for communication among ECG devices, by proposing an adaptive data structure that can handle multiple vital signals, as well as data for positioning, allergies, and demographic information on patients. The definition of a standardized data structure is an

---

[1] HSPA technology is a precursor to LTE.

important step towards the integration of the medical data measured by the WBAN sensors with various e-health information systems for monitoring or administrative purposes, belonging to hospitals, individual caregivers, home care, etc. Second, they implement a test bed of an M2M healthcare application for the remote monitoring of patients suffering from heart problems. The patient is equipped with a WBAN formed by a number of wearable sensors, a Global Positioning System (GPS) device, and a PDA. The PDA aggregates the sensory and geolocation data, as well as any additional information inserted manually by the patient, and plays the role of the M2M gateway. On the one hand, it employs Bluetooth technology to communicate with the WBAN nodes, and, on the other hand, it has mobile ADSL capabilities to forward the data to the remote server located at a hospital facility. To the other end of the system, a portable data acquisition system is considered, consisting of a medical monitor device, a GPS, and a laptop with Internet connectivity. Finally, a software application has been developed for the processing and visualization of the data retrieved by the healthcare provider.

The pilot testing of the proposed solution on real patients has revealed some very interesting conclusions. From the doctors' perspective, the use of the M2M e-health system has been an overall positive experience, facilitating patient monitoring and the collection of data. The patients, on the other hand, have given a more neutral evaluation. Even though they have generally been satisfied by the experience, they have shown more concerns on the wearability of the sensors, the user-friendliness of the software application, and the data collection process. A two-tier architecture is considered in Ref. [32] to implement a remote monitoring application for patients with chronic obstructive pulmonary disease (COPD). Bluetooth is used as the WBAN technology for the communication between the sensors and the coordinating node (e.g., a PDA). Apart from the Bluetooth network interface, the coordinating node can support two additional long-range wireless technologies for Internet connectivity with a remote medical server: cellular GPRS and Wi-Fi. The authors perform an interesting experiment by measuring system performance metrics of these two upper-tier technologies. The study clearly shows that GPRS and WLAN have complementary power and delay profiles: the energy consumption of GPRS is low but high delays may be observed, whereas Wi-Fi has higher energy cost but lower delays.

Based on the observed results, the authors provide guidelines for the design of an adaptive protocol that switches between the two long-range technologies depending on the scenario: (i) Wi-Fi is the recommended technology, when available, especially in emergency situations due to low-data latency. In these cases, the GPRS interface should remain on but at an idle state, employed only to receive incoming calls if needed and (ii) If the Wi-Fi connection is not available, the Wi-Fi network interface should be switched off completely and GPRS should be employed for communication.

In some works, ambient sensor networks for environmental monitoring are employed in conjunction with WBANs, in order to provide additional information on the patient's environment, such as temperature, humidity, and light conditions. Along this line, a three-tier network architecture is proposed in Ref. [33], for the remote monitoring of elderly or chronic patients in their residence. The lower tier consists of two systems: (i) a patient-worn fabric belt, which integrates the medical sensors and is equipped with a Bluetooth transceiver, and (ii) the ambient wireless sensors that form a ZigBee network and are deployed in the patient's surroundings (e.g., in the patient's home or a nursing house). In the middle tier, an ad hoc network of powerful mobile computing devices (e.g., laptops and PDAs) gathers the medical and ambient sensory data and forwards them to the higher tier. The middle-tier devices must have multiple network interfaces: Bluetooth and ZigBee to communicate with the lower tier and WLAN or cellular capabilities for connection with the higher layer. Finally, the higher tier is structured on the Internet and includes the application data-bases and servers that are accessed by the healthcare providers. The study involves a real implementation of the proposed architecture and tackles several security issues that arise along the three tiers. The proposed framework offers a flexible and secure solution for the monitoring of multiple patients that can be applied to different sce-narios, including home, hospital, and nursing home environments.

Sensor networks can also be employed for patient localization purposes. In Ref. [34], the authors propose a system architecture based on two independent subsystems for the monitoring and location tracking of patients within hospital environments. The healthcare monitoring subsystem consists of smart shirts with integrated medical sen-sors, each equipped with a wireless IEEE 802.15.4 module. The location subsystem has two components: (i) a deployment of wireless IEEE 802.15.4 nodes that are installed in known locations within the hospital infrastructure and broadcast periodic beacon frames and (ii) IEEE 802.15.4 end devices, held by the patients, that collect signal strength information from the received beacons. Both subsystems transmit their respective data (i.e., medical sensory data and signal strength information) to a gate-way through an IEEE 802.15.4-based ad hoc distribution network. The gateway has wired Internet connectivity and forwards the data to the management server and the monitoring e-health application. The proposed system has been tested with success in a hospital, achieving high reliability, sufficient battery lifetime of the sensors, and real-time data reconstruction. In terms of usability, the obtained feedback of the medical personnel has been taken into account to improve the software interface.

20.4.3   Security and privacy issues

Despite the great potential for improving the quality of life, the introduction of M2M e-health solutions raises considerable security and privacy challenges, mainly due to the confidential nature of the medical data exchanged in healthcare environments. The remaining of this section discusses the main security challenges, along with the most representative solutions proposed in the literature.

20.4.3.1   Challenges

In order to preserve the confidentiality of sensitive medical data, international orga-nizations oblige healthcare providers to follow specific privacy rules [35,36], by set-ting strict civil and criminal penalties to punish any sensitive data leakage. Hence, a plethora of research surveys has been recently released [37–42], focusing mainly on the M2M area network. In this context, the security framework is divided into two slightly overlapping parts: (i) *data security* and (ii) *data privacy*. The former deals with the secure data storage and transfer through the (usually wireless) medium, while the latter tackles access rights and secrecy issues related to the patient's private information, including among others identity, time, and location.

The requirements for the schemes of the first class do not substantially differ from the general requirements of wireless networks. However, the application of traditional security methods in M2M area networks is not straightforward, due to the intrinsic characteristics of wireless sensor devices, such as small size, restricted computational capabilities, and limited energy resources. In particular, the main issues concerning security are as follows:

- Data integrity: The broadcast nature of M2M area networks enables adversary users to intervene in the transmissions, posing high risks in emergency life-critical events. Therefore, proper data integrity mechanisms have to be put in place to ensure that the received data have not been altered.
- Data authentication: In healthcare environments, the authentication of the transmitted data is essential to guarantee and verify whether the received packets come from a trusted source.
- Data availability: The medical data and information have to be available upon any-time request, without being hampered by denial-of-service (DoS) attacks.
- Data freshness: In the e-health domain, data freshness is of crucial importance, since it indicates that the received physiological patient signals are up-to-date and not simply replayed by malicious users.

On the other hand, on ensuring privacy, the following fundamental requirements have to be considered:

- Data confidentiality: As mentioned above, health data are generally subject to ethical and legal obligations of confidentiality, thus offering substantial protection from malicious eavesdropping, which is further facilitated by the broadcast wireless nature.
- User authentication: User authentication is a strong prerequisite in healthcare environments so as to prevent unauthorized users from gaining access to sensitive medical information.
- User localization: The location privacy is one very important aspect in M2M area networks. Concealing the patient's location precludes malicious users to claim legitimate coordinates in the network while hindering any false signals that create confusion with regard to the patient's real physical location.

Apart from the aforementioned commonplace security requirements, new challenges are posed by the M2M concept that includes the interconnection of different technologies, as well as the data storage in multiple different physical locations. Considering the importance and the criticality of medical data, we will provide a brief an overview of the security-related research conducted in the e-health domain, providing useful insights on issues that arise either in the M2M area network or in the whole end-to-end system.

20.4.3.2 Approaches

Several cryptographic schemes have been already proposed in the literature, trying to provide effective solutions for data integrity, authentication, and confidentiality. Most schemes provide encrypting capabilities based on a secret key shared among nodes, either in software level [43] or in hardware level [44,45]. In addition, notable advances have been observed in the implementation of elliptic curve cryptography (ECC) [46,47], which has emerged as a promising alternative to RSA-based algorithms, guaranteeing the same level of security while employing a much smaller key size. To overcome the limitations of traditional symmetrical cryptography in wireless networks, biometrics [48,49] exploit particular physiological values of the patient's body in order to provide efficient cryptographic techniques in WBANs, thus gaining significant ground in healthcare environments.

Apart from the important technological achievements with regard to cryptography, several other works deal with privacy issues, motivated by the challenges that arise on the higher tiers of the network architecture due to the involvement of many different entities in end-to-end approaches. Narayanan and Gunes [50] present an information protection framework against unauthorized access in cloud-provisioned multitenant healthcare systems. The term multitenant explicitly refers to the plethora of authorities that need access to the sensitive medical data, including hospitals, clinics, pharmacies, professionals, and, in some cases, the patients themselves. The authors propose an improved access control scheme for cloud instances by extending the well-known task–role-based access control model [51] to include adaptive user roles and tasks in order to support multitenant cloud applications.

Similar issues about guaranteeing user privacy across several providers and organizations are also addressed in Refs. [52,53]. In particular, the work in Ref. [52] aims at extending the traditional service-oriented architecture framework to define a flexible policy-based approach for defining and monitoring streaming event data based on a general publish/subscribe model in business-to-business healthcare networks. The policy-based framework presented in this paper is a specific part of a comprehensive information system, named palliative information system, designed to support day-to-day home-care delivery for palliative care patients.

A preliminary approach to address security and privacy issues is presented in Ref. [53]. In emergency scenarios, the "on-the-fly" network integration and the information exchange among different entities are of paramount importance. However, the achievement of these goals is complicated, since each domain may correspond to a different authority. The proposed solution supports medical device integration and authentication among networks of different providers, dealing with interoperability challenges by using open standards like ISO/IEEE 11073–20601 [54], Device Profile for Web Services, and Bluetooth Health Device Profile for medical data transmission. Their solution claims to ensure cost-effectiveness, simplicity, and emergency support through sharing devices among authorities and dynamically reintegrating them in case of network alteration. Moreover, a new DPWS-based security model is described, without, however, considering scope crucial parameters such as trusting and key distribution among the authorities. Finally, it is worth noting the importance of guaranteeing an undisclosed location for the nodes (and consequently for the patient) in healthcare scenarios. Although location information is necessary in sensor networks, it can evolve in a serious threat in case that security and privacy restrictions are not met. To this end, many works in the literature have set the focus on designing effective secure localization methods [55,56], and the interested reader may further refer to Ref. [57] for a complete guide to secure localization schemes in sensor networks.

20.5 Existing projects

The aim of this section is to present an overview of the most recent and relevant research-funded projects for e-health applications.

HEALTH@HOME (Health at Home) (www.aal-europe.eu/projects/healthhome) aims to provide an end-to-end solution for the remote monitoring of cardiovascular and respiratory patient parameters. The data are continuously gathered through an automatic processing system and are accessed by the responsible medical personnel. A typical client/server architecture is adopted, where the client side is a residential gateway located at the patient's home, able to collect data from the biomedical sensors through wireless Bluetooth links. The most significant measured signals are ECG, SpO2, weight, blood pressure, chest impedance, respiration, and body posture. The measured data are sent through the gateway to a server located at the health service facilities that is integrated with the hospital information system. The gateway communicates with the server through ADSL as the primary transmission channel or mobile broadband (i.e., GSM/GPRS/UMTS) as the secondary (backup) data channel. Alarms are sent by short message service (SMS) directly to the physicians, the patients' relatives, and their caregivers. The http addresses the security issues in the communication between the gateway and the server through a certificate validation process. The proposed solution has been tested on 30 patients during monitoring periods of at least one month and has received positive feedback by both patients and medical personnel, as a reliable, user-friendly means of remote control and management of acute conditions.

IS-ACTIVE (Inertial Sensing Systems for Advanced Chronic Condition Monitoring and Risk Prevention) (www.is-active.eu) provides a person-centric healthcare solution for elderly people that suffer from chronic obstructive pulmonary disease (COPD). The project aims to provide real-time support to users in order to monitor, self-manage, and improve their physical condition by encouraging physical activity through visual feedback and real-time motivational cues. Since motion sensing is one of the main goals of IS-ACTIVE, inertial sensor nodes (accelerometers and gyroscopes) and sensor nodes that measure physiological data (heart rate, oxygen saturation, etc.) are employed. The nodes form a WBAN and report the sensory data to a central gateway, connected by cable to a computer. The node communication takes place at the 2.4-GHz ISM band, while two operation modes are adopted: (i) low-power, low-data-rate IEEE 802.15.4-compatible implementation, for long-term sensing and monitoring, for example, for activity-level monitoring applications, and (ii) high-data-rate, real-time motion capture via the proprietary FastMac networking protocol, for short-term, detailed sensory data acquisition, for example, for algorithm design and evaluation.

HELP: Home-based Empowered living for Parkinson Disease Patients (www.aaleurope.eu/projects/help) targets at designing a health monitoring system able to control disease progression and to mitigate Parkinson disease (PD) symptoms, thus improving the quality of life of affected elderly people. Although it provides an end-to-end solution that employs M2M communication for monitoring patients with PD, its aim is to design a control system for a subcutaneous infusion pump that administers the exact required drug dose according to the patients' level of activity without focusing on communication issues. This system is composed of the following components: (i) an intraoral electronic drug delivery device with miniaturized, noninvasive, and removable design, (ii) an external pump that delivers higher amounts of drug, (iii) a WBAN to gather information on the user environment to detect blockades, (iv) a telecommunication and service infrastructure to analyze and transfer data exchanged between the user and the automated system, and (v) a remote care unit for patient supervision.

WiserBAN (Smart miniature low-power wireless microsystem for Body Area Networks) (www.wiserban.eu) objective is to improve personal sensing capabilities by using tiny, unobtrusive, long-lifetime radio microsystems for WBAN sensor nodes, such as hearing aids, cardiac implants, insulin pumps, and cochlear implants. Particular emphasis is given on (i) sensor miniaturization for both implantable and wearable-based WBANs, (ii) the sensor data processing efficiency, and (iii) the development of a flexible/reconfigurable and low-power radio baseband system. To meet the aforementioned requirements, WiserBAN proposes a highly integrated microsystem that includes radio and antenna and data processing units. Two RF blocks in 65 nm CMOS technology that operate in the 2.4 GHz and the 402–405 MHz MICS band are implemented, along with reconfigurable PHY and MAC-layer protocols. Apart from the key medical-related use cases, WiserBAN has an ambitious exploitation plan with possible applications in home energy management, smart grids, and even military communication scenarios.

The goal of *CAALYX-MV: Complete Ambient Assisted Living Experiment—Market Validation* (www.caalyx-mv.eu) is to provide an end-to-end solution that is focused on improving the quality of life of elderly people. The proposed solution is composed of (i) a home system capable of monitoring and controlling social and health status of elder people and providing them with some tools and services to support their daily activities; (ii) a roaming system that comprises a smart textile shirt able to measure specific vital signs, detect falls, and communicate emergencies; and (iii) a care system for the monitoring of individuals by family, caretakers, and health services. All sensors in the WBAN are wearable; measure different parameters such as motion, blood pressure, and heart rate; and communicate using Bluetooth links with a mobile phone. The sensory data are sent through standard low-cost networking equipment to a GPS-enabled smart phone (3G/UMTS) that runs a completely autonomous software application. The application continuously analyzes sensor data in order to identify problematic conditions and promptly alert the care system.

The proposed system will be validated through three pilot programs that will test the usability and acceptability of the system by the users (both patients and caregivers) and will evaluate the reliability and detection accuracy of health problems in the monitored patients.

Help4Mood (http://www.help4mood.info/site/default.aspx) aims at developing an end-to-end system to help the recovery of people with major depression. The system is designed to be used together with other forms of therapy, such as self-help, counseling, or medication. The main components include: (i) a personal monitoring system to keep track of important behavior aspects, composed of sensors for both user activity and sleep monitoring, (ii) an interactive virtual agent asking patients about their health and well-being and providing a portal to trusted health information, and (iii) a decision support system handling the virtual agent to allow its customization to the individual needs of the person with depression.

The sensor devices communicate by using a proprietary low-power RF network protocol named SimpliciTI [58] over Bluetooth. To increase energy efficiency and reliability, the system adopts the idea of cooperation between nodes, achieved through the slight modification of the MAC protocol.

EXALTED (EXpAnding LTE for Devices) (http://www.ict-exalted.eu/) is a project that focuses on developing a new scalable network architecture to support the most challenging requirements for future wireless communication systems. Its aim is to provide secure, energy-efficient, and cost-effective M2M communications for low-end devices. Motivated by the inability of LTE Releases 8, 9, and 10 to serve a multitude of low-data-rate devices in an energy-, spectrum-, and cost-efficient way, EXALTED proposes improvements that can be easily integrated in the new LTE-M backbone. The LTE-M extension aims to fulfill the specific energy, spectrum, cost, and efficiency constraints of M2M communications, by proposing improvements on the PHY, MAC, Radio Link Control, Packet Data Convergence Protocol, and the Radio Resource Control layers of LTE. Security issues for LTE-M networks are also addressed. Finally, proof of concept of the proposed techniques will be provided through the implementation of a realistic test bed.

WSN4QoL: Wireless Sensor Networks for Quality of Life (http://www.wsn4qol.eu/) is a project focused on wireless communication technologies for e-health applications. The main objectives of WSN4QoL are (i) to provide a protocol stack architecture, which can accommodate a variety of protocols, algorithms, and sensor devices for healthcare applications; (ii) to develop reliable, energy-efficient, interference-robust communication protocols and algorithms; (iii) to develop distributed localization protocols that meet the constraints imposed by WBANs in healthcare scenarios; and (iv) to propose effective and efficient security solutions for the proposed communication protocols. The proposed protocols and algorithms will be integrated in healthcare commercial devices, in order to evaluate the performance improvements in realistic environments. To conclude, Table 20.2 provides a comparison among the main characteristics of the projects presented here.

Table 20.2 Summary of existing projects on e-health applications

Project	Main application	M2M area net. technology	M2M com. net. technology	End to end	Real-world validation	Security issues
HEALTH@HOME	CVD	Bluetooth	ADSL			
IS-ACTIVE	Fall detection	Bluetooth	UMTS	✗		✗
HELP	Parkinson's disease	Bluetooth	–			✗
WiserBAN	Healthcare sensors	IEEE 802.15.6	–	✗	✗	✗
CAALYX-MV	Independent living	Bluetooth	3G UMTS/Wi-Fi			
Help4Mood	Depression management	Bluetooth/SimpliciTI	ADSL			
EXALTED	Technology-oriented	–	LTE-M	✗		
WSN4QoL	Communication-oriented	IEEE 802.15.6	–	✗		

20.6   Concluding remarks

This chapter has provided an overview of M2M systems for e-health applications from a wireless communication perspective. After describing the high-level ETSI M2M system architecture, we presented the key candidate technologies that can be employed at different parts of the system to provide short-range interconnection of the sensor devices and long-range Internet connectivity. We, then, focused on end-to-end connectivity in M2M systems. After discussing the integration challenges between diverse communication technologies, we have highlighted different design approaches for end-to-end connectivity through examples of practical test bed implementations for healthcare services. Security and privacy challenges pertinent to e-health have also been addressed. Finally, a list of recent research projects in the context of e-health has been given, with emphasis on the different technical solutions adopted in each project.

In summary, the presented works and projects study different aspects of M2M systems for healthcare delivery, ranging from solving technical communication problems to the implementation of close to market solutions. Despite their differences, several common goals can be identified that can serve as guidelines for the design of successful end-to-end e-health applications: (i) miniaturization and enhanced wearability of the sensor devices, to provide unobtrusive monitoring that will not interfere with normal life activities of the patients; (ii) reliable two-way communication protocols that guarantee the prompt and successful delivery of data from the medical sensors to the medical personnel, as well as the reception of medical feedback at the patient; (iii) accurate detection of emergency situations to ensure timely medical intervention in life-threatening events. In addition, it is important to maintain a low probability of false-positive alarms, to avoid unnecessary hospitalizations and interventions, (iv) advanced security mechanisms to guarantee confidentiality and privacy of the medical data, and (v) having user-friendly and easy to learn application interfaces, to ensure the successful adoption of the e-health solutions, given that patients are often elderly people not familiar with the use of technology. Furthermore, the visualization of the monitored data must be done in a clear and helpful way for both the patient and the healthcare providers.

Acknowledgments

This work has been funded by the research projects WSN4QoL (286047), ESEE (324284), SGR (2014 SGR 1551) and GEOCOM (TEC2011-27723-C02-01).

References

[1] OECD Report. Machine-to-machine communications: connecting billions of devices. OECD Digital Economy Papers 192, (2012). Available online at http://dx.doi.org/10.1787/5k9gsh2gp043-en.

[2] K. Chang, A. Soong, M. Tseng, Z. Xiang, Global wireless machine-to-machine standardization, IEEE Internet Comput. 15 (2011) 64–69.

[3] oneM2M global Partnership Project. http://www.onem2m.org/, 2012.
[4] Machina Research. Machine-to-Machine (M2M) Communications in Healthcare 2010–20. 2011.
[5] ETSI. Machine to Machine Communications (M2M): Use Cases of M2M Applications for eHealth. Draft TR 102732 v0.4.1, 2011.
[6] G. Wu, S. Talwar, K. Johnsson, N. Himayat, K. Johnson, M2M: from mobile to embedded internet, IEEE Commun. Mag. 49 (2011) 36–43.
[7] K.-C. Chen, Machine-to-machine communications for healthcare, J. Comput. Sci. Eng. 6 (2012) 119–126.
[8] X. Shen, Emerging technologies for e-healthcare, IEEE Netw. 26 (2012) 2–3 [Editor's Note].
[9] A. Aragues, J. Escayola, I. Martinez, P. del Valle, P. Munoz, J. Trigo, J. Garcia, Trends and challenges of the emerging technologies toward interoperability and standardization in e-health communications, IEEE Commun. Mag. 49 (2011) 182–188.
[10] J. Ko, C. Lu, M. Srivastava, J. Stankovic, A. Terzis, M. Welsh, Wireless sensor networks for healthcare, Proc. IEEE 98 (2010) 1947–1960.
[11] M. Chen, S. Gonzalez, A. Vasilakos, H. Cao, V.C. Leung, Body area networks: a survey, Mobile Netw. Appl. 16 (2011) 171–193.
[12] B. Latre, B. Braem, I. Moerman, C. Blondia, P. Demeester, A survey on wireless body area networks, Wirel. Netw. 17 (2011) 1–18.
[13] H. Alemdar, C. Ersoy, Wireless sensor networks for healthcare: a survey, Comput. Netw. 54 (2010) 2688–2710.
[14] M. Patel, J. Wang, Applications, challenges, and prospective in emerging body area networking technologies, IEEE Wirel. Commun. 17 (2010) 80–88.
[15] ETSI. Machine to Machine Communications (M2M): Functional Architecture. Technical Specification TS 102690 v1.1.1, 2011.
[16] Enrico Scarrone, The ETSI M2M standard as enabler of a global market, in: In 11th edition of the M2M Forum, 2012. Available online at http://www.m2mforum.com/eng/images/stories//scarroneopening.pdf.
[17] H. Cao, V. Leung, C. Chow, H. Chan, Enabling technologies for wireless body area networks: a survey and outlook, IEEE Commun. Mag. 47 (2009) 84–93.
[18] IEEE Standard for Information technology—Telecommunications and information exchange between systems—Local and metropolitan area networks—Specific requirements—Part 15.1: Wireless Medium Access Control (MAC) and Physical Layer (PHY) specifications for Wireless Personal Area Networks (WPANs). IEEE Std 802.15.1-2005 (Revision of IEEE Std 802.15.1-2002), 2005.
[19] IEEE Standard for Local and metropolitan area networks—Part 15.4: Low-Rate Wireless Personal Area Networks (LR-WPANs). IEEE Std 802.15.4–2011 (Revision of IEEE Std 802.15.4–2006), 2011.
[20] Wireless Universal Serial Bus Specification 1.1, 2010.
[21] IEEE Standard for Information technology—Telecommunications and information exchange between systems—Local and metropolitan area networks—Specific requirements—Part 11: Wireless LAN Medium Access Control (MAC) and Physical Layer (PHY) specifications. IEEE Std 802.11-2007 (Revision of IEEE Std 802.11-2007), 2012.
[22] IEEE Standard for Air Interface for Broadband Wireless Access Systems. IEEE Std 802.16–2012 (Revision of IEEE Std 802.16–2009), 2012, pp. 1–2542.
[23] S. Baker, D. Hoglund, Medical-grade, mission-critical wireless networks [designing an enterprise mobility solution in the healthcare environment], IEEE Eng. Med. Biol. Mag. 27 (2008) 86–95.

[24] J. Misic, V. Misic, Bridging between IEEE 802.15.4 and IEEE 802.11b networks for multiparameter healthcare sensing, IEEE J. Sel. Areas Commun. 27 (2009) 435–449.

[25] J. Misic, V. Misic, Bridge performance in a multitier wireless network for healthcare monitoring, IEEE Wirel. Commun. 17 (2010) 90–95.

[26] Y. Zhang, N. Ansari, H. Tsunoda, Wireless telemedicine services over integrated IEEE 802.11/WLAN and IEEE 802.16/WiMAX networks, IEEE Wirel. Commun. 17 (2010) 30–36.

[27] D. Niyato, E. Hossain, S. Camorlinga, Remote patient monitoring service using heterogeneous wireless access networks: architecture and optimization, IEEE J. Sel. Areas Commun. 27 (2009) 412–423.

[28] R.S.H. Istepanaian, Y.-T. Zhang, Guest editorial introduction to the special section: 4G health—the long-term evolution of m-Health, IEEE Trans. Inf. Technol. Biomed. 16 (2012) 1–5.

[29] L. Skorin-Kapov, M. Matijasevic, Analysis of QoS requirements for e-health services and mapping to evolved packet system QoS classes, Int. J. Telemed. Appl. 9 (2010) 1–18 [Special issue on healthcare applications and services in converged networking environments].

[30] A. Alinejad, N. Philip, R. Istepanian, Cross-layer ultrasound video streaming over mobile WiMAX and HSUPA networks, IEEE Trans. Inf. Technol. Biomed. 16 (2012) 31–39.

[31] G.J. Mandellos, M.N. Koukias, I.S. Styliadis, D.K. Lymberopoulos, e-SCP-ECG+ protocol: an expansion on SCP-ECG protocol for health telemonitoring—pilot implementation, Int. J. Telemed. Appl. 2010 (2010) 1–17 [Special issue on healthcare applications and services in converged networking environments].

[32] K. Wac, M. Bargh, B.J. Van Beijnum, R. Bults, P. Pawar, A. Peddemors, Power- and delay-awareness of health telemonitoring services: the mobihealth system case study, IEEE J. Sel. Areas Commun. 27 (2009) 525–536.

[33] Y. Huang, M. Hsieh, H. Chao, S. Hung, J. Park, Pervasive, secure access to a hierarchical sensor-based healthcare monitoring architecture in wireless heterogeneous networks, IEEE J. Sel. Areas Commun. 27 (2009) 400–411.

[34] G. López, V. Custodio, J. Moreno, LOBIN: E-textile and wireless-sensor-network-based platform for healthcare monitoring in future hospital environments, IEEE Trans. Inf. Technol. Biomed. 14 (2010) 1446–1458.

[35] Office for Civil Rights, United State Department of Health and Human Services. Medical privacy—National standards to protect the privacy of personal health information. Available online at http://www.hhs.gov/ocr/privacy/hipaa/administrative/privacyrule/index.html.

[36] The Data Protection Directive. EU Directive 95/46/EC. Available online at http://www.dataprotection.ie/viewdoc.asp?m=&fn=/documents/legal/6aii-1c.htm#1.

[37] P. Kumar, H.J. Lee, Security issues in healthcare applications using wireless medical sensor networks: a survey, Sensors 12 (1) (2011) 55–91.

[38] T. Dimitriou, I. Krontiris, Security issues in biomedical wireless sensor networks, in: Proc. of the 1st International Symposium on Applied Sciences on Biomedical and Communication Technologies (ISABEL 2008), 2008, pp. 1–5.

[39] D. Halperin, T.S. Heydt-Benjamin, K. Fu, T. Kohno, W.H. Maisel, Security and privacy for implantable medical devices, IEEE Pervasive Comput. 7 (2008) 30–39.

[40] E. Weippl, A. Holzinger, A.M. Tjoa, Security aspects of ubiquitous computing in health care, e & i Elektrotech. Inf.tech. 123 (2006) 156–161, http://dx.doi.org/10.1007/s00502-006-0336.

[41] D. Kotz, A threat taxonomy for mHealth privacy, in: Proc. of the 3rd International Conference on Communication Systems and Networks (COMSNETS 2011), 2011, pp. 1–6.

[42] S. Avancha, A. Baxi, D. Kotz, Privacy in mobile technology for personal healthcare, ACM Comput. Surv. 45 (2012) 3:1–3:54.

[43] C. Karlof, N. Sastry, D. Wagner, TinySec: a link layer security architecture for wireless sensor networks, in: Proc. of the 2nd International Conference on Embedded Networked Sensor Systems (SenSys 2004), ACM, New York, NY, USA, 2004, pp. 162–175.

[44] M. Healy, T. Newe, E. Lewis, Efficiently securing data on a wireless sensor network, J. Phys. Conf. Ser. 76 (1) (2007) 012063.

[45] A. Wood, G. Virone, T. Doan, Q. Cao, L. Selavo, Y. Wu, L. Fang, Z. He, S. Lin, J. Stankovic, ALARM-NET: Wireless sensor networks for assisted-living and residential monitoring, Tech. rep., Department of Computer Science, University of Virginia, Tech. Rep. CS-2006–1, 2006.

[46] A. Liu, P. Ning, TinyECC: a configurable library for elliptic curve cryptography in wireless sensor networks, in: Proc. of the International Conference on Information Processing in Sensor Networks, (IPSN 2008), 2008, pp. 245–256.

[47] P. Szczechowiak, L.B. Oliveira, M. Scott, M. Collier, R. Dahab, NanoECC: testing the limits of elliptic curve cryptography in sensor networks, in: Proc. of the 5th European Conference on Wireless sensor networks (EWSN 2008), Springer-Verlag, Berlin, Heidelberg, 2008, pp. 305–320.

[48] F.M. Bui, D. Hatzinakos, Biometric methods for secure communications in body sensor networks: resource-efficient key management and signal-level data scrambling, EURASIP J. Adv. Signal Process. 2008 (2008) 1–16.

[49] K.K. Venkatasubramanian, E.K.S. Gupta, Security for pervasive health monitoring sensor applications, in: Proc. of the 4th International Conference on Intelligent Sensing and Information Processing (ICISIP 2006), 2006, pp. 197–202.

[50] H. Narayanan, M. Gunes, Ensuring access control in cloud provisioned healthcare systems, in: Proc. of IEEE Consumer Communications and Networking Conference (CCNC 2011), 2011, pp. 247–251.

[51] S. Oh, S. Park, Task-role-based access control model, Inf. Syst. 28 (2003) 533–562.

[52] B. Eze, C. Kuziemsky, L. Peyton, G. Middleton, A. Mouttham, Policy-based data integration for e-Health monitoring processes in a B2B environment: experiences from Canada, J. Theor. Appl. Electron. Commer. Res. 5 (2010) 56–70.

[53] A. Kliem, M. Hovestadt, O. Kao, Security and communication architecture for networked medical devices in mobility-aware ehealth environments, in: Proc. of IEEE First International Conference on Mobile Services (MS 2012), 2012, pp. 112–114.

[54] ISO/IEC/IEEE Health informatics–Personal health device communication–Part 20601: Application profile–Optimized exchange protocol. ISO/IEEE 11073–20601:2010(E), 2010, pp. 1–208.

[55] N. Labraoui, M. Gueroui, Secure range-free localization scheme in wireless sensor networks, in: Proc. of the 10th International Symposium on Programming and Systems (ISPS 2011), 2011, pp. 1–8.

[56] J. Jiang, G. Han, L. Shu, H.-C. Chao, S. Nishio, A novel secure localization scheme against collaborative collusion in wireless sensor networks, in: Proc. of the 7th International Wireless Communications and Mobile Computing Conference (IWCMC 2011), 2011, pp. 308–313.

[57] A. Srinivasan, J. Wu, A survey on secure localization in wireless sensor networks, in: B. Furht (Ed.), Encyclopedia of Wireless and Mobile Communications, CRC Press, Taylor and Francis Group, Boca Raton, London, 2008.

[58] Simpliciti protocol stack, www.ti.com/tool/simpliciti.

Index

Printed in the United States
By Bookmasters